Feature Papers in Drug Toxicity

Feature Papers in Drug Toxicity

Editor

Youssef Sari

MDPI

Basel • Beijing • Wuhan • Barcelona • Belgrade • Novi Sad • Cluj • Manchester

Editor
Youssef Sari
Pharmacology and
Experimental Therapeutics
The University of Toledo,
College of Pharmacy and
Pharmaceutical Sciences
Toledo
United States

Editorial Office
MDPI
St. Alban-Anlage 66
4052 Basel, Switzerland

This is a reprint of articles from the Special Issue published online in the open access journal *Toxics* (ISSN 2305-6304) (available at: www.mdpi.com/journal/toxics/special_issues/IJ1L3CYE8E).

For citation purposes, cite each article independently as indicated on the article page online and as indicated below:

Lastname, A.A.; Lastname, B.B. Article Title. *Journal Name* **Year**, *Volume Number*, Page Range.

ISBN 978-3-7258-0442-9 (Hbk)
ISBN 978-3-7258-0441-2 (PDF)
doi.org/10.3390/books978-3-7258-0441-2

Contents

About the Editor

Youssef Sari

Dr. Youssef Sari is a Professor of Pharmacology at the University of Toledo, College of Pharmacy and Pharmaceutical Sciences, Department of Pharmacology and Experimental Therapeutics. Dr. Sari's research has contributed significantly to the field of drugs of abuse, including alcohol, nicotine, cocaine, opioids, and methamphetamine. He has over 122 publications in good- to high-quality, peer-reviewed research journals and over 8 published book chapters. Based on his outstanding achievement in his field, he is now listed as Top #1 Expert Worldwide in the area of Excitatory Amino Acid Transporter 2 in the brain by Expertscape Inc. He has received several awards and honors in recognition of his excellent contributions to the scientific literature in the field of drugs of abuse. Among these awards are the Sigma Xi Young Investigator Award, the President's Research and Scholarship Award, Shining Star Award, and the Outstanding Faculty Research and Scholarship Award. Dr. Sari's research programs have been funded for several years by prestigious awards from the National Institutes of Health. Based on his meritorious scientific achievements, he has secured several NIH funding for the last 15 years. He is currently serving as ad hoc reviewer at the National Institute of Health, USA. He has an incredible record of mentoring and tirelessly interacting with his mentees and other junior colleagues in and outside the laboratory. Most of his mentees are currently holding the title of Assistant Professor or Assistant Professor/Head of the Department at National and International Universities.

Preface

This preface introduces the Special Issue "Feature Papers in Drug Toxicity". Drug toxicity is a major problem involving neurotoxicity and organ injury. There are many drugs which have toxic effects as a result of acute or chronic exposure as well as prenatal exposure. These drugs include abused drugs and drugs that have toxic effects in the brain and other organs of the body. Understanding the pharmacological and toxicological nature of these drugs expands research in the field of toxicology. In this Special Issue, several abused drugs were tested in animal models, including ethanol, hydrocodone, and heroin. In addition, the toxicological effects of other drugs are reported in this Special Issue, including pentylenetetrazol, metronidazole, paracetamol, paclitaxel, and tilmicosin. Importantly, several protective agents were tested by researchers and are reported in this Special Issue. These agents protect against the toxic effects induced by abused and other drugs. In addition, studies report the protective effects of certain plant extracts against drug-induced toxicity.

This Special Issue is for the benefit of scientists and readers in the field of toxicity involving drugs of abuse as well as medications that may have toxic side effects. The Special Issue is also for the benefit of scientists and readers who are interested in research involving the protective effects of plant extracts to prevent toxic effects of other chemicals.

Youssef Sari
Editor

toxics

MDPI

Article

Carbetocin Inhibits Behavioral Sensitization to Ethanol in Male and Female Mice, Independent of Corticosterone Levels

Beatriz Yamada Costa [1], Luana Gasparini Santos [2], Priscila Marianno [1], Mariana Rae [1], Marina Gomes de Almeida [1], Malcon Carneiro de Brito [1], Rosângela Eichler [1] and Rosana Camarini [1,*]

[1] Department of Pharmacology, Institute of Biomedical Sciences, Universidade de São Paulo, São Paulo 05508-900, Brazil; beatriz.cvs@usp.br (B.Y.C.); primarianno@usp.br (P.M.); marianarae@hotmail.com (M.R.); malconbrito@usp.br (M.C.d.B.); eichler@usp.br (R.E.)

[2] School of Pharmaceutical Sciences, Universidade de São Paulo, São Paulo 05508-900, Brazil; luuanagasparini@gmail.com

* Correspondence: camarini@icb.usp.br

Abstract: Oxytocin (OXT), a pro-social peptide, is increasingly recognized as a potential protective substance against drug addiction. In the context of ethanol, previous research has shown OXT's properties in reducing self-administration, alleviating motor impairment in rodents, and reducing craving in humans. However, its role in behavioral sensitization, a neuroadaptive response resulting from repeated drug exposure linked to an increased drug incentive, remains unexplored. OXT is recognized for its role in regulating the hypothalamic–pituitary–adrenal (HPA) axis, in which corticosterone is acknowledged as a significant factor in the development of behavioral sensitization. This study aimed to investigate the effects of carbetocin (CBT), an analogue of OXT, on the expression of behavioral sensitization to ethanol and the concurrent alterations in plasma corticosterone levels in male and female Swiss mice. We also aimed to confirm previous studies on OXT's impact on ethanol consumption in male mice, but with a focus on CBT, using the two-bottle choice model and the drinking in the dark (DID) methodology. For the sensitization study, the mice received either ethanol (1.8 g/kg, i.p.) or saline treatments daily for 15 consecutive days, followed by treatment with carbetocin (0.64 mg/kg, i.p.) or a vehicle for 6 days. Subsequently, on day 22, all the animals underwent an ethanol challenge to assess the expression of behavioral sensitization. The plasma corticosterone levels were measured on days 21 and 22. The CBT effectively prevented the expression of ethanol-induced behavioral sensitization in both male and female subjects, with no alterations having been detected in their corticosterone levels. In the ethanol consumption study, following an initial phase of ethanol acquisition, the male mice underwent a 6-day treatment with CBT i.p. or saline before being re-exposed to ethanol. We also found a reduction in their ethanol consumption due to the CBT treatment. In conclusion, carbetocin emerges as a promising and effective intervention for mitigating ethanol-induced behavioral sensitization and reducing ethanol intake, highlighting its potential significance in alcohol addiction treatment.

Keywords: behavioral sensitization; ethanol; oxytocin; addiction; estrous cycle; ethanol self-administration

Citation: Costa, B.Y.; Santos, L.G.; Marianno, P.; Rae, M.; de Almeida, M.G.; de Brito, M.C.; Eichler, R.; Camarini, R. Carbetocin Inhibits Behavioral Sensitization to Ethanol in Male and Female Mice, Independent of Corticosterone Levels. *Toxics* **2023**, *11*, 893. https://doi.org/10.3390/toxics11110893

Academic Editor: Youssef Sari

Received: 23 September 2023
Revised: 17 October 2023
Accepted: 27 October 2023
Published: 31 October 2023

1. Introduction

Alcohol, a globally consumed psychoactive substance, has a myriad of adverse consequences for individuals and society [1]. It significantly contributes to the development or exacerbation of over 200 different diseases and health conditions classified in the ICD-10 system [1]. The likelihood of mortality from any of these causes, as well as the risk of developing cancers, escalates as alcohol consumption levels increase, and the level of consumption for minimizing health-related harm is zero [2]. Ethanol consumption is linked to a spectrum of hepatic disorders, encompassing liver inflammation, fatty liver disease, and cirrhosis [3]. Beyond hepatotoxicity, alcohol-derived metabolites contribute to oxidative stress and impaired cognitive function and exert systemic repercussions on multiple organ systems, encompassing the cardiovascular and gastrointestinal systems [4,5].

Alcohol use disorder (AUD) is a health condition that affects both men and women, albeit with notable differences in terms of prevalence, manifestation, and consequences between the genders. The detrimental consumption of alcohol contributes to 7.1% of the global disease burden in males and 2.2% in females [6]. Men exhibit higher rates of alcohol consumption and AUD compared to women [7]. This has often been attributed to cultural and societal factors, including gender-specific expectations that may encourage men to engage in alcohol-related risky behaviors [8]. However, recent studies suggest that the gender gap in AUD is narrowing, with an increasing number of women experiencing alcohol-related issues [9].

Currently, only a few medications are licensed for treating AUD, including disulfiram, naltrexone, acamprosate, and nalmefene. Ongoing research is necessary due to the limitations of the existing drugs for the treatment of AUD [10]. Moreover, tailored treatment approaches that consider gender-specific factors have become increasingly important in addressing AUD. Gender-sensitive interventions, support groups, and healthcare services have been shown to enhance the effectiveness of treatment and recovery strategies for both men and women [11].

One of the emerging treatment options is oxytocin (OXT), a neuropeptide involved in the modulation of different behaviors, such as mood, social interaction, couple formation, and stress [12]. The effect of this neuropeptide on addiction has received great attention [13], with studies indicating that OXT administration decreases alcohol self-administration [14] and reduces cue-reactivity to ethanol in rats and humans [15]. Notably, alcohol-dependent rats exhibited significant changes in their OXT systems, whereas female rats showed no alterations [16]. Furthermore, genetic disruption of the OXT receptor using knockout mice influenced alcohol consumption in female mice, resulting in an increased intake before and after their exposure to stress, while male mice showed no significant genotypic differences [17], indicating sex-specific responses to OXT.

Although there are several studies demonstrating the effects of OXT on alcohol consumption in both males and females, the role of OXT in the behavioral sensitization to ethanol and potential sex-specific responses, in particular, still remain unknown. While models of self-administration address the rewarding effects of drug abuse [18], behavioral sensitization focuses on neuroadaptive processes, as it is described as the psychomotor manifestation of sensitization in neuronal pathways [19]. Studies using mice have shown that females are more sensitive to ethanol-induced behavioral sensitization than males, pointing to a sex-dependent criteria for this phenomenon [20,21].

Although sensitization can be associated with several behaviors, increased locomotor activity is the most commonly studied phenomenon. Nonetheless, sensitization can affect not only behavioral but also neurochemical or neuroendocrine processes, an effect which can be observed in an increase in neurotransmitters' release or hormonal secretions such as corticosterone [22], for instance. Corticosterone plays an important role in the development of behavioral sensitization, since the activation of the hypothalamic–pituitary–adrenal (HPA) axis has been described to increase drug use [23]. Specifically, regarding behavioral sensitization to alcohol, a cross-sensitization between alcohol and stress has also been reported [24,25]. In this context, OXT contributes to the regulation of stress responses through its interaction with the HPA axis [26]. This interaction involves a cascade of neuroendocrine processes wherein OXT may influence the release of hormones, including glucocorticoids [27].

In our research, we chose to investigate the effects of carbetocin (CBT), a synthetic analogue of OXT. Despite its structural similarity to OXT, it exhibits a longer half-life. CBT has been investigated for its potential to prevent the priming-induced reinstatement of morphine-seeking behaviors [28,29] and has the advantage of not inducing any alterations in the plasma corticosterone levels [28]. In this study, we aimed to assess the impact of CBT on the expression of behavioral sensitization to ethanol and its effects on plasma corticosterone levels in male and female mice. Additionally, we investigated the influence of CBT on ethanol intake in male mice.

2. Materials and Methods

2.1. Animals

Thirty-two male and thirty-two female Swiss mice were housed in groups of four, with food and water ad libitum, in an experimental room, with controlled temperature (24 ± 2 °C) and light conditions (light/dark cycle of 12 h; lights on at 7:00 a.m.). Swiss mice were used owing to their sensitivity to the stimulant effects of ethanol, making them useful for studying behavioral sensitization in male and female mice [30,31].

For the ethanol consumption experiment, twenty-eight adult male C57BL/6 mice, 8–10 weeks old, were housed in groups of four, with food and water ad libitum, with controlled temperature and light conditions (light/dark cycle of 12 h; lights off at 7:00 a.m.). C57BL/6 mice were chosen for their genetic predisposition to voluntarily consume significant amounts of ethanol, but their low sensitivity to behavioral sensitization, making them a good choice for alcohol intake studies [32–34]. The animals were acclimatized to the reverse cycle at least 2 weeks before the experiments. Red incandescent lights were utilized during the dark phase to facilitate mice handling by the investigators.

All the procedures were approved by the Ethics Committee on the Use of Animals of the Institute of Biomedical Sciences (University of Sao Paulo) (CEUA—ICB/USP), under CEUA numbers 9998280518 and 4512140222 and protocol 25/2016, in accordance with Law 11,794 of 8 October 2008, Decree 6899 of 15 July 2009, as well as with the rules issued by the National Council for Control of Animal Experimentation (CONCEA). Efforts were made to minimize pain and suffering and reduce the use of animals. Two male Swiss mice were excluded from this cohort due to their aggressive behavior and two male C57BL/6 mice died from unknown causes.

2.2. Drugs

Ethanol (95%; Labsynth, Diadema, SP, Brazil) was administered intraperitoneally (i.p.) in a 20% (*v*/*v*) solution, prepared with a saline solution (NaCl 0.9%), at a dose of 1.8 g/kg. The saline solution was used as a control solution and injected i.p. For the voluntary ethanol intake, ethanol was diluted to 20% (*v*/*v*) in tap water. Carbetocin (CBT) (Sigma-Aldrich, St. Louis, MO, USA), a synthetic analog of OXT, was dissolved in a saline solution and administered i.p. for 6 days, following the locomotor sensitization protocol, and, before the re-exposure to ethanol in the DID paradigm = for 6 consecutive days, at a dose of 6.4 mg/kg. CBT was chosen due to its longer half-life (85–100 min) compared to OXT (3–5 min) and its stability, facilitating its handling for a longer period of time [35].

2.3. Identification of the Estrous Cycle Phase

In female mice, the phases of the estrous cycle were identified with a fresh cytological analysis of vaginal lavage. The animals were properly restrained, and a careful vaginal wash was performed with 15 µL of 0.9% saline solution. The saline was injected, aspirated, and then placed on a histological slide for microscope viewing and phase identification [36]. This procedure ensured a consistent timing for blood collection and measurement of corticosterone.

2.4. Experimental Design

2.4.1. Effects of Carbetocin on Ethanol-Induced Behavioral Sensitization

The experimental design is depicted in Figure 1. The locomotor activity was evaluated in an open field, a plexiglass arena measuring 40 cm in diameter, and a wall measuring 50 cm in height. The animals received the saline or ethanol injections according to the experimental group and, after five minutes, were placed into the center of the apparatus. The total horizontal locomotor activity was evaluated for a period of 5 min, as the peak in the locomotor activation induced by ethanol occurs between 5 and 10 min after ethanol administration [37,38]. All the experiments were performed between 9:00 a.m. and 10:00 a.m. The activity was recorded using a digital camera and a video-capture system. The EthoVision® software version 11.5.1026 (Noldus, The Netherlands) was used to quantify the distance covered by each animal, as previously described [39].

Figure 1. Experimental design 1: The experimental design involved the initial administration of saline in H1 and H2 (habituation days), followed by a subsequent treatment with either saline or 1.8 g/kg of ethanol (from days D1 to D15). From days D16 to D21, the mice received injections of either saline or 6.4 g/kg of carbetocin (CBT). On day 22 (D22), all the mice were challenged with 1.8 g/kg of ethanol, resulting in the following four groups: SAL-SAL, SAL-CBT, ETOH-SAL, and ETOH-CBT.

For the first two days (H1 and H2), all the animals received an intraperitoneal saline solution and had their locomotor activity evaluated in the open field, to familiarize them with the experimenter's handling and minimize the novelty effect of the apparatus.

Following the habituation period, the mice were randomly assigned to either the saline (SAL) or 1.8 g/kg ethanol (EtOH) groups. During the 15 days of treatment (D1–D15), the mice received daily injections of either SAL or 1.8 g/kg of ETOH i.p., based on their group assignment. From D16 to D21, the animals underwent a period of ethanol withdrawal, during which half of the animals in each group received 6.4 mg/kg of CBT i.p. and the other half received isovolumetric injections of saline as a control. On D22, all the animals were challenged with an injection of 1.8 g/kg of ethanol. For each sex, we established four distinct groups, as follows: SAL-SAL, SAL-CBT, ETOH-SAL, and ETOH-CBT (n = 7–8/group).

The locomotor activity was evaluated on days H1, H2, D1, D8, D15, and D22. In addition, blood samples from the caudal vein were collected for the subsequent measurement of plasma corticosterone on D21, between 1:00 p.m. and 3:00 p.m., and D22, after the behavioral test, for the subsequent corticosterone and ethanol measurements.

2.4.2. Effects of Carbetocin on Ethanol Consumption

The experimental design is shown in Figure 2. We employed the drinking in the dark (DID) protocol, using the two-bottle choice method, to assess the effects of CBT on ethanol consumption. Three hours after the onset of the dark phase, the animals were single-housed with free access to two bottles: one contained ethanol (95% v/v; Labsynth, SP, Brazil) diluted to 20% (v/v) in tap water, and the other contained tap water. The mice were allowed to freely consume both solutions for a 2 h period. Subsequently, the bottles were removed, and the mice were returned to their respective home cages. The bottles were weighed both before and immediately after the consumption sessions, and the differences in weight were converted into the volumes of ethanol and water solutions consumed. The ethanol consumption in grams per kilogram (g/kg) was determined by taking into account the density of the ethanol, the concentration of the solution, the quantity of the solution consumed, and the body weight of each subject. The solutions were replaced daily, and the positions of the bottles were regularly interchanged to eliminate potential side preferences.

Throughout the DID procedure, a separate cage with two bottles was employed as a control to account for any liquid loss from handling or evaporation. The volume lost in these control bottles was subtracted from the measured volume of ethanol or water consumed by each animal.

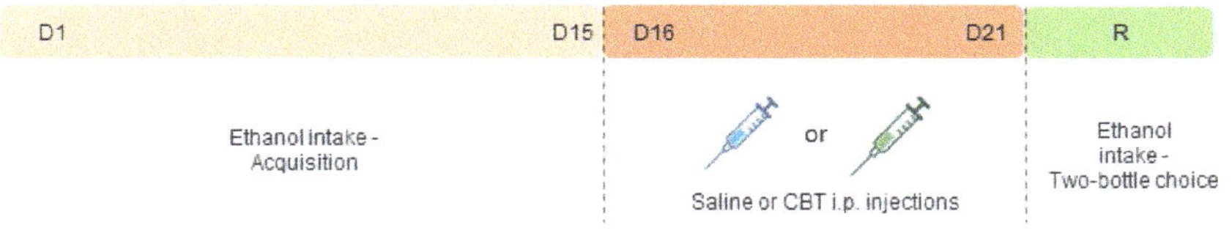

Figure 2. Experimental design 2. After an initial 15-day period of alcohol acquisition (D1–D15) using the DID paradigm, the mice underwent a 6-day treatment phase with either saline or 6.4 mg/kg of carbetocin (CBT) i.p (from days D16 to D21). The administration of the CBT occurred at different time points, either 1 h or 24 h prior to re-exposure (R). The R phase involved the re-exposure to the two-bottle choice test (water vs. ethanol). Three groups were formed, as follows: SAL, CBT-1H, and CBT-24H.

Since previous studies showed no OXT-specific changes in female mice after ethanol exposure [16], the DID protocol was employed to evaluate voluntary ethanol consumption only in male mice, with modifications adapted from a previous study [40,41]. With this study we sought to investigate whether CBT would yield results consistent with previous research conducted with OXT. This investigation involved an experimental paradigm comprising distinct phases, including an initial period of acquisition, followed by a withdrawal phase, and, ultimately, re-exposure to ethanol. The mice were exposed to the DID paradigm for 15 days to ensure the stabilization of their ethanol consumption (acquisition phase). Following the acquisition phase, the animals were randomly distributed into three groups—CTL (n = eight), CBT-1H (n = nine), and CBT-24H (n = nine)—and treated accordingly, with either saline or CBT (6.4 mg/kg), for six consecutive days during a period of ethanol deprivation. The CBT was administered at two different time points: either 1 h (CBT-1H) or 24 h (CBT-24H) prior to re-exposure (R). Subsequently, the animals were given two bottles, with free access to ethanol (20%) and water for 24 h, and their consumption was measured at both the 2 h and 24 h marks from the onset of drinking, following the protocol described in Marianno et al., 2017 [41].

2.5. Blood Collection for Biochemical Analysis

Blood collection for a subsequent corticosterone measurement was taken on D21, between 1:00 p.m. and 3:00 p.m. Approximately 100 μL of blood was collected from the caudal vein and placed in microcentrifuge tubes containing heparin (100 U/mL, in the volume of 10% of the total volume of the blood collected). The samples (n = 7/group) were centrifuged at 2000× g at 4 °C for 10 min and the plasma was transferred to a clean tube and stored at −80 °C. The corticosterone levels were determined using the IBL Corticosterone Enzyme Immunoassay Kit (Tecan Trading AG, Männedorf, Switzerland), following the manufacturer's procedures.

Blood samples were collected on D22, after the animals were euthanized, for a subsequent corticosterone and ethanol measurement, between 9:00 a.m. and 11:00 a.m. Approximately 250 μL of blood were placed in microcentrifuge tubes containing heparin (250 U/mL, in the volume of 10% of the total volume of the blood collected). The samples were processed as described above. The corticosterone levels (n = 5–7/group) and the blood ethanol concentration (BEC) (n = 5/group) were assayed. The BEC was analyzed using the Ethanol Assay Kit Abcam (Abcam plc, Cambridge, UK), following the manufacturer's procedures. Some of the samples underwent hemolysis, resulting in a reduction in the number of samples.

2.6. Statistical Analysis

The results were submitted for statistical analysis using the Statistica program, version 7.0. Levene's test was employed to assess the homogeneity of variances. A four-way analysis of variance (ANOVA) for repeated measures was performed to analyze the data related to the mice's locomotor activity in the behavioral sensitization experiment, with "pretreatment" (SAL or ETOH), "treatment" (SAL or CBT), and "sex" (MALE or FEMALE) as the between-group statistical factors, and "time" as the repeated measure. Follow-up three-way ANOVAs for repeated measures were performed for each sex. The locomotor response to the ethanol challenge (D22) was analyzed with two-way ANOVAs, using "pretreatment" and "treatment" as the between-group factors. For the analysis of the corticosterone levels, a three-way ANOVA (pretreatment X treatment X sex) was followed up with two-way ANOVAs using "pretreatment" and "treatment" as the between-group factors for males. For females, we included the diestrus vs. non-diestrus phases as a factor to control for hormonal variations (pretreatment X treatment X estrous phase). For the analysis of the blood ethanol concentration, we used a three-way ANOVA (pretreatment X treatment X sex).

As for the data from the mice's ethanol intake during the acquisition phase, we conducted a one-way ANOVA for repeated measures, with "time" as the repeated measure. The analysis of the re-exposure (R) data (2 h) was conducted using a two-way ANOVA for repeated measures, considering time as the repeated measure (with two levels: mean of the last 5 days of acquisition and re-exposure) and group (CTL, CBT-1H, CBT-24H) as the between-subjects factor. The analysis of re-exposure to ethanol for 24 h was performed using a one-way ANOVA. A Newman–Keuls post hoc test was used to compare the means when statistical significance was found in the repeated measures, and, for the non-repeated measures, the Tukey test was used. The values of $p < 0.05$ were considered significant. Statistical details other than those explicitly mentioned in the main text can be found in the Supplementary Materials.

3. Results

3.1. CBT Inhibited the Expression of Behavioral Sensitization in Male and Female Mice

A four-way ANOVA used to analyze the locomotor activity on H1 and H2 (Figure 3A,B) revealed significant effects of sex [$F(1,54) = 5.28$, $p < 0.05$] and time [$F(1,54) = 18.57$, $p < 0.001$]. The female mice exhibited a reduced motor activity compared to the males, which may be attributed to differences in sensitivity to novelty. The decrease in locomotor activity from H2 to H1 indicates a typical habituation response to the apparatus. Following that, we performed a four-way ANOVA to analyze the locomotor activity on D1, D8, and D15. Given the absence of significant sex differences [$F(1,54) = 0.06$, $p = 0.80$], we continued the analysis with separate three-way ANOVAs for each sex.

The statistical analysis of the locomotor activity in the male mice on D1, D8, and D15 revealed significant effects of the pretreatment [$F(1,26) = 8.81$, $p < 0.01$] and of the time [$F(2,52) = 16.22$, $p < 0.001$], as well as an interaction between the pretreatment and the time [$F(2,52) = 9.22$, $p < 0.01$]. As depicted in Figure 3A, the locomotor activity of the male mice who had been subjected to ethanol treatment was higher on D8 and D15 compared to D1, as detected with the post hoc Newman–Keuls test. These results suggest that the male mice exhibited behavioral sensitization starting from D8, which was further confirmed by their increased locomotor activity on D15.

Similar results were observed in the female mice (Figure 3B). A three-way ANOVA for repeated measures also detected significant effects of the pretreatment [$F(1,28) = 63.70$, $p < 0.001$] and of the time [$F(2,56) = 11.20$, $p < 0.001$], as well as an interaction between the pretreatment and the treatment [$F(1,28) = 9.22$, $p < 0.01$] and an interaction between the pretreatment and the time [$F(2,56) = 8.56$, $p < 0.001$]. The post hoc Newman–Keuls test showed significant differences on D8 and D15 compared to D1 in the ethanol-pretreated group.

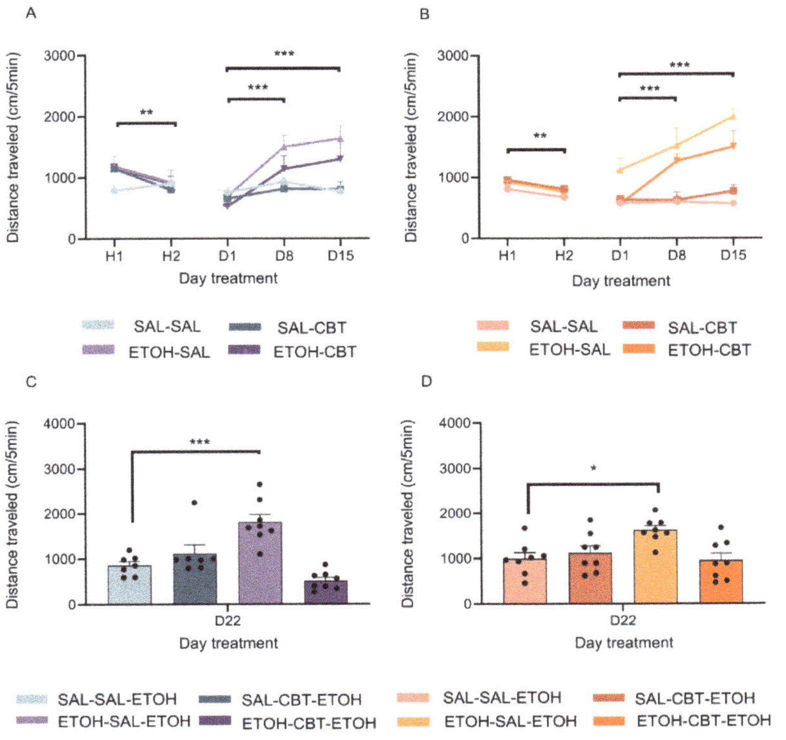

Figure 3. Effects of CBT on the expression of locomotor sensitization. The figure displays locomotor activity (cm) measured over 5 min time-periods. After 2 days of saline injections (H1 and H2), the mice received either SAL or ETOH i.p. injections daily for 15 days (D1–D15) based on their group assignment [Figures (**A**) (male) and (**B**) (female)]. From days D16 to D21, the animals underwent a period of ethanol withdrawal, during which half of the animals in each group received 6.4 mg/kg of CBT i.p. and the other half received injections of saline as a control. On the following day (D22) [Figures (**C**) (male) and (**D**) (female)], all the animals were challenged with an injection of 1.8 g/kg of ethanol. For each sex, four distinct groups were established, as follows: SAL-SAL (n = 7–8/group), SAL-CBT (n = 7–8/group), EtOH-SAL (n = 8/group), and EtOH-CBT (n = 8/group). The locomotor activity in H2 was lower than in H1. The locomotor activity of the mice subjected to ethanol treatment was higher on D8 and D15 compared to D1 in both the male and female mice. The activity of the ETOH-SAL-ETOH group on D22 differed from that of the SAL-SAL-ETOH; * $p < 0.05$, ** $p < 0.01$, *** $p < 0.001$. The data represent the mean ± SEM. Figure 3C,D feature individual data points.

Figure 3C shows the locomotor activity of the male mice measured on D22, when all the mice received a challenge injection of 1.8 g/kg of ethanol. A two-way ANOVA revealed significant effects of the treatment [F(1,26) = 13.94, $p < 0.001$] and an interaction between the pretreatment and the treatment [F(1,26) = 30.86, $p < 0.001$]. The animals previously exposed to repeated ethanol treatment followed by saline during the abstinence period (ETOH-SAL-ETOH) exhibited a heightened locomotor activity compared to the animals who had been pre-exposed to saline and were administered saline during this phase (SAL-SAL-ETOH), indicating a more pronounced response in the mice subjected to repeated ethanol administration as opposed to those receiving a single, acute ethanol injection. These data confirmed the expression of ethanol-induced behavioral sensitization in the ETOH-SAL group. No significant difference was detected in the locomotor activity of the ETOH-CBT-ETOH group compared to that of the SAL-SAL-ETOH group, suggesting that CBT was effective in reversing ethanol sensitization. No significant differences were found between the SAL-SAL-ETOH and SAL-CBT-ETOH groups, showing that CBT did not affect the locomotion of these animals.

Likewise, a two-way ANOVA applied to the data from the female mice on D22 revealed a significant treatment effect [$F(1,28) = 4.12$, $p = 0.05$] and an interaction between the pretreatment and the treatment [$F(1,28) = 8.77$, $p < 0.01$]. The post hoc test indicated that the ETOH-SAL-ETOH group exhibited a greater locomotor activity than the SAL-SAL-ETOH (Figure 3D) group. No significant differences were found in the locomotor activity between the ETOH-CBT-ETOH group and the SAL-SAL-ETOH or SAL-CBT-ETOH groups. The results confirm the efficacy of CBT in reversing ethanol sensitization in the female mice as well. In alignment with the previous results found in the male mice, no significant differences were found between the SAL-SAL-ETOH and SAL-CBT-ETOH groups, underscoring that CBT did not exert an impact on the locomotion of these animals.

3.2. CBT Influence on Behavioral Sensitization Is Not Mediated by Alterations in the Stress Hormone Corticosterone

The analysis of the plasma corticosterone levels from D21 (Figure 4A) with a three-way ANOVA revealed a significant effect associated with the sex factor [$F(1,48) = 5.92$, $p < 0.05$], with the female mice exhibiting higher corticosterone levels than the male mice. No other effects or interactions were found. A follow-up analysis of the data from the male mice using a two-way ANOVA confirmed no significant main effects or interactions. For the female mice, we considered the diestrus and non-diestrus phases as a variable (cycle). In this case, we found an effect of the cycle [$F(1,20) = 73.51$, $p < 0.001$]. While ANOVAs unveiled effects related to the pretreatment [$F (1,20) = 6.16$, $p = 0.02$] and the treatment [$F (1,20) = 8.64$, $p = 0.01$], interpreting these findings is challenging due to the influence of hormonal variations and the unequal distribution of female mice across different phases within each group. A post hoc analysis revealed higher corticosterone levels in the non-diestrus phases compared to the diestrus.

The analysis of the plasma corticosterone levels from D22 (Figure 4B) using a three-way ANOVA revealed significant effects of the sex [$F(1,36) = 22.27$, $p < 0.01$] and pretreatment [$F(1,36) = 18.72$, $p < 0.001$] factors, as well as interactions between the sex and the pretreatment [$F(1,36) = 5.68$, $p < 0.05$], and among the sex, the pretreatment, and the treatment [$F(1,36) = 6.38$, $p < 0.05$]. No other main effects or significant interactions were observed. A post hoc analysis revealed higher corticosterone levels in the female mice compared to the male mice, as observed in D21.

Subsequently, we conducted a two-way ANOVA to analyze the data from the male mice, revealing a significant effect of the pretreatment [$F(1,16) = 5.02$, $p < 0.05$]. The mice who had been previously exposed to ethanol exhibited lower corticosterone levels compared to the ethanol-naïve mice when both groups were challenged with ethanol. Additionally, a significant treatment effect [$F(1,16) = 8.66$, $p < 0.05$] demonstrated that CBT increased the corticosterone levels in the male mice, irrespective of the pretreatment.

The analysis of the corticosterone levels in the female mice using a 3-way ANOVA revealed an effect of the estrous cycle [$F (1,16) = 12.26$, $p < 0.01$], indicating higher hormone levels during the non-diestrus phases compared to the diestrus. We also found an effect of the pretreatment, similar to that observed in the male mice [$F(1,16) = 22.49$, $p < 0.01$]. However, differently from the male mice, CBT treatment did not alter the corticosterone levels in the female mice on D22, when they were challenged with ethanol [$F(1,16) = 0.12$, $p = 0.73$].

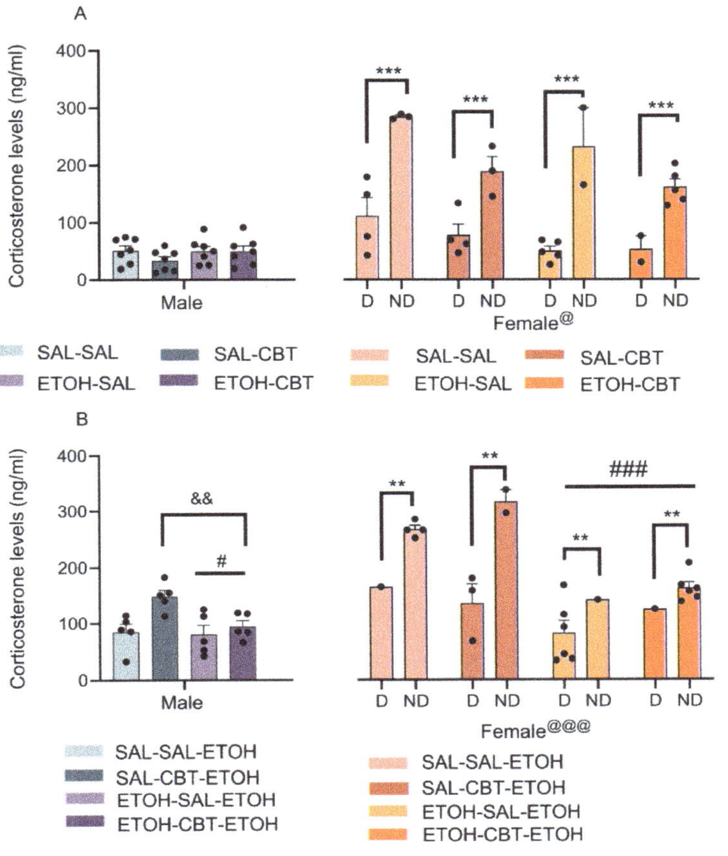

Figure 4. Effects of carbetocin (CBT) and ethanol on plasma corticosterone levels. The male and female animals were pretreated with either saline or ethanol for 15 days, followed by treatment with either saline or CBT during a 6-day abstinence period, and they were challenged with 1.8 g/kg of ethanol on the following day. The corticosterone concentrations were measured on D21 (n = 7/group, Figure (**A**)) and D22 (n = 5–7/group, Figure (**B**)). D: diestrus; ND: non-diestrus. @ $p < 0.05$ and @@@ $p < 0.001$: the female mice showed higher corticosterone levels compared to the male mice. # $p < 0.05$ and ### $p < 0.001$: the mice pretreated with ethanol (ETOH-) exhibited lower levels of corticosterone than those pretreated with saline (SAL-). && $p < 0.01$: the mice treated with CBT (SAL-CBT and ETOH-CBT) displayed higher corticosterone levels than those treated with saline (SAL-SAL and ETOH-SAL). ** $p < 0.01$ and *** $p < 0.001$: the corticosterone levels in the non-diestrus are higher than in the diestrus. The data represent the mean ± SEM. Figure 4A,B feature individual data points.

3.3. CBT Does Not Alter Ethanol Metabolism

The analysis of the blood ethanol concentration (BEC) (Figure 5) using a three-way ANOVA revealed statistically significant effects related to the sex [$F(1,32) = 5.77$, $p < 0.05$] and the pretreatment [$F(1,32) = 17.95$, $p < 0.001$], but it did not show any statistically significant differences associated with the treatment [$F(1,32) = 0.22$, $p = 0.64$]. The female mice showed higher BECs compared to the male mice. Moreover, the mice with prior ethanol exposure exhibited a lower BEC compared to the alcohol-naïve mice. No other main effects or interactions were observed.

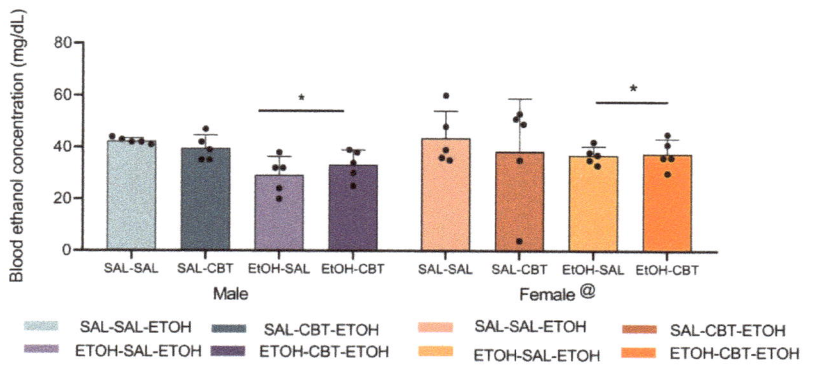

Figure 5. Effects of CBT and ethanol on blood ethanol concentration (BEC). The male and female animals were pretreated with either saline or ethanol for 15 days (D1 to D15), followed by treatment with either saline or CBT during a 6-day abstinence period (D16 to D21). On day 22, all the animals received an injection of 1.8 g/kg of ethanol. The BEC was measured on D22 (n = 5/group). @ $p < 0.05$: the female mice showed higher BECs compared to the male mice. * $p < 0.05$: the ethanol pretreatment resulted in a lower BEC compared to the saline pretreatment. Figure 5 features individual data points.

3.4. CBT Decreases Ethanol Intake in Male Mice

The analysis of the data from the last 5 days of the 15-day acquisition phase using a one-way ANOVA for repeated measures revealed no differences among them ($F(4,180) = 0.56$, $p = 0.69$; Figure 6A). This indicates that, at the end of the acquisition phase, the mice had reached stable levels of ethanol consumption.

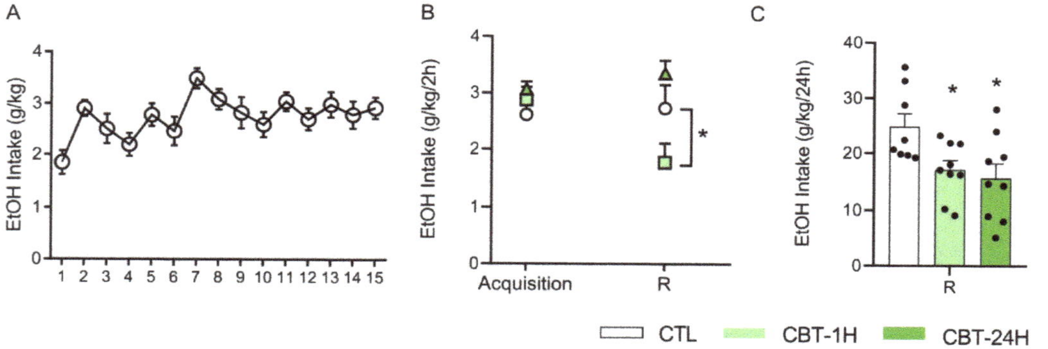

Figure 6. CBT decreases ethanol intake (g/kg) in male mice. The male mice were exposed to the DID paradigm for 15 days to ensure the stabilization of ethanol consumption (acquisition phase) (**A**). Following the acquisition phase, the animals were randomly distributed into three groups—control (CTL, n = eight), CBT-1H (n = nine), and CBT-24H (n = nine)—and treated accordingly for 6 consecutive days. After a six-day period of ethanol deprivation, the mice were re-exposed to the two-bottle choice (R) method, with free access to ethanol and water for 24 h. The consumption was measured at both 2 (**B**) and 24 h (**C**) from the initiation of drinking. * $p < 0.05$: differs from CTL. DID = drinking in the dark. Figure 6C features individual data points.

Figure 6B shows the ethanol intake averaged over the final 5 days of the acquisition phase and the consumption after 2 h from re-exposure. The ANOVA revealed significant effects of the treatment ($F(2,23) = 4.11$, $p < 0.05$) and a treatment X time interaction ($F(2,23) = 5.14$, $p < 0.05$), with no significant effect of the time ($F(2,23) = 1.45$, $p = 0.24$). During the 2 h re-exposure period, the CBT-1H group exhibited a significant reduction in their ethanol intake compared to both the acquisition phase and the other two groups.

Ethanol consumption was additionally measured at the end of the 24 h period of free access to both ethanol and water, during the re-exposure phase (Figure 6C). The one-way ANOVA revealed a significant treatment effect ($F(2,23) = 4.90$, $p < 0.05$), with the groups that received CBT exhibiting a reduction in their ethanol consumption compared to the control group.

4. Discussion

To our knowledge, this study is the first to demonstrate that CBT can reverse ethanol-induced behavioral sensitization in both male and female Swiss mice. CBT treatment alone failed to induce changes in the mice's locomotor activity; hence, it can be inferred that the reduction in behavioral sensitization is not contingent upon alterations in locomotor activity induced by CBT.

The concept of sensitization, initially described by Segal and Mandell in 1974 [42], entails the gradual and persistent amplification of specific behaviors after repeated exposure to stimulant drugs. It serves as a well-studied model of neuroplasticity. Following intermittent stimulant drug treatment, such as amphetamine or cocaine, or after administering stimulant ethanol doses, sensitized behaviors may manifest with increased intensity, faster onset, or at lower doses than before sensitization [43,44]. There is strong evidence linking behavioral sensitization to changes in limbic neurochemical systems, which play a role in various psychiatric and substance use disorders [45].

In a recent study in our laboratory [46], we observed that CBT enhanced the rewarding effects of ethanol, as measured with conditioned place preference (CPP), which primarily reflects the rewarding aspects of a substance and relies on Pavlovian learning mechanisms [47]. In the present study, we adopted the behavioral sensitization paradigm to model a different facet of addiction, aligned with the incentive-sensitization theory proposed by Robinson and Berridge in 1993 [48]. Our findings suggest that, although CBT may enhance the rewarding effect of ethanol in certain contexts, it can also exert protective effects against the expression of ethanol-induced behavioral sensitization. These results highlight the complex role of CBT in modulating different aspects of ethanol addiction. It is important to emphasize that, while sensitization may contribute to the development of addiction by enhancing the incentive salience of drugs, it does not imply an inescapable cycle of substance use. One can experience sensitization without progressing to a chronic substance dependence, and, likewise, individuals can develop substance use disorders without exhibiting sensitization.

Regarding our findings about corticosterone levels, CBT did not change corticosterone levels on D21, in agreement with previous studies [28]. On D22, following a challenge injection with ethanol, the levels of this hormone decreased in the mice who had been previously exposed to ethanol compared to those who had been pre-exposed to saline. This reduction in corticosterone levels in the mice with prior ethanol exposure, as opposed to those receiving an acute ethanol dose, suggests a potential development of tolerance or adaptation to the stress-inducing effects of ethanol. While CBT treatment increased the corticosterone levels in the male mice, the female mice did not exhibit the same response, indicating that the results may be affected by estrous cycle variations. Our study did not reveal any specific effect of CBT on the corticosterone levels of the male or female mice who had undergone ethanol sensitization. This suggests that the reversal of sensitization using CBT is not contingent upon corticosterone levels and involves a distinct pathway or mechanism for modulating sensitization. The OXT system contributes to the regulation of stress responses through its interaction with the HPA axis [49], but its potential to alleviate addiction-related behaviors exacerbated by stress has yet to be explored [50]. When it comes to intranasal OXT administration in individuals with AUD, for example, its effects on alcohol craving appear to vary, influenced by the individuals' anxiety levels [51].

The higher levels of corticosterone in the female mice compared to the male mice can likely be attributed to hormonal variations, given that approximately 50% of the female mice were in non-diestrus phases. When we analyzed the corticosterone levels considering the estrous cycle, the non-diestrus phases differed from the diestrus, exhibiting higher levels of corticosterone. These findings underscore the significance of identifying specific phases of the estrous cycle, considering the associated hormonal variations. Diestrus, known for its longer duration, is characterized by a period of quiescence and lower estradiol levels [51–53]. In contrast, the pre-ovulatory period, considered as non-diestrus, is characterized by an increased estradiol secretion [54]. Glucocorticoids and estradiol can mutually influence each other [55], as evidenced by studies demonstrating the enhancement of corticosterone secretion with estradiol administration in female rats [56]. The observed interplay between estradiol and corticosterone highlights the complexity of hormonal regulation during different phases of the estrous cycle.

While we observed no sex-specific differences in ethanol sensitization or the influence of CBT treatment on this response, previous research by Hansson et al. (2018) [15] revealed significant alterations in the OXT system among male, dependent rats and in the post-mortem brains of human individuals with alcohol addiction, but not in the female subjects [16]. The divergent responses in the OXT system between male and female subjects highlight the need for gender-specific considerations in addiction research and treatment approaches.

This study also revealed that the female mice exhibited higher BEC levels compared to the male mice, likely attributed to differences in metabolism, body composition, and alcohol absorption rates between the sexes. In fact, in humans, men typically exhibit higher gastric alcohol dehydrogenase (ADH) activity compared to women, leading to a lower, peak BEC in men compared to women [42]. Similar results have been described for mice [43]. It is also worth noting that the mice with prior ethanol exposure exhibited a lower BEC compared to the alcohol-naïve mice, as previously demonstrated. This phenomenon is likely attributable to the development of pharmacokinetic tolerance. Of greater significance, the treatment with CBT did not result in any changes in the BEC levels, suggesting that the CBT mechanism for reducing ethanol sensitization is unlikely to be linked to alterations in ethanol metabolism.

By using CBT, our research substantiated prior findings seen in studies employing OXT which demonstrated its effectiveness in reducing cue-induced reinstatement response in male, dependent rats [15] and ethanol consumption in various self-administration models in male mice [14]. More recently, King et al. (2021) [57] demonstrated the involvement of endogenous OXT in the hypothalamus in controlling ethanol consumption and suggested that signaling through OXT receptors plays a role in reducing ethanol consumption in a binge-like drinking model. In this study, we further elucidated the effects of CBT administration both 1 h and 24 h before ethanol consumption in a two-bottle choice model. Our results revealed that CBT effectively reduced ethanol intake when the bottles were available for the 24 h session. However, in the 2 h session, only the CBT injection administered 1 h prior to the session demonstrated a significant decrease in ethanol intake. It is important to consider the context of ethanol withdrawal effects on anxiety.

The observed reduction in ethanol intake might indicate that CBT has the potential to alleviate anxiety or craving associated with ethanol withdrawal, contributing to a decrease in consumption. In fact, OXT has been shown to modulate stress, anxiety, and craving behaviors (see Rae et al. 2022 [58] as review). The fact that CBT effectively reduced ethanol intake during the 24 h session, regardless of whether it was administered 1 h or 24 h before the session, suggests that CBT appears to have a sustained efficacy in reducing ethanol intake during prolonged access to ethanol, irrespective of the timing of its administration.

It is important to highlight that the effects of OXT have been tested on other drugs, such as methamphetamine, and that the results are promising, showing a dose-dependent attenuation of motor hyperactivity through OXT administration, an effect blocked by an OXT antagonist [59]. Furthermore, with regard to opioids, OXT has been shown to decrease the acquisition and maintenance of heroin self-administration [60].

Behavioral sensitization to ethanol can also be accompanied by changes in other behaviors, such as reactivity to stress, reward sensitivity, and cognitive function. More research is needed to assess the effects of CBT on other aspects of behavioral sensitization to ethanol.

We can conclude that CBT attenuates the neuroplastic events underlying behavioral sensitization in male and female mice and decreases ethanol intake in male mice. It should be emphasized that the relationship between behavioral sensitization and dependence is still debatable. Although there are studies showing that sensitized animals are more vulnerable to increased ethanol consumption [61], this agreement is not unanimous [62]. Nevertheless, the role of sensitization in the neuroadaptive processes that occur with repeated drug exposure should be considered as a phenomenon related to the psychological desire or "wanting" for the drug [48]. These findings point to CBT as a potential therapeutic tool for addressing alcohol use disorders and reducing the health-toxic risks associated with excessive ethanol consumption.

Supplementary Materials: The following supporting information can be downloaded at https://www.mdpi.com/article/10.3390/toxics11110893/s1: Table S1: Statistical analysis.

Author Contributions: Conceptualization: P.M., M.R. and R.C.; formal analysis, B.Y.C., P.M., R.E. and R.C.; investigation, B.Y.C., L.G.S., M.G.d.A., M.C.d.B. and R.E.; writing B.Y.C., P.M., M.R., M.C.d.B. and R.C.; original supervision, P.M., M.R. and R.C.; funding acquisition, R.C. All authors have read and agreed to the published version of the manuscript.

Funding: The funding for this study was provided with grant #2021/04816-1 from the Fundação de Amparo à Pesquisa do Estado de São Paulo (FAPESP). R.C. is a Research Fellow of the National Council for Scientific and Technological Development (CNPq). B.Y.C. was the recipient of a scholarship from the Fundação de Amparo à Pesquisa do Estado de São Paulo (FAPESP), with grant 2022/05101-9. L.G.S. was the recipient of a scholarship from the National Counsel of Technological and Scientific Development (CNPq/PIBIC), with grant 156686/2018-0. The sponsors had no involvement in the design of this study and in the collection, analyses, and interpretation of the data, nor in the writing of this manuscript and in the decision to submit this article for publication.

Institutional Review Board Statement: All the procedures were approved by the Ethics Committee on the Use of Animals of the Institute of Biomedical Sciences (University of Sao Paulo) (CEUA—ICB/USP), under CEUA numbers 9998280518 and 4512140222 and protocol 25/2016, in accordance with Law 11,794 of 8 October 2008, Decree 6899 of 15 July 2009, as well as with the rules issued by the National Council for Control of Animal Experimentation (CONCEA). Efforts were made to minimize pain and suffering and reduce the use of animals.

Informed Consent Statement: Not applicable.

Data Availability Statement: The data presented in this study are available on request from the corresponding author.

Conflicts of Interest: The authors declare no conflict of interest.

References

1. World Health Organization. *Global Status Report on Alcohol and Health 2018*; World Health Organization: Geneva, Switzerland, 2018.
2. GBD 2016 Alcohol and Drug Use Collaborators. The global burden of disease attributable to alcohol and drug use in 195 countries and territories, 1990–2016: A systematic analysis for the Global Burden of Disease Study 2016. *Lancet Psychiatry* **2018**, *5*, 987–1012. [CrossRef] [PubMed]
3. Osna, N.A.; Donohue, T.M., Jr.; Kharbanda, K.K. Alcoholic Liver Disease: Pathogenesis and Current Management. *Alcohol. Res.* **2017**, *38*, 147–161. [PubMed]
4. El-Mas, M.M.; Abdel-Rahman, A.A. Role of Alcohol Oxidative Metabolism in Its Cardiovascular and Autonomic Effects. *Adv. Exp. Med. Biol.* **2019**, *1193*, 1–33. [PubMed]

5. Vijayraghavan, S.; Porcher, L.; Mieczkowski, P.A.; Saini, N. Acetaldehyde makes a distinct mutation signature in single-stranded DNA. *Nucleic Acids Res.* **2022**, *50*, 7451–7464. [CrossRef]
6. World Health Organization. *World Health Statistics 2023: Monitoring Health for the SDGs, Sustainable Development Goals*; World Health Organization: Geneva, Switzerland, 2023.
7. Goh, C.M.J.; Asharani, P.V.; Abdin, E.; Shahwan, S.; Zhang, Y.; Sambasivam, R.; Vaingankar, J.A.; Ma, S.; Chong, A.S.; Subramaniam, M. Gender Differences in Alcohol Use: A Nationwide Study in a Multiethnic Population. *Int. J. Ment. Health Addict.* **2022**, 1–15. [CrossRef]
8. Sudhinaraset, M.; Wigglesworth, C.; Takeuchi, D.T. Social and Cultural Contexts of Alcohol Use: Influences in a Social-Ecological Framework. *Alcohol. Res.* **2016**, *38*, 35–45.
9. White, A.M. Gender Differences in the Epidemiology of Alcohol Use and Related Harms in the United States. *Alcohol. Res.* **2020**, *40*, 1. [CrossRef]
10. Camarini, R. Mesenchymal stem cells as new perspective for the treatment of alcohol use disorder. *Gene Ther.* **2019**, *27*, 471–473. [CrossRef]
11. Green, C.A. Gender and use of substance abuse treatment services. *Alcohol. Res. Health* **2006**, *29*, 55–62.
12. Gimpl, G.; Fahrenholz, F. The oxytocin receptor system: Structure, function, and regulation. *Physiol. Rev.* **2001**, *81*, 629–683. [CrossRef]
13. Ryabinin, A.E.; Fulenwider, H.D. Alcohol and oxytocin: Scrutinizing the relationship. *Neurosci. Biobehav. Rev.* **2021**, *127*, 852–864. [CrossRef] [PubMed]
14. King, C.E.; Griffin, W.C.; Luderman, L.N.; Kates, M.M.; McGinty, J.F.; Becker, H.C. Oxytocin reduces ethanol self-administration in mice. *Alcohol. Clin. Exp. Res.* **2017**, *41*, 955–964. [CrossRef] [PubMed]
15. Hansson, A.C.; Koopmann, A.; Uhrig, S.; Bühler, S.; Domi, E.; Kiessling, E.; Ciccocioppo, R.; Froemke, R.C.; Grinevich, V.; Kiefer, F.; et al. Oxytocin Reduces Alcohol Cue-Reactivity in Alcohol-Dependent Rats and Humans. *Neuropsychopharmacology* **2018**, *43*, 1235–1246. [CrossRef] [PubMed]
16. Hansson, A.C.; Spanagel, R. No changes in the oxytocin system in alcohol-dependent female rodents and humans: Towards a sex-specific psychopharmacology in alcoholism. *Addict. Biol.* **2021**, *26*, e12945. [CrossRef]
17. Rodriguez, K.M.; Smith, B.L.; Caldwell, H.K. Voluntary alcohol consumption is increased in female, but not male, oxytocin receptor knockout mice. *Brain Behav.* **2020**, *10*, e01749. [CrossRef]
18. Bardo, M.T.; Bevins, R.A. Conditioned place preference: What does it add to our preclinical understanding of drug reward? *Psychopharmacology* **2000**, *153*, 31–43. [CrossRef]
19. Carrara-Nascimento, P.F.; Hoffmann, L.B.; Flório, J.C.; Planeta, C.S.; Camarini, R. Effects of Ethanol Exposure During Adolescence or Adulthood on Locomotor Sensitization and Dopamine Levels in the Reward System. *Front. Behav. Neurosci.* **2020**, *14*, 31. [CrossRef]
20. Quigley, J.A.; Logsdon, M.K.; Turner, C.A.; Gonzalez, I.L.; Leonardo, N.B.; Becker, J.B. Sex differences in vulnerability to addiction. *Neuropharmacology* **2021**, *187*, 108491. [CrossRef]
21. Masur, J.; Boerngen, R. The excitatory component of ethanol in mice: A chronic study. *Pharmacol. Biochem. Behav.* **1980**, *13*, 777–780. [CrossRef]
22. Camarini, R.; Marianno, P.; Rae, M. Chapter three—Social Factors in Ethanol Sensitization. *Int. Rev. Neurobiol.* **2018**, *140*, 53–80.
23. Morley-Fletcher, S.; Rea, M.; Maccari, S.; Laviola, G. Environmental enrichment during adolescence reverses the effects of prenatal stress on play behavior and HPA axis reactivity in rats. *Eur. J. Neurosci.* **2003**, *18*, 3367–3374. [CrossRef] [PubMed]
24. Santos-Rocha, J.B.; Rae, M.; Teixeira, A.M.A.; Teixeira, S.A.; Munhoz, C.D.; Muscará, M.N.; Marcourakis, T.; Szumlinski, K.K.; Camarini, R. Involvement of neuronal nitric oxide synthase in cross-sensitization between chronic unpredictable stress and ethanol in adolescent and adult mice. *Alcohol* **2018**, *68*, 71–79. [CrossRef] [PubMed]
25. dos Santos, J.R.B.; Rae, M.; Teixeira, S.A.; Muscará, M.N.; Szumlinski, K.K.; Camarini, R. The effect of MK-801 on stress-ethanol cross-sensitization is dissociable from its effects on nNOS activity. *Alcohol* **2023**, *112*, 31–39. [CrossRef] [PubMed]
26. Neumann, I.D.; Krömer, S.A.; Toschi, N.; Ebner, K. Brain oxytocin inhibits the (re)activity of the hypothalamo-pituitary-adrenal axis in male rats: Involvement of hypothalamic and limbic brain regions. *Regul. Pept.* **2000**, *96*, 31–38. [CrossRef] [PubMed]
27. Windle, R.J.; Shanks, N.; Lightman, S.L.; Ingram, C.D. Central oxytocin administration reduces stress-induced corticosterone release and anxiety behavior in rats. *Endocrinology* **1997**, *138*, 2829–2834. [CrossRef]
28. Georgiou, P.; Zanos, P.; Garcia-Carmona, J.A.; Hourani, S.; Kitchen, I.; Kieffer, B.L.; Laorden, M.L.; Bailey, A. The oxytocin analogue carbetocin prevents priming-induced reinstatement of morphine-seeking: Involvement of dopaminergic, noradrenergic and MOPr systems. *Eur. Neuropsychopharmacol.* **2015**, *25*, 2459–2464. [CrossRef]
29. Zanos, P.; Georgiou, P.; Wright, S.R.; Hourani, S.M.; Kitchen, I.; Winsky-Sommerer, R.; Bailey, A. The oxytocin analogue carbetocin prevents emotional impairment and stress-induced reinstatement of opioid-seeking in morphine-abstinent mice. *Neuropsychopharmacology* **2014**, *39*, 855–865. [CrossRef]
30. Carrara-Nascimento, P.F.; Griffin III, W.C.; Pastrello, D.M.; Olive, M.F.; Camarini, R. Changes in extracellular levels of glutamate in the nucleus accumbens after ethanol-induced behavioral sensitization in adolescent and adult mice. *Alcohol* **2011**, *45*, 451–460. [CrossRef]
31. Didne, V.; van Ingelgom, T.; Tirelli, E.; Quertemont, E. Long-term exposure to daily ethanol injections in DBA/2J and Swiss mice: Lessons for the interpretation of ethanol sensitization. *PLoS ONE* **2019**, *14*, e0214696. [CrossRef]

32. Le, A.D.; Ko, J.; Chow, S.; Quan, B. Alcohol consumption by C57BL/6, BALB/c, and DBA/2 mice in a limited access paradigm. *Pharmacol. Biochem. Behav.* **1994**, *47*, 375–378. [CrossRef]
33. Camarini, R.; Hodge, C.W. Ethanol preexposure increases ethanol self-administration in C57BL/6J and DBA/2J mice. *Pharmacol. Biochem. Behav.* **2004**, *79*, 623–632. [CrossRef] [PubMed]
34. Crabbe, J.C.; Phillips, T.J. Pharmacogenetic studies of alcohol self-administration and withdrawal. *Psychopharmacology* **2004**, *174*, 539–560. [CrossRef] [PubMed]
35. Passoni, I.; Leonzino, M.; Gigliucci, V.; Chini, B.; Busnelli, M. Carbetocin is a Functional Selective Gq Agonist That Does Not Promote Oxytocin Receptor Recycling After Inducing β-Arrestin-Independent Internalisation. *J. Neuroendocrinol.* **2016**, *28*, 12363. [CrossRef] [PubMed]
36. Caligioni, C.S. Assessing reproductive status/stages in mice. *Curr. Protoc. Neurosci.* **2009**. Appendix 4: Appendix 4I. [CrossRef]
37. Phillips, T.J.; Huson, M.; Gwiazdon, C.; Burkhart-Kasch, S.; Shen, E.H. Effects of acute and repeated ethanol exposures on the locomotor activity of BXD recombinant inbred mice. *Alcohol. Clin. Exp. Res.* **1995**, *19*, 269–278. [CrossRef]
38. Legastelois, R.; Botia, B.; Naassila, M. Sensitization to the stimulant motor effects of ethanol is not dependent on tolerance to ataxic or sedative properties of ethanol in female mice. *Drug Alcohol. Depend.* **2015**, *3*, 4. [CrossRef]
39. Rueda, A.V.L.; Teixeira, A.M.A.; Yonamine, M.; Camarini, R. Environmental enrichment blocks ethanol-induced locomotor sensitization and decreases BDNF levels in the prefrontal cortex in mice. *Addict. Biol.* **2012**, *17*, 736–745. [CrossRef]
40. Rhodes, J.S.; Best, K.; Belknap, J.K.; Finn, D.A.; Crabbe, J.C. Evaluation of a simple model of ethanol drinking to intoxication in C57BL/6J mice. *Physiol. Behav.* **2005**, *84*, 53–63. [CrossRef]
41. Marianno, P.; Abrahao, K.P.; Camarini, R. Environmental enrichment blunts ethanol consumption after restraint stress in C57BL/6 mice. *PLoS ONE* **2017**, *12*, e0170317. [CrossRef]
42. Segal, D.S.; Mandell, A.J. Long-Term Administration of d-Amphetamine: Progressive Augmentation of Motor Activity and Stereotypy. *Pharmacol. Biochem. Behav.* **1974**, *2*, 249–255. [CrossRef]
43. Segal, D.S.; Geyer, M.A.; Schuckit, M.A. Stimulant-induced psychosis: An evaluation of animal methods. *Essays Neurochem. Neuropharmacol.* **1981**, *5*, 95–129. [PubMed]
44. Phillips, T.J.; Roberts, A.J.; Lessov, C.N. Behavioral sensitization to ethanol: Genetics and the effects of stress. *Pharmacol. Biochem. Behav.* **1997**, *57*, 487–493. [CrossRef] [PubMed]
45. Richtand, N. Behavioral Sensitization, Alternative Splicing, and D3 Dopamine Receptor-Mediated Inhibitory Function. *Neuropsychopharmacollogy* **2006**, *31*, 2368–2375. [CrossRef]
46. Rae, M.B.; Zanos, P.; Georgiou, P.; Chivers, P.; Bailey, A.; Camarini, R. Environmental enrichment enhances conditioned place preference to ethanol via an oxytocinergic-dependent mechanism in male mice. *Neuropharmacology* **2018**, *138*, 267–274. [CrossRef] [PubMed]
47. Kuhn, B.N.; Kalivas, P.W.; Bobadilla, A.C. Understanding Addiction Using Animal Models. *Front. Behav. Neurosci.* **2019**, *13*, 262. [CrossRef] [PubMed]
48. Robinson, T.E.; Berridge, K.C. The neural basis of drug craving: An incentive-sensitization theory of addiction. *Brain Res. Rev.* **1993**, *18*, 247–391. [CrossRef]
49. Stephens, M.A.; Wand, G. Stress and the HPA axis: Role of glucocorticoids in alcohol dependence. *Alcohol Res.* **2012**, *34*, 468–483.
50. McGinty, G.; Hyland, P.; Shevlin, M. Trauma Response and Psychosis: Investigating the Association between PTSD Symptomology and Psychotic Experiences. *Eur. J. Psychotraumatology* **2019**, *10*, 2–4.
51. Mitchell, J.M.; Arcuni, P.A.; Weinstein, D.; Woolley, J.D. Intranasal Oxytocin Selectively Modulates Social Perception, Craving, and Approach Behavior in Subjects With Alcohol Use Disorder. *J. Addict. Med.* **2016**, *10*, 182–189. [CrossRef]
52. Greaves, P. Chapter 12—Female Genital Tract. In *Histopathology of Preclinical Toxicity Studies*, 4th ed.; Academic Press: Boston, MA, USA, 2012; pp. 667–723.
53. Wangikar, P.; Ahmed, T.; Vangala, S. Chapter 76: Toxicology Pathology of reproductive system. In *Reproductive and Developmental Toxicology*, 1st ed.; Gupta, R.C., Ed.; Elsevier: Amsterdam, The Netherlands, 2011; pp. 1003–1026.
54. Seligowski, A.V.; Hurly, J.; Mellen, E.; Ressler, K.J.; Ramikie, T.S. Translational studies of estradiol and progesterone in fear and PTSD. *Eur. J. Psychotraumatol.* **2020**, *11*, 1723857. [CrossRef]
55. Ycaza Herrera, A.; Mather, M. Actions and interactions of estradiol and glucocorticoids in cognition and the brain: Implications for aging women. *Neurosci. Biobehav. Rev.* **2015**, *55*, 36–52. [CrossRef]
56. Seale, J.V.; Wood, S.A.; Atkinson, H.C.; Harbuz, M.S.; Lightman, S.L. Gonadal steroid replacement reverses gonadectomy-induced changes in the corticosterone pulse profile and stress-induced hypothalamic-pituitary-adrenal axis activity of male and female rats. *J. Neuroendocrinol.* **2004**, *16*, 989–998. [CrossRef] [PubMed]
57. King, C.E.; Griffin, W.C.; Lopez, M.F.; Becker, H.C. Activation of hypothalamic oxytocin neurons reduces binge-like alcohol drinking through signaling at central oxytocin receptors. *Neuropsychopharmacology* **2021**, *46*, 1950–1957. [CrossRef] [PubMed]
58. Rae, M.; Lemos Duarte, M.; Gomes, I.; Camarini, R.; Devi, L.A. Oxytocin and vasopressin: Signalling, behavioural modulation and potential therapeutic effects. *Br. J. Pharmacol.* **2022**, *179*, 1544–1564. [CrossRef]
59. Qi, J.; Yang, J.Y.; Song, M.; Li, Y.; Wang, F.; Wu, C.F. Inhibition by oxytocin of methamphetamine-induced hyperactivity related to dopamine turnover in the mesolimbic region in mice. *Naunyn Schmiedebergs Arch. Pharmacol.* **2008**, *376*, 441–448. [CrossRef] [PubMed]

60. Kovacs, G.L.; Sarnyai, Z.; Babarczi, E.; Szabo, G.; Telegdy, G. The role of oxytocin-dopamine interactions in cocaine-induced locomotor hyperactivity. *Neuropharmacology* **1990**, *29*, 365–366. [CrossRef]
61. Abrahao, K.P.; Ariwodola, O.J.; Butler, T.R.; Rau, A.R.; Skelly, M.J.; Carter, E.; Alexander, N.P.; McCool, B.A.; Souza-Formigoni, M.L.; Weiner, J.L. Locomotor sensitization to ethanol impairs NMDA receptor-dependent synaptic plasticity in the nucleus accumbens and increases ethanol self-administration. *J. Neurosci.* **2013**, *33*, 4834–4842. [CrossRef]
62. Ribeiro, A.F.; Pigatto, G.; Goeldner, F.O.; Lopes, J.F.; de Lacerda, R.B. Lack of relation between drug-seeking behavior in an addiction model and the expression of behavioral sensitization in response to ethanol challenge in mice. *J. Neural Transm.* **2008**, *115*, 43–54. [CrossRef]

toxics

MDPI

Article

Effects of Chronic Hydrocodone Exposure and Ceftriaxone on the Expression of Astrocytic Glutamate Transporters in Mesocorticolimbic Brain Regions of C57/BL Mice

Woonyen Wong and Youssef Sari *

Department of Pharmacology and Experimental Therapeutics, College of Pharmacy and Pharmaceutical Sciences, The University of Toledo, Toledo, OH 43614, USA; woonyen.wong@rockets.utoledo.edu
* Correspondence: youssef.sari@utoledo.edu; Tel.: +1-419-383-1507

Abstract: Exposure to opioids can lead to the alteration of several neurotransmitters. Among these neurotransmitters, glutamate is thought to be involved in opioid dependence. Glutamate neurotransmission is mainly regulated by astrocytic glutamate transporters such as glutamate transporter 1 (GLT-1) and cystine/glutamate antiporter (xCT). Our laboratory has shown that exposure to lower doses of hydrocodone reduced the expression of xCT in the nucleus accumbens (NAc) and the hippocampus. In the present study, we investigated the effects of chronic exposure to hydrocodone, and tested ceftriaxone as a GLT-1 upregulator in mesocorticolimbic brain regions such as the NAc, the amygdala (AMY), and the dorsomedial prefrontal cortex (dmPFC). Eight-week-old male mice were divided into three groups: (1) the saline vehicle control group; (2) the hydrocodone group; and (3) the hydrocodone + ceftriaxone group. Mice were injected with hydrocodone (10 mg/kg, i.p.) or saline for 14 days. On day seven, the hydrocodone/ceftriaxone group was injected with ceftriaxone (200 mg/kg, i.p.) for last seven days. Chronic exposure to hydrocodone reduced the expression of GLT-1, xCT, protein kinase B (AKT), extracellular signal-regulated kinases (ERK), and c-Jun N-terminal Kinase (JNK) in NAc, AMY, and dmPFC. However, hydrocodone exposure increased the expression of G-protein-coupled metabotropic glutamate receptors (mGluR5) in the NAc, AMY, and dmPFC. Importantly, ceftriaxone treatment normalized the expression of mGluR5, GLT-1, and xCT in all these brain regions, except for xCT in the AMY. Importantly, ceftriaxone treatment attenuated hydrocodone-induced downregulation of signaling pathways such as AKT, ERK, and JNK expression in the NAc, AMY, and dmPFC. These findings demonstrate that ceftriaxone has potential therapeutic effects in reversing hydrocodone-induced downregulation of GLT-1 and xCT in selected reward brain regions, and this might be mediated through the downstream kinase signaling pathways such as AKT, ERK, and JNK.

Keywords: opioids; ceftriaxone; glutamate; GLT-1; xCT; kinases; drug exposure; mesocorticolimbic

Citation: Wong, W.; Sari, Y. Effects of Chronic Hydrocodone Exposure and Ceftriaxone on the Expression of Astrocytic Glutamate Transporters in Mesocorticolimbic Brain Regions of C57/BL Mice. *Toxics* **2023**, *11*, 870. https://doi.org/10.3390/toxics11100870

Academic Editor: Guido Cavaletti

Received: 22 August 2023
Revised: 26 September 2023
Accepted: 18 October 2023
Published: 20 October 2023

1. Introduction

Opioid use disorder (OUD) has been considered as a major health issue in the United States [1]. Opioids are commonly used to treat chronic pain; however, their misuses are associated with the development of dependence, and overdose leading to deaths. There are several classes of opioids, and hydrocodone is considered a semi-synthetic opioid, which is widely used in the management of chronic pain associated with surgery procedures and musculoskeletal injuries [2]. Hydrocodone exerts its analgesic effect by activating the mu-opioid receptor, a G-protein coupled receptors (GPCR), which can inhibit the production of cyclic adenosine monophosphate (cAMP) leading to the activation of a G-protein-gated inwardly rectifying potassium channel (GIRK) [3]. Furthermore, opioids dysregulate several neurotransmitters, including glutamate [4]. Indeed, the activation of opioid receptors (mainly mu receptor) induced release of glutamate in the nucleus accumbens (NAc) core, and this effect was mainly observed in astrocytes [5], which highly express the major glutamate transporter type 1 (GLT-1) and the cystine/glutamate antiporter (xCT) [6–9].

Glutamate homeostasis is dysregulated by exposure to substances of abuse, including alcohol, nicotine, cocaine, and methamphetamine, and this effect has been associated with the downregulation of GLT-1 expression in several reward brain regions such as the NAc, dorsomedial prefrontal cortex (dmPFC), amygdala (AMY), and hippocampus [10–15]. In addition, xCT was also found to be downregulated during NAc exposure to substances of abuse such as cocaine, alcohol, and nicotine [16–19]. Importantly, chronic exposure to opioids alters glutamate transport and glutamate clearance. For example, chronic exposure to morphine downregulated several glutamate transporters, including GLT-1 [20,21]; this might be associated with an increase in extracellular glutamate concentrations in the brain [20,22]. Regarding xCT, a study from our laboratory showed that hydrocodone administered at a lower dose decreased the expression of this protein in the NAc and hippocampus in an animal model of conditioned place preference (CPP) [23]. Importantly, ceftriaxone, a beta-lactam antibiotic known to upregulate GLT-1 and xCT [9,24], was shown to attenuate the effect of hydrocodone-induced downregulation in xCT in these brain regions [23]. These studies and others clearly demonstrate that chronic exposure to drugs of abuse downregulated the expression of GLT-1 and xCT and increased extracellular glutamate concentrations in the central reward brain regions, and CEF and other beta-lactams have the potential to attenuate these effects.

In this study, we investigated the effects of chronic exposure to a higher dose of hydrocodone (10 mg/kg) on the expression of GLT-1 and xCT in certain central reward brain regions such as the NAc, AMY, and dmPFC. Importantly, we determined whether ceftriaxone treatment would normalize the expression of these glutamate transporters. We further investigated whether chronic exposure to hydrocodone affects the expression of mGluR5, and determined whether ceftriaxone would attenuate this effect. Finally, we aimed to investigate the signaling pathways involved in hydrocodone-induced changes in GLT-1 and xCT expression. We focused on the expression of signaling pathways such as ERK, JNK, and Akt since some of these kinases are suggested to be involved in the regulatory effect of ceftriaxone in GLT-1 expression in the brain [25–27].

2. Materials and Methods

2.1. Animal

Male C57BL/6 mice (Jackson Laboratory, 25–30 g, 8 weeks of age, $n = 7$) were used in this study. This study tested a total of 21 male mice. Mice were housed in a room that was maintained at 21 °C on a 12/12 h light/dark cycle. Mice had free access to water and food. All experimental procedures were approved by the Institutional Animal Care and Use Committee (IACUC), University of Toledo. This is in accordance with the guidelines set by the National Institutes of Health for the use of animals in research as described in the Guide for the Care and Use of Laboratory Animals under approved protocol number 400155 (2 August 2022), The University of Toledo.

2.2. Drugs and Dosing

Male mice were handled three days before starting the experiment for acclimation. Mice were then divided into three groups: (1) saline group, mice were intraperitoneal (i.p.) injected with saline vehicle from day 1–14 ($n = 7$); (2) hydrocodone group, mice were i.p. injected hydrocodone (10 mg/kg) from day 1–14 ($n = 7$); and (3) hydrocodone + ceftriaxone group, mice were i.p. injected hydrocodone (10 mg/kg) from day 1–14, and ceftriaxone (200 mg/kg) was i.p. injected from day 7–14 ($n = 7$) (Figure 1). Hydrocodone (Sigma-Aldrich, St. Louis, MO, USA) was dissolved in saline at 10 mg/kg. Ceftriaxone was purchased from (Pfizer, Lake Forest, IL, USA) and was dissolved in saline at 200 mg/kg. Note that equal volumes of saline (control group) and hydrocodone (10 mg/kg, i.p., hydrocodone group) were injected from day 1 through 14; ceftriaxone (200 mg/kg, i.p.) was administered from day 7 through 14 for the ceftriaxone/hydrocodone group. Mice were euthanized on day 15 by CO_2 inhalation as approved by UT-IACUC (Figure 1).

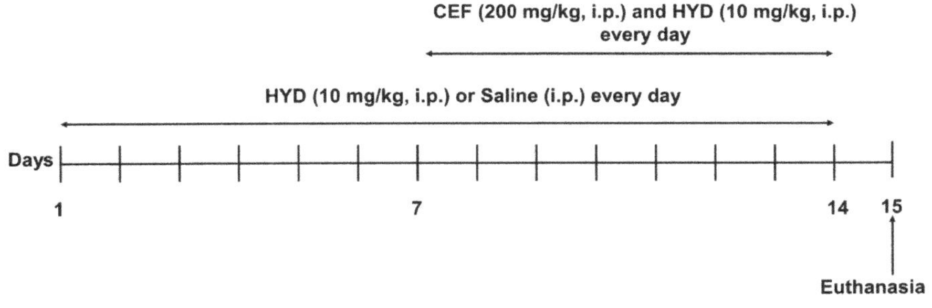

Figure 1. Timeline of the experimental procedure. CEF: ceftriaxone; HYD: hydrocodone.

2.3. Brain Tissue Extraction

Animals were euthanized by CO_2 inhalation after 7 h of fasting on day 15. Fasting was applied for further study, which aims to investigate potential changes in liver tissues. For the present study, brains were isolated and frozen immediately on dry ice and stored at −80 °C. NAc (core and shell), dmPFC (cingulate cortex and prelimbic cortex), HIP (cornu ammonis, CA, subfield: CA1, CA2, and CA3), and AMY (central amygdala, basomedial amygdala and basolateral amygdala) were extracted using a cryostat machine (Leica CM1950). All brain regions were selected using the Brain Mice Atlas [28]. Brain samples were stored at −80 °C for subsequent Western blot analyses.

2.4. Western Blot Analysis

Western blot was used to determine protein expression of phospho-ERK, ERK, phospho-Akt, Akt, phospho-JNK, JNK, xCT, GLT-1, mGluR5, and β-tubulin in the NAc (core and shell), AMY, and dmPFC. Samples were lysed using a lysis buffer (50 mM Tris–HCl, 150 mM NaCl, 1 mM EDTA, 0.5% NP-40, 1% Triton, 0.1% SDS) with phosphatase and protease inhibitors. The amount of protein in each tissue sample was quantified using a detergent compatible protein assay (Bio-Rad, Hercules, CA, USA). An equal amount of protein from each sample was mixed with laemmili dye, and the mixtures were loaded onto 10% Tris-glycerine gel to separate the protein using electrophoresis. Then, proteins were transferred from gels to a polyvinylidene difluoride (PVDF) membrane. Subsequently, the PVDF membranes were blocked with 5% fat-free milk in Tris-buffered saline with Tween 20 (TBST) at room temperature for 30 min. Membranes were incubated overnight at 4 °C with primary antibodies: rabbit anti-*p*-ERK (1:1000, Abcam, Cambridge, UK, ab201015), rabbit anti-ERK (1:1000, Abcam, ab17942), rabbit anti-*p*-JNK (1:1000, Cell Signaling, Danvers, MA, USA, 9251), rabbit anti-JNK (1:1000, Cell Signaling, 9252), rabbit anti-*p*-Akt (1:1000, Cell Signaling, 4060), rabbit anti-Akt (1:1000, Cell Signaling, 4691), rabbit anti-GLT-1 (1:5000, Abcam ab205248), rabbit anti-xCT (1:1000, Abcam ab125186), and rabbit anti-mGluR5 (1:1000, Abcam ab76316). Mouse anti-β-tubulin (1:1000, BioLeagend, San Diego, CA, USA) was used as a control loading protein. On the following day, membranes were washed five times with TBST and incubated with the match secondary antibody (1:4000) for 60 min. The membranes were then washed with TBST and dried for further analysis. The dried membranes were incubated with chemiluminescent reagents (Super Signal West Pico, Perce Inc., Appleton, WI, USA) for 1–2 min. Digitized blot images were developed using the GeneSys imaging system. Quantification and analysis of the expression of *p*-ERK, ERK, *p*-JNK, JNK, *p*-Akt, Akt, GLT-1, xCT, mGluR5, and β-tubulin blots were performed using ImageJ software (Version 1.53t 24). The control group was reported as 100% to measure the changes in the expression of proteins of interest in the NAc (core and shell), AMY, and dmPFC as described in our previous studies [23,29].

2.5. Statistical Analysis

All statistical analyses were performed using GraphPad Prism software (Version 10). One-way ANOVA with Newman–Keuls as a post-hoc multiple comparison test was used to analyze the Western blot data as a percentage (relative to control values) ratio to the loading protein, β-tubulin. The data are reported for a $p < 0.05$ level of significance.

3. Results

3.1. Effect of Chronic Hydrocodone Exposure and Ceftriaxone on GLT-1 Protein Expression in the NAC, AMY, and dmPFC

Data analyses revealed a significant difference in the expression of GLT-1 in NAc among all tested groups ($F_{2,13} = 11.16$, $p < 0.01$, Figure 2). Newman–Keuls post-hoc analyses showed a significant decrease in GLT-1 expression in the NAc in the hydrocodone group compared to the control group ($p < 0.05$), and ceftriaxone ($p < 0.01$) significantly increased GLT-1 expression in the NAc as compared to the hydrocodone group (Figure 2A). In addition, statistical analysis revealed a significant difference in the expression of GLT-1 in the AMY ($F_{2,13} = 11.16$, $p < 0.01$, Figure 2B) and the dmPFC ($F_{2,15} = 29.24$, $p < 0.0001$, Figure 2C) among all tested groups. Newman–Keuls post-hoc analyses showed a significant decrease in GLT-1 expression in the AMY ($p < 0.05$, Figure 2B) and the dmPFC ($p < 0.01$, Figure 2C) of the hydrocodone group compared to the control group. Importantly, ceftriaxone significantly increased GLT-1 expression in the AMY ($p < 0.01$, Figure 2B) and dmPFC ($p < 0.0001$, Figure 2C) as compared to the hydrocodone group. The hydrocodone–ceftriaxone groups showed significantly increased GLT-1 expression compared to the control group in the AMY ($p < 0.05$, Figure 2B) and the dmPFC ($p < 0.05$, Figure 2C). However, no significant changes were observed between the control and hydrocodone-ceftriaxone groups in the NAc (Figure 2).

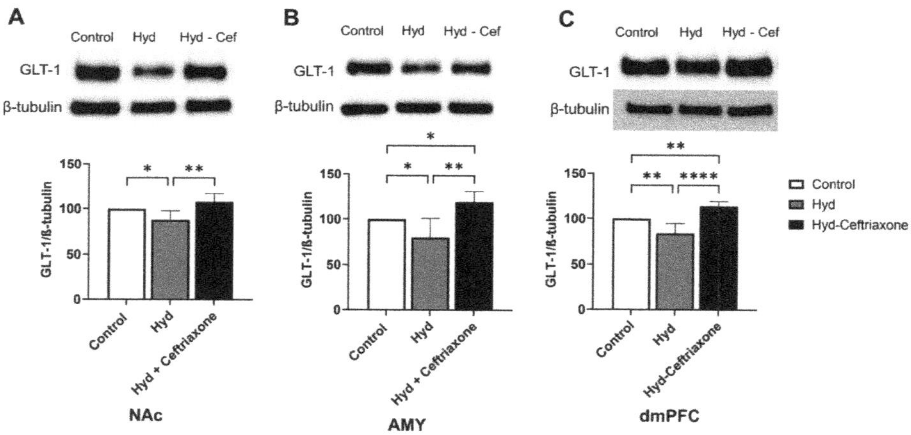

Figure 2. Effect of chronic hydrocodone exposure on the expression of GLT-1 in the NAc, AMY, and dmPFC. (**A**) Immunoblots for GLT-1 and β-tubulin in the NAc. Quantitative analysis using one-way ANOVA followed by Newman–Keuls post-hoc test indicated that GLT-1 was significantly downregulated in the hydrocodone group compared to the control group, while post-treatment with ceftriaxone (200 mg/kg) upregulated GLT-1 expression compared to the hydrocodone group in the NAc. (**B**) Immunoblots for GLT-1 and β-tubulin in the AMY. Quantitative analysis using one-way ANOVA followed by Newman–Keuls post-hoc test showed that GLT-1 was significantly downregulated in the hydrocodone group compared to the control group, while post-treatment with ceftriaxone (200 mg/kg) upregulated GLT-1 expression compared to the hydrocodone group in the AMY. (**C**) Immunoblots for GLT-1 and β-tubulin in the dmPFC. Quantitative analysis using one-way ANOVA followed by Newman–Keuls post-hoc test indicated that GLT-1 expression was significantly

downregulated in the hydrocodone group compared to the control group, while ceftriaxone (200 mg/kg) upregulated GLT-1 expression compared to the hydrocodone group in the dmPFC. Control group data were represented as 100%. Each column is expressed as mean ± S.E.M ($n = 7$/group), (* $p < 0.05$, ** $p < 0.01$, and **** $p < 0.0001$).

3.2. Effect of Chronic Hydrocodone Exposure and Ceftriaxone on xCT Protein Expression in the NAc, AMY, and dmPFC

We further investigated the effects of chronic hydrocodone exposure and ceftriaxone treatment on the expression of xCT in mesocorticolimbic brain regions. There were significant differences in xCT expression in the NAc ($F_{2,12} = 8.364$, $p < 0.01$, Figure 3A) and dmPFC ($F_{2,15} = 15.07$, $p < 0.001$, Figure 3C) among all tested groups. However, there were no significant changes in xCT expression between all tested groups in the AMY ($F_{2,18} = 0.03247$, $p > 0.05$, Figure 3B). Newman–Keuls post-hoc analyses revealed significant decreases in xCT expression in the NAc ($p < 0.05$, Figure 3A) and the dmPFC ($p < 0.001$, Figure 3C) in the hydrocodone group as compared to the control group. Importantly, ceftriaxone attenuated hydrocodone-induced downregulation in the NAc ($p < 0.01$, Figure 3A) and dmPFC ($p < 0.001$, Figure 3C). Quantitative analysis revealed non-significant differences in xCT expression among control and hydrocodone–ceftriaxone groups in the NAc, AMY, and dmPFC (Figure 3).

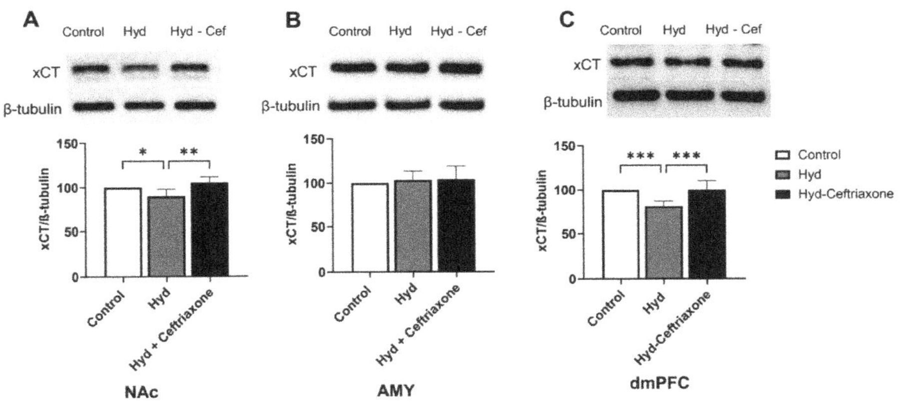

Figure 3. Effect of chronic hydrocodone exposure on xCT expression in the NAc, AMY, and dmPFC. (**A**) Immunoblots for xCT and β-tubulin in the NAc. Quantitative analysis using one-way ANOVA followed by Newman–Keuls post-hoc test indicated that xCT expression was significantly decreased in the hydrocodone group compared to the control group, while ceftriaxone (200 mg/kg) normalized GLT-1 expression compared to the hydrocodone group in the NAc. (**B**) Immunoblots for xCT and β-tubulin in the AMY. Quantitative analysis using one-way ANOVA followed by Newman–Keuls post-hoc test indicated that there were no significant differences in xCT expression among all tested groups in the AMY. (**C**) Immunoblots for xCT and β-tubulin in dmPFC. Quantitative analysis using one-way ANOVA followed by Newman–Keuls post-hoc test showed that xCT was significantly decreased in the hydrocodone group compared to the control group, while post-treatment with ceftriaxone (200 mg/kg) normalized xCT expression compared to the hydrocodone group in the dmPFC. Control group data were represented as 100%. Each column is expressed as mean ± S.E.M ($n = 7$/group), (* $p < 0.05$, ** $p < 0.01$ and *** $p < 0.001$).

3.3. Effects of Chronic Hydrocodone Exposure and Ceftriaxone on p-ERK Protein Expression in the NAc, AMY, and dmPFC

We also investigated the effects of ceftriaxone on kinase signaling pathways such as *p*-ERK in mesocorticolimbic brain regions. Western blot data analyses revealed significant differences in *p*-ERK expression in the NAc ($F_{2,14} = 19.73$, $p < 0.0001$, Figure 4A), AMY

($F_{2,13}$ = 7.141, $p < 0.01$, Figure 4B), and dmPFC ($F_{2,14}$ = 12.44, $p < 0.001$, Figure 4C) among all groups. Newman–Keuls post hoc analyses revealed that chronic hydrocodone exposure decreased *p*-ERK expression in the NAc ($p < 0.0001$, Figure 4A), AMY ($p < 0.05$, Figure 4B), and dmPFC ($p < 0.001$, Figure 4C) compared to the control group. Importantly, ceftriaxone attenuated hydrocodone-induced downregulation of *p*-ERK in the NAc ($p < 0.001$, Figure 4A), AMY ($p < 0.01$, Figure 4B) and the dmPFC ($p < 0.01$, Figure 4C). No significant changes were detected between the hydrocodone and hydrocodone–ceftriaxone groups.

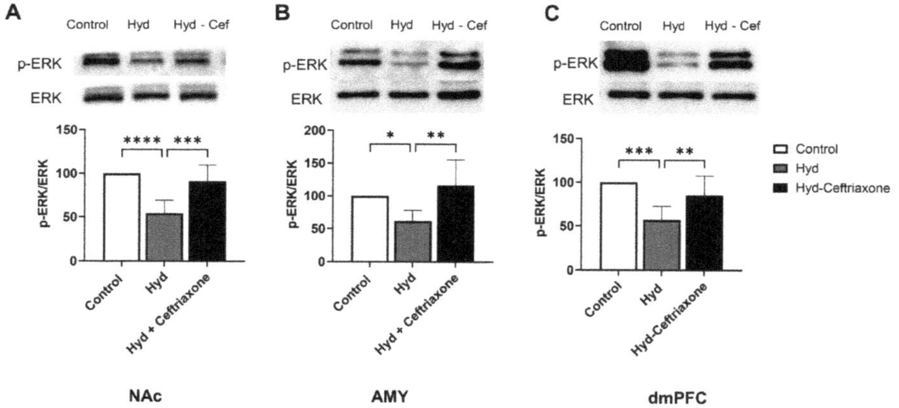

Figure 4. Effect of chronic hydrocodone exposure on *p*-ERK expression in the NAc, AMY, and dmPFC. (**A**) Immunoblots for *p*-ERK and ERK in the NAc. Quantitative analysis using one-way ANOVA followed by Newman–Keuls post hoc test revealed that chronic hydrocodone exposure downregulated *p*-ERK expression in the NAc compared to the control group, while ceftriaxone (200 mg/kg) upregulated *p*-ERK expression compared to the hydrocodone group. (**B**) Immunoblots for *p*-ERK and ERK in the AMY. Quantitative analysis using one-way ANOVA followed by Newman–Keuls post hoc test showed that *p*-ERK was significantly downregulated in the hydrocodone group compared to the control group, while post-treatment with ceftriaxone (200 mg/kg) upregulated *p*-ERK expression in the AMY compared to the hydrocodone group. (**C**) Immunoblots for *p*-ERK and ERK in the dmPFC. Quantitative analysis using one-way ANOVA followed by Newman–Keuls post hoc test showed that *p*-ERK expression was significantly downregulated in the hydrocodone group compared to the control group, while ceftriaxone (200 mg/kg) upregulated *p*-ERK expression in the dmPFC compared to the hydrocodone group. Control group data were represented as 100%. Each column is expressed as mean ± S.E.M (n = 7/group), (* $p < 0.05$, ** $p < 0.01$, *** $p < 0.001$ and **** $p < 0.0001$).

3.4. Effect of Chronic Hydrocodone Exposure and Ceftriaxone on p-AKT Protein Expression in the NAc, AMY, and dmPFC

The effect of chronic hydrocodone exposure on *p*-AKT expression was also measured in mesocorticolimbic brain regions. Data analyses revealed significant differences in *p*-Akt expression in the NAc ($F_{2,13}$ = 5.970, $p < 0.05$, Figure 5A), AMY ($F_{2,13}$ = 18.37, $p < 0.001$, Figure 5B), and dmPFC ($F_{2,18}$ = 27.26, $p < 0.0001$, Figure 5C) among all groups. Chronic hydrocodone exposure decreased *p*-Akt expression in the NAc ($p < 0.05$, Figure 5A), AMY ($p < 0.01$, Figure 5B), and dmPFC ($p < 0.0001$, Figure 5C) compared to the control group. Importantly, ceftriaxone attenuated hydrocodone-induced downregulation of *p*-Akt expression ($p < 0.05$, Figure 5A) in the NAc, AMY ($p < 0.001$, Figure 5B), and dmPFC ($p < 0.0001$, Figure 5C). Significant differences were observed between control and hydrocodone-ceftriaxone groups in the AMY ($p < 0.05$, Figure 5B) and dmPFC ($p < 0.05$, Figure 5C); no significant difference was detected in the NAc (Figure 5A).

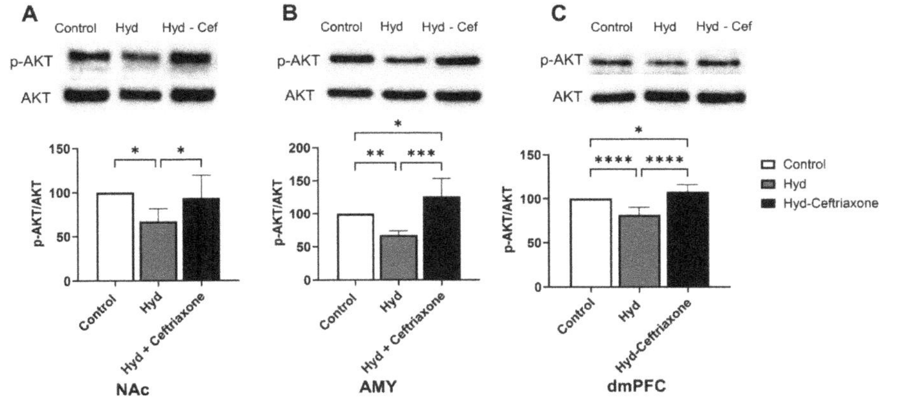

Figure 5. Effect of chronic hydrocodone exposure on *p*-AKT expression in the NAc, AMY, and dmPFC. (**A**) Immunoblots for *p*-AKT and AKT in the NAc. Quantitative analysis using one-way ANOVA followed by Newman–Keuls post hoc test indicated that *p*-AKT expression was significantly decreased in the hydrocodone group as compared to the control group, while ceftriaxone (200 mg/kg) normalized *p*-AKT expression in the NAc compared to the hydrocodone group. (**B**) Immunoblots for *p*-AKT and AKT in the AMY. Quantitative analysis using one-way ANOVA followed by Newman–Keuls post hoc test showed that *p*-AKT was significantly decreased in the hydrocodone group compared to the control group, while post-treatment with ceftriaxone (200 mg/kg) normalized the expression of *p*-AKT expression in the AMY compared to the hydrocodone group. (**C**) Immunoblots for *p*-AKT and AKT in the dmPFC. Quantitative analysis using one-way ANOVA followed by Newman–Keuls post hoc test showed that *p*-AKT expression was significantly reduced in the hydrocodone group compared to the control group, while ceftriaxone (200 mg/kg) normalized *p*-AKT expression in the dmPFC compared to the hydrocodone group. Control group data were represented as 100%. Each column is expressed as mean ± S.E.M ($n = 7$/group), (* $p < 0.05$, ** $p < 0.01$, *** $p < 0.001$ and **** $p < 0.0001$).

3.5. Effect of Chronic Hydrocodone Exposure and Ceftriaxone on p-JNK Protein Expression in the NAc, AMY, and dmPFC

We further investigated the effect of chronic exposure to hydrocodone on *p*-JNK expression in mesocorticolimbic brain regions. Data analyses revealed a significant difference in *p*-JNK expression in the NAc ($F_{2,14} = 7.577$, $p < 0.01$, Figure 6A), AMY ($F_{2,15} = 11.82$, $p < 0.001$, Figure 6B), and dmPFC ($F_{2,18} = 18.36$, $p < 0.0001$, Figure 6C) among all tested groups. Hydrocodone exposure decreased *p*-JNK expression in the NAc ($p < 0.05$, Figure 6A), AMY ($p < 0.01$, Figure 6B), and dmPFC ($p < 0.001$, Figure 6C) compared to the control group. Importantly, ceftriaxone attenuated hydrocodone-induced downregulation in *p*-JNK expression in the NAc ($p < 0.01$, Figure 6A), AMY ($p < 0.001$, Figure 6B), and dmPFC ($p < 0.0001$, Figure 6C). No significant difference was found in *p*-JNK expression between the control and hydrocodone–ceftriaxone groups in all three brain regions (Figure 6).

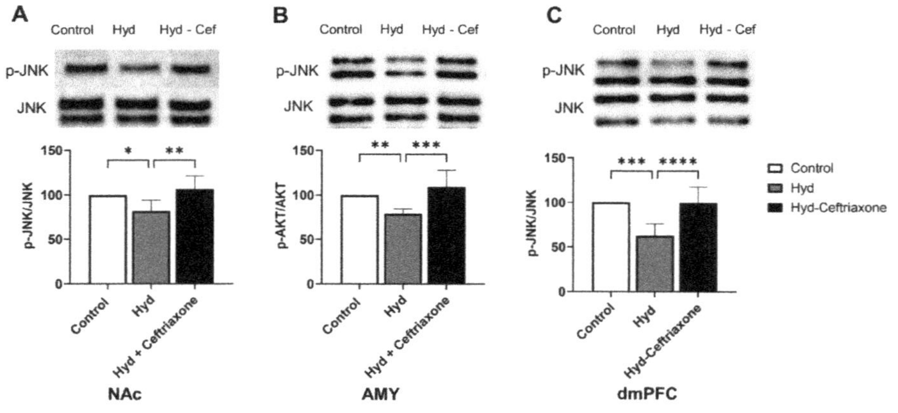

Figure 6. Effect of chronic hydrocodone exposure on *p*-JNK expression in the NAc, AMY, and dmPFC. (**A**) Immunoblots for *p*-JNK and JNK in the NAc. Quantitative analysis using one-way ANOVA followed by Newman–Keuls post hoc test revealed that *p*-JNK expression in the NAc was significantly downregulated in the hydrocodone group compared to the control group, while ceftriaxone (200 mg/kg) upregulated *p*-AKT expression compared to the hydrocodone group. (**B**) Immunoblots for *p*-JNK and JNK in the AMY. Quantitative analysis using one-way ANOVA followed by Newman–Keuls post hoc test showed that *p*-JNK was significantly downregulated in the hydrocodone group compared to the control group, while post-treatment with ceftriaxone (200 mg/kg) upregulated *p*-JNK expression in the AMY compared to the hydrocodone group. (**C**) Immunoblots for *p*-JNK and JNK in the dmPFC. Quantitative analysis using one-way ANOVA followed by Newman–Keuls post hoc test indicated that *p*-JNK expression was significantly downregulated in the hydrocodone group compared to the control group, while ceftriaxone (200 mg/kg) upregulated *p*-JNK expression in the dmPFC compared to the hydrocodone group. Each column is expressed as mean ± S.E.M (n = 7/group), (* $p < 0.05$, ** $p < 0.01$, *** $p < 0.001$ and **** $p < 0.0001$).

3.6. Effect of Chronic Hydrocodone Exposure and Ceftriaxone on mGluR5 Protein Expression in the NAc, AMY, and dmPFC

We finally determined the effect of hydrocodone exposure on mGluR5 expression in mesocorticolimbic brain regions. Statistical analyses revealed a significance difference in mGluR5 expression in the NAc ($F_{2,15} = 17.81$, $p < 0.001$, Figure 7A), AMY ($F_{2,12} = 8.018$, $p < 0.01$, Figure 7B), and dmPFC ($F_{2,12} = 18.59$, $p < 0.001$, Figure 7C). Furthermore, chronic hydrocodone exposure increased mGluR5 expression in the NAc ($p < 0.05$, Figure 7A), AMY ($p < 0.01$, Figure 7B), and dmPFC ($p < 0.01$, Figure 7C). Importantly, ceftriaxone attenuated hydrocodone-induced upregulation in mGluR5 expression in the NAc ($p < 0.0001$, Figure 7A), AMY ($p < 0.01$, Figure 7B), and dmPFC ($p < 0.001$, Figure 7C) compared to the control and hydrocodone groups. No significant differences were observed between the control and hydrocodone–ceftriaxone groups in the AMY (Figure 7B). However, there were significant differences between the control and hydrocodone–ceftriaxone groups in the NAc ($p < 0.01$, Figure 7A) and the dmPFC ($p < 0.05$, Figure 7C).

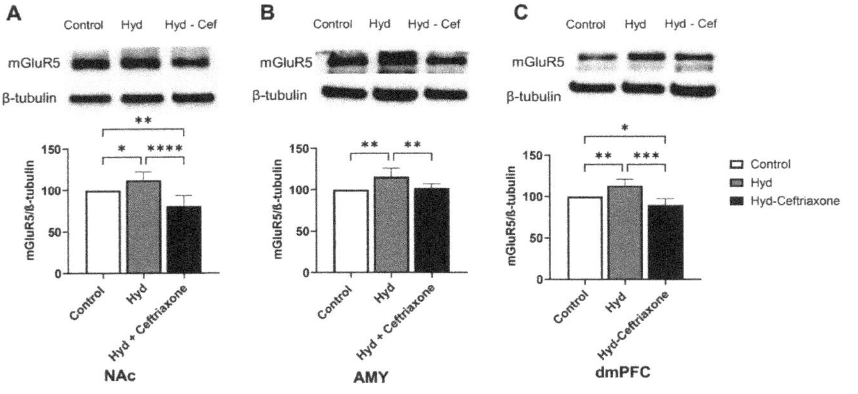

Figure 7. Effect of chronic hydrocodone exposure on mGluR5 expression in the NAc, AMY, and dmPFC. (**A**) Immunoblots for mGluR5 and β-tubulin in the NAc. Quantitative analysis using one-way ANOVA followed by Newman–Keuls post hoc test showed that mGluR5 expression was significantly upregulated in the hydrocodone group compared to the control group, while ceftriaxone (200 mg/kg) normalized mGluR5 expression in the NAc compared to the hydrocodone group. (**B**) Immunoblots for mGluR5 and β-tubulin in the AMY. Quantitative analysis using one-way ANOVA followed by Newman–Keuls post hoc test showed that mGluR5 was significantly upregulated in the hydrocodone group compared to the control group, while post-treatment with ceftriaxone (200 mg/kg) normalized mGluR5 expression in the AMY compared to the hydrocodone group. (**C**) Immunoblots for mGluR5 and β-tubulin in the dmPFC. Quantitative analysis using one-way ANOVA followed by Newman–Keuls post hoc test showed that mGluR5 expression was significantly upregulated in the hydrocodone group compared to the control group, while ceftriaxone (200 mg/kg) normalized mGluR5 expression in the dmPFC compared to hydrocodone group. Control group data were represented as 100%. Each column is expressed as mean ± S.E.M ($n = 7$/group), (* $p < 0.05$, ** $p < 0.01$, *** $p < 0.001$ and **** $p < 0.0001$).

4. Discussion

The hyperglutamatergic state is a major neurochemical unbalance that might be the cause of many neurological diseases and psychiatric disorders, including drug addiction [30,31]. GLT-1 plays an important role in regulating the majority of extracellular glutamate concentrations in the brain, and a reduction in GLT-1 expression is often associated with relapse to drugs of abuse. Therefore, restoring glutamate homeostasis may have a therapeutic, beneficial effect against neuroexcitotoxicity caused by chronic exposure to drugs of abuse. The present study demonstrated that chronic hydrocodone exposure alters the expression of GLT-1, xCT, mGluR5, *p*-ERK/ERK, *p*-Akt/Akt, and *p*-JNK/JNK in mesocorticolimibic brain regions. Ceftriaxone, known to upregulate GLT-1, attenuated hydrocodone-induced alteration of the expression of these target proteins. This study focused on three brain regions: the dmPFC, AMY, and NAc. These brain regions are reciprocally connecting glutamatergic projections, which are involved in drug seeking and drug dependence [6,9]. GLT-1 is a major glutamate transporter that regulates most of the extracellular glutamate concentrations in these key reward brain regions and others [8]. Previous studies from our laboratory demonstrated that chronic exposure to ethanol induced downregulation of GLT-1, as well xCT expression in central reward brain regions, including the NAc and ceftriaxone, attenuated this effect [6,16,32]. It is important to note that xCT is colocalized with GLT-1 in astrocytes to regulate basal extracellular glutamate concentrations [7]. Thus, both GLT-1 and xCT are critical in regulating the excess of extracellular glutamate that is mediated through astrocytes. In this study, we focused on the effect of chronic exposure of hydrocodone in the mesocorticolimbic brain regions involved drug dependence.

Exposure to hydrocodone (10 mg/kg, i.p.) for 14 days downregulated GLT-1 expression in the dmPFC, AMY, and NAc. In contrast to our current findings, previous studies from our laboratory reported no change in GLT-1 expression with hydrocodone exposure (5 mg/kg, i.p.) in the NAc, dmPFC, hippocampus, and AMY [23]. However, this study showed that hydrocodone exposure downregulated xCT expression in the NAc and the hippocampus, and ceftriaxone attenuated this effect. This difference may be attributable to the duration of hydrocodone exposure and the dose of hydrocodone tested. In addition, it has been suggested that a reduction in GLT-1 expression in the NAc is associated with chronic exposure to drugs of abuse, including alcohol, nicotine, heroin, and amphetamine [12,16,33,34]. Hence, our study showed that chronic exposure to higher doses of hydrocodone resulted in a decrease in GLT-1 expression in the dmPFC, AMY, and NAc. Furthermore, we found that xCT expression was downregulated in the dmPFC and NAc, but not in the AMY. The result from this current study is in accordance with our previous study which showed that exposure to a lower dose of hydrocodone (5 mg/kg, i.p.) in alcohol-preferring (P) rats reduced xCT expression in the NAc in a CPP model [23]. In the present study, ceftriaxone restored GLT-1 expression in the brains of mice exposed to hydrocodone. We suggest that the upregulatory effect of ceftriaxone in GLT-1 expression may decrease extracellular glutamate concentrations and increase glutamate uptake in models of drugs abuse, including ethanol and opioids [6,24]. Treatment with ceftriaxone reversed the effects of hydrocodone-induced downregulation of GLT-1 and xCT expression in the dmPFC and NAc. This is consistent with previous findings indicating the downregulation of GLT-1 and xCT expression in the NAc of animals exposed to drugs of abuse, including hydrocodone, ethanol, cocaine, nicotine, and methamphetamine, and these involved different behavioral paradigms such as drug seeking, self-administration, and reinstatement [11,15,23,35–37]. Our present findings demonstrated clearly that β-lactams (e.g., ceftriaxone and other β-lactams) have the potential to attenuate the effects of chronic exposure to opioids and normalize glutamate transporters to prevent neuroexcitotoxicity that might be mediated through excess of extracellular glutamate concentrations at the synaptic cleft.

Furthermore, this study explored the signaling pathways involved in GLT-1 and xCT upregulation. Thus, we focused on investigating the effect of hydrocodone exposure in Akt expression and its phosphorylated form. The Akt pathway is involved in synaptic and structural neuroadaptations, and *p*-Akt expression was decreased in the NAc after exposure to drug abuse, including morphine, heroin, and nicotine [38–40]. These findings are in accordance with our present findings showing that *p*-Akt is downregulated in the dmPFC, AMY, and NAc after chronic exposure to hydrocodone. Several studies reported that the Akt signaling pathway, which functions downstream of phosphatidylinositol 3-kinase (PI-3K) and the nuclear transcription factor-κB (NF-κB), is involved in the upregulation of GLT-1 expression [41]. Additionally, our studies and others confirmed the association between the Akt pathway and the upregulation of GLT-1 expression [41–44]. Our results showed that *p*-Akt expression was upregulated after ceftriaxone treatment, suggesting that the Akt signaling pathway is involved in a ceftriaxone-mediated increase in GLT-1 (Figure 8).

We further investigated the ERK signaling pathway as it is involved in neuroplasticity, and signal transduction [45]. Our study showed that exposure to hydrocodone for 14 days downregulated *p*-ERK expression in the dmPFC, AMY, and NAc, and this effect was attenuated with ceftriaxone treatment. Downregulation of ERK in the NAc is in accordance with previous studies showing that chronic morphine exposure reduced both ERK and Akt in the NAc of male Sprague-Dawley rats and CD-1 mice [38,46,47]. In addition, *p*-ERK is also known to initiate the transcription factors such as NF-kB and CREB, which in turn regulate GLT-1 transcription [48]. Therefore, our current study further showed that ceftriaxone activated the ERK signaling pathway, and consequently modulated GLT-1 expression (Figure 8).

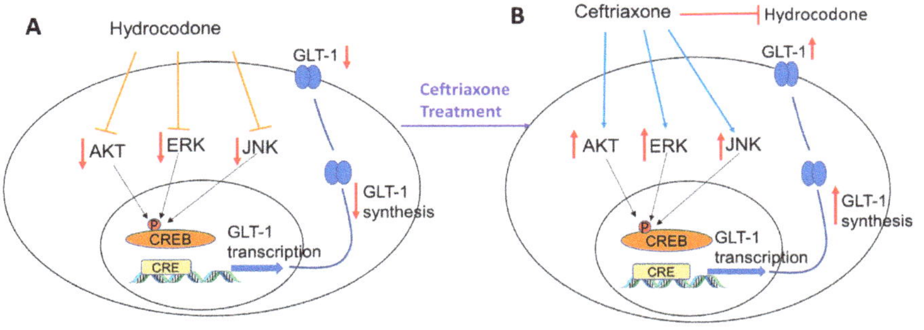

Figure 8. Schematic representation summarizing the effects of chronic hydrocodone exposure on GLT-1 expression in mesocorticolimbic brain regions. (**A**) Chronic exposure to hydrocodone reduced ERK, AKT, and JNK signaling kinases leading to GLT-1 downregulation. (**B**) Ceftriaxone treatment attenuated hydrocodone-induced GLT-1 downregulation by upregulating signaling kinases such as ERK, AKT, and JNK.

Furthermore, we investigated the JNK pathway as an important signaling pathway in the mitogen-activated protein kinases (MAPK) family. A study has reported that CREB activation is dependent on JNK [49]. Additionally, CREB phosphorylation is required for synaptic plasticity and memory consolidation [50–52]. In this study, we showed that chronic exposure to hydrocodone reduced *p*-JNK expression in the dmPFC, AMY, and NAC. Hydrocodone-induced downregulation of *p*-JNK expression was attenuated after ceftriaxone treatment. Here, we suggest that upregulation of GLT-1 expression in the dmPFC, AMY, and NAc might be associated with activation of CREB through JNK phosphorylation [53] (Figure 8). However, studies are warranted to validate this assumption.

In addition to signaling pathways, mGluRs have also been implicated in opioid reward. mGluR5 is known to play a facilitative role in mediating the potentiating effects of opioids and is highly expressed in reward-related brain regions, including the NAc and dmPFC [54]. It is important to note that mGluR5 is selectively increased in the NAc under a morphine-CPP paradigm and repeated exposure to cocaine [55,56]. This is in accordance with our present study demonstrating that chronic exposure to hydrocodone increased mGluR5 expression in the dmPFC, AMY, and NAc. Hydrocodone-induced upregulation in mGluR5 expression was attenuated with ceftriaxone treatment.

5. Conclusions

This study revealed that chronic exposure to hydrocodone induced dysregulation of glutamatergic system in the NAc, AMY, and dmPFC. Treatment with ceftriaxone successfully attenuated hydrocodone-induced dysfunction in this glutamatergic system. This was associated with the reversal of hydrocodone-induced changes in mGluR5, GLT-1, xCT, ERK, AKT, and JNK expression in the NAc, AMY, and dmPFC. We revealed that the upregulatory or normalizing effect of ceftriaxone in GLT-1 expression was mediated in part through the kinase signaling pathways such as the ERK, AKT, and JNK.

This study was proof of a concept to determine the effects of exposure to hydrocodone for 14 days on the expression of astrocytic glutamate transporters (GLT-1 and xCT). The study was limited to determine the signaling pathways involved in the upregulatory effects of ceftriaxone on GLT-1 and xCT expression. Further studies are warranted to investigate the beneficial preclinical effects of ceftriaxone and other beta-lactams in a model of self-administration or CPP of hydrocodone and other highly potent opioids (e.g., fentanyl). In addition, further studies are warranted to investigate the beneficial preclinical effects of ceftriaxone-induced upregulation of GLT-1 and xCT on a model of opioid overdose.

Author Contributions: W.W. participated in study design and conceptualization, drafted and revised the manuscript, performed animal injections, performed brain dissection, protein assay, and Western blot assay, and collected the data. Y.S. conceptualized and designed the study, critically revised the manuscript for intellectual content, and approved the final version of the manuscript. All authors have read and agreed to the published version of the manuscript.

Funding: This work was supported in part by the National Institutes of Health (AA029674 to YS). The content is solely the responsibility of the authors and does not necessarily represent the official views of the National Institutes of Health.

Institutional Review Board Statement: Animal research was approved by the Institutional Animal Care and Use Committee (IACUC) of The University of Toledo, following the guidelines governing the use of animals in research of the National Institutes of Health as described in the Guide for the Care and Use of Laboratory Animals (protocol# 400155; 2 August 2022).

Data Availability Statement: The data presented in this study are available in this research article.

Acknowledgments: The authors would like to thank the National Institutes of Health and the University of Toledo for their support.

Conflicts of Interest: The authors declare no conflict of interest.

References

1. Rudd, R.A.; Seth, P.; David, F.; Scholl, L. Increases in Drug and Opioid-Involved Overdose Deaths—United States, 2010–2015. *MMWR Morb. Mortal. Wkly. Rep.* **2016**, *65*, 1445–1452. [CrossRef]
2. Singla, A.; Sloan, P. Pharmacokinetic evaluation of hydrocodone/acetaminophen for pain management. *J. Opioid Manag.* **2013**, *9*, 71–80. [CrossRef]
3. Gould, H.J., III; Paul, D. Hydrocodone extended-release: Pharmacodynamics, pharmacokinetics and behavioral pharmacology of a controversy. *Pharmacol. Res.* **2015**, *91*, 99–103. [CrossRef]
4. Reeves, K.C.; Shah, N.; Munoz, B.; Atwood, B.K. Opioid Receptor-Mediated Regulation of Neurotransmission in the Brain. *Front. Mol. Neurosci.* **2022**, *15*, 919773. [CrossRef]
5. Corkrum, M.; Rothwell, P.E.; Thomas, M.J.; Kofuji, P.; Araque, A. Opioid-Mediated Astrocyte-Neuron Signaling in the Nucleus Accumbens. *Cells* **2019**, *8*, 586. [CrossRef]
6. Alasmari, F.; Goodwani, S.; McCullumsmith, R.E.; Sari, Y. Role of glutamatergic system and mesocorticolimbic circuits in alcohol dependence. *Prog. Neurobiol.* **2018**, *171*, 32–49. [CrossRef]
7. Baker, D.A.; Xi, Z.X.; Shen, H.; Swanson, C.J.; Kalivas, P.W. The origin and neuronal function of in vivo nonsynaptic glutamate. *J. Neurosci.* **2002**, *22*, 9134–9141. [CrossRef]
8. Danbolt, N.C. Glutamate uptake. *Prog. Neurobiol.* **2001**, *65*, 1–105. [CrossRef]
9. Rao, P.S.; Sari, Y. Glutamate transporter 1: Target for the treatment of alcohol dependence. *Curr. Med. Chem.* **2012**, *19*, 5148–5156. [CrossRef]
10. Alasmari, F.; Crotty Alexander, L.E.; Nelson, J.A.; Schiefer, I.T.; Breen, E.; Drummond, C.A.; Sari, Y. Effects of chronic inhalation of electronic cigarettes containing nicotine on glial glutamate transporters and alpha-7 nicotinic acetylcholine receptor in female CD-1 mice. *Prog. Neuropsychopharmacol. Biol. Psychiatry* **2017**, *77*, 1–8. [CrossRef]
11. Alhaddad, H.; Alasmari, F.; Alhamadani, B.; Wong, W.; Bell, R.L.; Sari, Y. Effects of chronic ethanol consumption on the expression of GLT-1 and neuroplasticity-related proteins in the nucleus accumbens of alcohol-preferring rats. *Brain Res. Bull.* **2020**, *165*, 272–280. [CrossRef] [PubMed]
12. Alhaddad, H.; Wong, W.; Sari, A.T.; Crotty Alexander, L.E.; Sari, Y. Effects of 3-Month Exposure to E-Cigarette Aerosols on Glutamatergic Receptors and Transporters in Mesolimbic Brain Regions of Female C57BL/6 Mice. *Toxics* **2020**, *8*, 95. [CrossRef] [PubMed]
13. Althobaiti, Y.S.; Alshehri, F.S.; Almalki, A.H.; Sari, Y. Effects of Ceftriaxone on Glial Glutamate Transporters in Wistar Rats Administered Sequential Ethanol and Methamphetamine. *Front. Neurosci.* **2016**, *10*, 427. [CrossRef]
14. Das, S.C.; Yamamoto, B.K.; Hristov, A.M.; Sari, Y. Ceftriaxone attenuates ethanol drinking and restores extracellular glutamate concentration through normalization of GLT-1 in nucleus accumbens of male alcohol-preferring rats. *Neuropharmacology* **2015**, *97*, 67–74. [CrossRef] [PubMed]
15. Knackstedt, L.A.; Melendez, R.I.; Kalivas, P.W. Ceftriaxone restores glutamate homeostasis and prevents relapse to cocaine seeking. *Biol. Psychiatry* **2010**, *67*, 81–84. [CrossRef]
16. Alhaddad, H.; Das, S.C.; Sari, Y. Effects of ceftriaxone on ethanol intake: A possible role for xCT and GLT-1 isoforms modulation of glutamate levels in P rats. *Psychopharmacology* **2014**, *231*, 4049–4057. [CrossRef]
17. Das, S.C.; Althobaiti, Y.S.; Hammad, A.M.; Alasmari, F.; Sari, Y. Role of suppressing GLT-1 and xCT in ceftriaxone-induced attenuation of relapse-like alcohol drinking in alcohol-preferring rats. *Addict. Biol.* **2022**, *27*, e13178. [CrossRef]

18. Knackstedt, L.A.; LaRowe, S.; Mardikian, P.; Malcolm, R.; Upadhyaya, H.; Hedden, S.; Markou, A.; Kalivas, P.W. The role of cystine-glutamate exchange in nicotine dependence in rats and humans. *Biol. Psychiatry* **2009**, *65*, 841–845. [CrossRef]
19. Sondheimer, I.; Knackstedt, L.A. Ceftriaxone prevents the induction of cocaine sensitization and produces enduring attenuation of cue-and cocaine-primed reinstatement of cocaine-seeking. *Behav. Brain Res.* **2011**, *225*, 252–258. [CrossRef]
20. Mao, J.; Sung, B.; Ji, R.R.; Lim, G. Chronic morphine induces downregulation of spinal glutamate transporters: Implications in morphine tolerance and abnormal pain sensitivity. *J. Neurosci.* **2002**, *22*, 8312–8323. [CrossRef]
21. Ozawa, T.; Nakagawa, T.; Shige, K.; Minami, M.; Satoh, M. Changes in the expression of glial glutamate transporters in the rat brain accompanied with morphine dependence and naloxone-precipitated withdrawal. *Brain Res.* **2001**, *905*, 254–258. [CrossRef]
22. Wang, X.F.; Zhao, T.Y.; Su, R.B.; Wu, N.; Li, J. Agmatine Prevents Adaptation of the Hippocampal Glutamate System in Chronic Morphine-Treated Rats. *Neurosci. Bull.* **2016**, *32*, 523–530. [CrossRef]
23. Alshehri, F.S.; Hakami, A.Y.; Althobaiti, Y.S.; Sari, Y. Effects of ceftriaxone on hydrocodone seeking behavior and glial glutamate transporters in P rats. *Behav. Brain Res.* **2018**, *347*, 368–376. [CrossRef] [PubMed]
24. Abulseoud, O.A.; Alasmari, F.; Hussein, A.M.; Sari, Y. Ceftriaxone as a Novel Therapeutic Agent for Hyperglutamatergic States: Bridging the Gap Between Preclinical Results and Clinical Translation. *Front. Neurosci.* **2022**, *16*, 841036. [CrossRef] [PubMed]
25. Han, X.; Yang, L.; Du, H.; Sun, Q.; Wang, X.; Cong, L.; Liu, X.; Yin, L.; Li, S.; Du, Y. Insulin Attenuates Beta-Amyloid-Associated Insulin/Akt/EAAT Signaling Perturbations in Human Astrocytes. *Cell. Mol. Neurobiol.* **2016**, *36*, 851–864. [CrossRef] [PubMed]
26. Karki, P.; Hong, P.; Johnson, J., Jr.; Pajarillo, E.; Son, D.S.; Aschner, M.; Lee, E.Y. Arundic Acid Increases Expression and Function of Astrocytic Glutamate Transporter EAAT1 Via the ERK, Akt, and NF-kappaB Pathways. *Mol. Neurobiol.* **2018**, *55*, 5031–5046. [CrossRef]
27. Li, L.B.; Toan, S.V.; Zelenaia, O.; Watson, D.J.; Wolfe, J.H.; Rothstein, J.D.; Robinson, M.B. Regulation of astrocytic glutamate transporter expression by Akt: Evidence for a selective transcriptional effect on the GLT-1/EAAT2 subtype. *J. Neurochem.* **2006**, *97*, 759–771. [CrossRef]
28. Paxinos, G.; Halliday, G.; Watson, C.; Kassem, M.S. *Atlas of the Developing Mouse Brain*; Academic Press: Cambridge, MA, USA, 2020.
29. Alshehri, F.S.; Althobaiti, Y.S.; Sari, Y. Effects of administered ethanol and methamphetamine on glial glutamate transporters in rat striatum and hippocampus. *J. Mol. Neurosci.* **2017**, *61*, 343–350. [CrossRef]
30. Karki, P.; Lee, E.; Aschner, M. Manganese neurotoxicity: A focus on glutamate transporters. *Ann. Occup. Environ. Med.* **2013**, *25*, 4. [CrossRef]
31. Pajarillo, E.; Rizor, A.; Lee, J.; Aschner, M.; Lee, E. The role of astrocytic glutamate transporters GLT-1 and GLAST in neurological disorders: Potential targets for neurotherapeutics. *Neuropharmacology* **2019**, *161*, 107559. [CrossRef]
32. Sari, Y.; Sakai, M.; Weedman, J.M.; Rebec, G.V.; Bell, R.L. Ceftriaxone, a beta-lactam antibiotic, reduces ethanol consumption in alcohol-preferring rats. *Alcohol Alcohol.* **2011**, *46*, 239–246. [CrossRef] [PubMed]
33. Shen, H.W.; Scofield, M.D.; Boger, H.; Hensley, M.; Kalivas, P.W. Synaptic Glutamate Spillover Due to Impaired Glutamate Uptake Mediates Heroin Relapse. *J. Neurosci.* **2014**, *34*, 5649–5657. [CrossRef] [PubMed]
34. Althobaiti, Y.S.; Alshehri, F.S.; Hakami, A.Y.; Hammad, A.M.; Sari, Y. Effects of Clavulanic Acid Treatment on Reinstatement to Methamphetamine, Glial Glutamate Transporters, and mGluR 2/3 Expression in P Rats Exposed to Ethanol. *J. Mol. Neurosci.* **2019**, *67*, 1–15. [CrossRef] [PubMed]
35. Logan, C.N.; Bechard, A.R.; Hamor, P.U.; Wu, L.; Schwendt, M.; Knackstedt, L.A. Ceftriaxone and mGlu2/3 interactions in the nucleus accumbens core affect the reinstatement of cocaine-seeking in male and female rats. *Psychopharmacology* **2020**, *237*, 2007–2018. [CrossRef] [PubMed]
36. Sari, Y.; Toalston, J.E.; Rao, P.S.S.; Bell, R.L. Effects of ceftriaxone on ethanol, nicotine or sucrose intake by alcohol-preferring (P) rats and its association with GLT-1 expression. *Neuroscience* **2016**, *326*, 117–125. [CrossRef]
37. Sari, Y.; Smith, K.D.; Ali, P.K.; Rebec, G.V. Upregulation of GLT1 attenuates cue-induced reinstatement of cocaine-seeking behavior in rats. *J. Neurosci.* **2009**, *29*, 9239–9243. [CrossRef]
38. Muller, D.L.; Unterwald, E.M. In vivo regulation of extracellular signal-regulated protein kinase (ERK) and protein kinase B (Akt) phosphorylation by acute and chronic morphine. *J. Pharmacol. Exp. Ther.* **2004**, *310*, 774–782. [CrossRef]
39. Hudson, R.; Green, M.; Wright, D.J.; Renard, J.; Jobson, C.E.L.; Jung, T.; Rushlow, W.; Laviolette, S.R. Adolescent nicotine induces depressive and anxiogenic effects through ERK 1-2 and Akt-GSK-3 pathways and neuronal dysregulation in the nucleus accumbens. *Addict. Biol.* **2021**, *26*, e12891. [CrossRef]
40. Zhu, H.; Zhuang, D.; Lou, Z.; Lai, M.; Fu, D.; Hong, Q.; Liu, H.; Zhou, W. Akt and its phosphorylation in nucleus accumbens mediate heroin-seeking behavior induced by cues in rats. *Addict. Biol.* **2021**, *26*, e13013. [CrossRef]
41. Rao, P.S.; Saternos, H.; Goodwani, S.; Sari, Y. Effects of ceftriaxone on GLT1 isoforms, xCT and associated signaling pathways in P rats exposed to ethanol. *Psychopharmacology* **2015**, *232*, 2333–2342. [CrossRef]
42. Zhang, X.; Shi, M.; Bjoras, M.; Wang, W.; Zhang, G.; Han, J.; Liu, Z.; Zhang, Y.; Wang, B.; Chen, J.; et al. Ginsenoside Rd promotes glutamate clearance by up-regulating glial glutamate transporter GLT-1 via PI3K/AKT and ERK1/2 pathways. *Front. Pharmacol.* **2013**, *4*, 152. [CrossRef] [PubMed]
43. Goodwani, S.; Rao, P.S.S.; Bell, R.L.; Sari, Y. Amoxicillin and amoxicillin/clavulanate reduce ethanol intake and increase GLT-1 expression as well as AKT phosphorylation in mesocorticolimbic regions. *Brain Res.* **2015**, *1622*, 397–408. [CrossRef] [PubMed]

44. Wu, X.; Kihara, T.; Akaike, A.; Niidome, T.; Sugimoto, H. PI3K/Akt/mTOR signaling regulates glutamate transporter 1 in astrocytes. *Biochem. Biophys. Res. Commun.* **2010**, *393*, 514–518. [CrossRef] [PubMed]
45. Lein, E.S.; Hawrylycz, M.J.; Ao, N.; Ayres, M.; Bensinger, A.; Bernard, A.; Boe, A.F.; Boguski, M.S.; Brockway, K.S.; Byrnes, E.J.; et al. Genome-wide atlas of gene expression in the adult mouse brain. *Nature* **2007**, *445*, 168–176. [CrossRef] [PubMed]
46. Rosas, M.; Porru, S.; Fenu, S.; Ruiu, S.; Peana, A.T.; Papale, A.; Brambilla, R.; Di Chiara, G.; Acquas, E. Role of nucleus accumbens mu opioid receptors in the effects of morphine on ERK1/2 phosphorylation. *Psychopharmacology* **2016**, *233*, 2943–2954. [CrossRef]
47. Rosas, M.; Porru, S.; Sabariego, M.; Piludu, M.A.; Giorgi, O.; Corda, M.G.; Acquas, E. Effects of morphine on place conditioning and ERK1/2 phosphorylation in the nucleus accumbens of psychogenetically selected Roman low-and high-avoidance rats. *Psychopharmacology* **2018**, *235*, 59–69.
48. Lee, E.; Sidoryk-Wegrzynowicz, M.; Wang, N.; Webb, A.; Son, D.S.; Lee, K.; Aschner, M. GPR30 regulates glutamate transporter GLT-1 expression in rat primary astrocytes. *J. Biol. Chem.* **2012**, *287*, 26817–26828. [CrossRef]
49. Zhi, Y.; Lu, C.; Zhu, G.; Li, Z.; Zhu, P.; Liu, Y.; Shi, W.; Su, L.; Jiang, J.; Qu, J.; et al. Positive regulation of the CREB phosphorylation via JNK-dependent pathway prevents antimony-induced neuronal apoptosis in PC12 cell and mice brain. *Neurotoxicology* **2020**, *81*, 101–108. [CrossRef]
50. Walton, M.R.; Dragunow, M. Is CREB a key to neuronal survival? *Trends Neurosci.* **2000**, *23*, 48–53. [CrossRef]
51. Walters, C.L.; Blendy, J.A. Different requirements for cAMP response element binding protein in positive and negative reinforcing properties of drugs of abuse. *J. Neurosci.* **2001**, *21*, 9438–9444. [CrossRef]
52. Valverde, O.; Mantamadiotis, T.; Torrecilla, M.; Ugedo, L.; Pineda, J.; Bleckmann, S.; Gass, P.; Kretz, O.; Mitchell, J.M.; Schutz, G.; et al. Modulation of anxiety-like behavior and morphine dependence in CREB-deficient mice. *Neuropsychopharmacology* **2004**, *29*, 1122–1133. [CrossRef] [PubMed]
53. Ramos-Miguel, A.; García-Fuster, M.; Callado, L.; La Harpe, R.; Meana, J.; García-Sevilla, J. Phosphorylation of FADD (Fas-associated death domain protein) at serine 194 is increased in the prefrontal cortex of opiate abusers: Relation to mitogen activated protein kinase, phosphoprotein enriched in astrocytes of 15 kDa, and Akt signaling pathways involved in neuroplasticity. *Neuroscience* **2009**, *161*, 23–38. [PubMed]
54. Ferraguti, F.; Shigemoto, R. Metabotropic glutamate receptors. *Cell Tissue Res.* **2006**, *326*, 483–504.
55. Ghasemzadeh, M.B.; Nelson, L.C.; Lu, X.Y.; Kalivas, P.W. Neuroadaptations in ionotropic and metabotropic glutamate receptor mRNA produced by cocaine treatment. *J. Neurochem.* **1999**, *72*, 157–165. [CrossRef]
56. Qi, C.; Wang, X.; Ge, F.; Li, Y.; Shen, F.; Wang, J.; Cui, C. mGluR 5 in the nucleus accumbens shell regulates morphine-associated contextual memory through reactive oxygen species signaling. *Addict. Biol.* **2015**, *20*, 927–940. [CrossRef] [PubMed]

toxics

Article

Natural Polyphenols—Resveratrol, Quercetin, Magnolol, and β-Catechin—Block Certain Aspects of Heroin Addiction and Modulate Striatal IL-6 and TNF-α

Shaimaa ElShebiney [1,*], Rania Elgohary [1], Marwa El-Shamarka [1], Noha Mowaad [1] and Osama A. Abulseoud [2,3,*]

1 Department of Narcotics, Ergogenics, and Poisons, National Research Centre, Dokki, Cairo 12622, Egypt
2 Department of Psychiatry and Psychology, Mayo Clinic, Phoenix, AZ 85001, USA
3 Department of Neuroscience, Graduate School of Biomedical Sciences, Mayo Clinic College of Medicine, Phoenix, AZ 85001, USA
* Correspondence: sa.elshebiney@nrc.sci.eg (S.E.); abulseoud.osama@mayo.edu (O.A.A.)

Abstract: We have examined the effects of four different polyphenols in attenuating heroin addiction using a conditioned place preference (CPP) paradigm. Adult male Sprague Dawley rats received heroin (alternating with saline) in escalating doses starting from 10 mg/kg, i.p. up to 80 mg/kg/d for 14 consecutive days. The rats were treated with distilled water (1 mL), quercetin (50 mg/kg/d), β-catechin (100 mg/kg/d), resveratrol (30 mg/kg/d), or magnolol (50 mg/kg/d) through oral gavage for 7 consecutive days, 30 min before heroin administration, starting on day 8. Heroin withdrawal manifestations were assessed 24 h post last heroin administration following the administration of naloxone (1 mg/kg i.p). Heroin CPP reinstatement was tested following a single dose of heroin (10 mg/kg i.p.) administration. Striatal interleukin 6 (IL-6) and tumor necrosis factor alpha (TNF-α) were quantified (ELISA) after naloxone-precipitated heroin withdrawal. Compared to the vehicle, the heroin-administered rats spent significantly more time in the heroin-paired chamber ($p < 0.0001$). Concomitant administration of resveratrol and quercetin prevented the acquisition of heroin CPP, while resveratrol, quercetin, and magnolol blocked heroin-triggered reinstatement. Magnolol, quercetin, and β-catechin blocked naloxone-precipitated heroin withdrawal and increased striatal IL-6 concentration ($p < 0.01$). Resveratrol administration was associated with significantly higher withdrawal scores compared to those of the control animals ($p < 0.0001$). The results of this study show that different polyphenols target specific behavioral domains of heroin addiction in a CPP model and modulate the increase in striatal inflammatory cytokines TNF-α and IL-6 observed during naloxone-precipitated heroin withdrawal. Further research is needed to study the clinical utility of polyphenols and to investigate the intriguing finding that resveratrol enhances, rather than attenuates naloxone-precipitated heroin withdrawal.

Keywords: heroin; conditioned place preference (CPP); naloxone precipitated withdrawal; polyphenols; resveratrol; magnolol; β-catechin

Citation: ElShebiney, S.; Elgohary, R.; El-Shamarka, M.; Mowaad, N.; Abulseoud, O.A. Natural Polyphenols—Resveratrol, Quercetin, Magnolol, and β-Catechin—Block Certain Aspects of Heroin Addiction and Modulate Striatal IL-6 and TNF-α. *Toxics* **2023**, *11*, 379. https://doi.org/10.3390/toxics11040379

Academic Editors: Jürgen Gailer and Guido Cavaletti

Received: 4 February 2023
Revised: 4 April 2023
Accepted: 14 April 2023
Published: 17 April 2023

1. Introduction

Heroin is a highly addictive illicit opiate that represents one of the main contributors to the global burden of illness [1]. In the United States alone, heroin use has reached epidemic proportions, affecting about 1.6% of the population aged 12 or older [2–4]. In Egypt, one study from the Poison Treatment Center including all cases of acute substance intoxication between 2015–2019 reported that opiate (tramadol) was the most common substance of exposure, and the greatest cause of fatality [5].

Current pharmacological treatment options for heroin addiction target μ opioid receptors as either full agonists, such as methadone, partial agonists, such as buprenorphine—which is also a κ-antagonist—or full antagonists, such as naltrexone [6–8]. Besides their suboptimal efficacy [9], the stigma around methadone [10] added to the complex logistics for dispensing buprenorphine [11] and the problem of diversion [12]; all these factors limit the patient availability of these medications and highlight the urgent need for novel, non-opioid, pharmacological agents to treat different stages of heroin addiction.

Illicit drug use, including opiates, is associated with dysregulated immune signaling [13,14], with reports of both activation and suppression of inflammatory cytokines. Morphine administration, in one study, increased microglial release of central inflammatory mediators, such as tumor necrosis factor (TNF)-alpha and interleukin (IL)-6 [15,16], and plasma IL-6 was significantly higher in methadone-maintained heroin users compared to healthy control subjects [17]. On the other hand, heroin and other opiates suppress the microglial secretion of TNF-α [14] This immunomodulatory effect of opiates is not mediated by all opioid receptors, and other signaling pathways play a significant role in opioid addiction behaviors [14,18,19].

Peroxisome proliferator-activated receptor-alpha (PPAR-α) is known to regulate cellular inflammatory response [20,21], and polyphenols, such as resveratrol, quercetin, magnolol, and β-catechin, found in many different plants [22–25], possess immune-modulatory properties, likely through the activation of PPAR-α [26–29], and their potential efficacy for treating opioid use disorder seems to be promising.

Yunusoglu et al. examined the effect of resveratrol on alcohol-induced conditioned place preference (CPP) in mice. Pretreatment with resveratrol, dose dependently, impaired ethanol preference acquisition, reinstating and facilitating the extinction of alcohol CPP [30]. Furthermore, Singh et al., showed that repeated administration of another polyphenol, quercetin, attenuated the development of tolerance to the analgesic effect of morphine and suppressed naloxone-precipitated withdrawal [31]. Quercetin pretreatment 30 min before ethanol administration in a CPP paradigm attenuated acquisition and reinstatement and accelerated the extinction of ethanol-CPP [32]. Moreover, the effects of quercetin and β-catechin on naloxone-precipitated withdrawal were tested in vitro. Both quercetin and catechin, injected into the guinea-pig ileum 10 min before morphine, were capable of blocking naloxone-induced contracture after exposure to morphine in a concentration-dependent fashion [33]. Taken together, there is evidence that polyphenols attenuate behavioral manifestations of ethanol, morphine, and methamphetamine administration and attenuate drug-induced activation of certain inflammatory cytokines. In this study, we aimed to examine the efficacy of four different natural polyphenols in blocking behavioral manifestations of heroin administration using a conditioned place preference (CPP) paradigm.

Conditioned place preference (CPP) is a valid and reliable method used to assess the rewarding properties of various drugs of abuse [34], such as drug-paired craving and relapse [35].

2. Methods

2.1. Animals

Adult inbred male Sprague Dawley rats weighing 180 to 210 g (n = 88) were used for the experiments. The study was conducted according to the guidelines of the Declaration of Helsinki and approved by the Animal Care and Use Committee of the Egyptian National Research Center (protocol #19-220/20/11/2019). The rats were housed in standard plastic cages, with 4 animals/cage, in a controlled environment (temperature, 25–26 °C, humidity, 45–65%, and 12 h dark: light cycle, with lights on at 7:00 a.m.) and food/water were provided ad libitum.

2.2. Groups

Rats were randomly assigned to two cohorts. The first cohort (n = 56) was used to establish CPP to test naloxone-precipitated withdrawal, and to assay striatal IL-6 and TNF-α concentration. The second cohort (n = 32) was established to test heroin-triggered reinstatement. Rats in the first cohort were assigned to one of five groups: (1) control (n = 8), (2) heroin + quercetin (n = 12), (3) heroin + resveratrol (n = 12), (4) heroin + β-catechin (n = 12), and (5) heroin + magnolol (n = 12). Rats in the second cohort were assigned to one of six groups (n = 8 each): (1) vehicle negative control, (2) heroin positive control, (3) quercetin, (4) resveratrol, (5) β-catechin, and (6) magnolol.

2.3. Drugs

Heroin was provided from the Criminal Justice Laboratories under the permission of the Ministry of Justice, Cairo, Egypt. Resveratrol (Doctor's best, CA, USA), quercetin (Naturebell, Chino, CA, USA), β-catechin (Puritan's Pride, Holbrook, NY, USA), and magnolol (Nutricrafters, Sparks, NV, USA) were obtained from a local pharmacy as dietary supplements. We discarded the capsule, which contains inactive ingredients such as cellulose, gelatin, rice flour, silica, and maltodextrin. The content of the capsule contains the active substance only. We dissolved the active ingredient in distilled water to the required concentration to be administered orally at 1 mL doses.

2.4. Behavioral Study

(A) Conditioned place preference (first cohort)

- Habituation: Before the start of the procedure, the rats were habituated to the place preference laboratory room for one hour. During the preconditioning phase (1 day), the animals were allowed to freely explore the whole apparatus for 15 min. The time spent in each chamber, while the door is open, was recorded (unconditioned preference), and then the animals were returned to their home cages.
- Establishing CPP: The following day, control rats received saline (0.5 mL/kg, i.p.), while heroin-primed rats received heroin in the least preferred chamber in escalating doses, starting from 10 mg/kg, i.p. up to 80 mg/kg daily for 14 consecutive days (10 mg/kg/d for 4 days, then 20 mg/kg/d for 4 days, then 40 mg/kg/d for 4 days, then 80 mg/kg/d for 2 days). The rats received treatments for heroin CPP starting on day 8. Animals were administered polyphenols (quercetin 50 mg/kg/d [36], or β-catechin 100 mg/kg/d [37], or resveratrol 30 mg/kg/d [38], p.o., or magnolol 50 mg/kg/d, p.o. [39]) or distilled water (1 mL) through oral gavage for 7 consecutive days starting at day 8 of heroin administration.
- Testing for heroin CPP and testing the efficacy of polyphenols in blocking heroin CPP: On day 15, the animals were tested for heroin preference during 10 min of free access to both chambers. The percentage of time spent in the drug-paired chamber was recorded manually by a blinded observer in real time.
- Testing for the efficacy of polyphenols in attenuating naloxone-precipitated heroin withdrawal: On day 16 after testing for heroin CPP, rats were challenged with naloxone (1 mg/kg, i.p.) after 24 h of the last heroin dose (between 8:00 a.m. and 12:00 p.m.) to precipitate withdrawal, and they were observed in a transparent cylinder arena for a 30 min test period to detect withdrawal symptoms; scores were recorded manually in real time by a blinded observer [40]. Specific withdrawal signs, including jumping, wet dog shakes, head shakes, teeth chattering, tremors, and rearing movements, were counted during every 5 min observation. Irritation, piloerection, salivation, diarrhea, and grooming were observed and scored on a four-point scale: 0 = absent; 1 = mild; 2 = moderate; 3 = severe. The scores for each time period were combined [1].
- Testing for the effect of naloxone-precipitated withdrawal on spontaneous locomotor activity using an open field: On day 16, the rats were screened in an open field at the end of the experiment for 5 min (Fernandes et al., 2012). Each rat was placed in the center of the field (100 × 100 cm white box), and the number of squares crossed, as well as the vertical rears, were monitored by a blind observer in real time.
- Testing for the effect of naloxone-precipitated withdrawal on anxiety using elevated plus maze (EPM): Following the open field experiment, the rats were tested

for anxiety using EPM. The maze was raised 40 cm off the floor with two equal crossed arms (10 cm wide and 100 cm long), and one arm was closed by 30 cm high walls. Animals were placed at the intersection facing one open arm and allowed to freely move; the time spent in the open or closed arm, in addition to the number of entries into any of the arms, was recorded in real time by a blinded observer. Maze sessions of 5 min each were held after 90 min of naloxone-induced withdrawal [41].

- Testing for the effect of naloxone-precipitated withdrawal on sucrose preference: On day 17 (next day of naloxone-precipitated withdrawal), the rats were deprived of food for 12 h, starting at 8:00 p.m. and continuing until 8:00 a.m. the next day, and were placed in individual cages and provided two regular 200 mL bottles: one containing 3% sucrose solution and the other containing tap water. On the next day (8:00 a.m.), the volumes of sucrose-containing water and plain water were recorded after 24 h. Sucrose intake was calculated: sucrose preference = sucrose intake/total intake (sucrose + water intake) × 100 (28).
- Euthanasia and brain tissue collection: The rats were euthanized by decapitation under light anesthesia after the end of the sucrose preference test on day 17. Brain tissues were dissected and stored at −80 °C for molecular assay (Figure 1A).

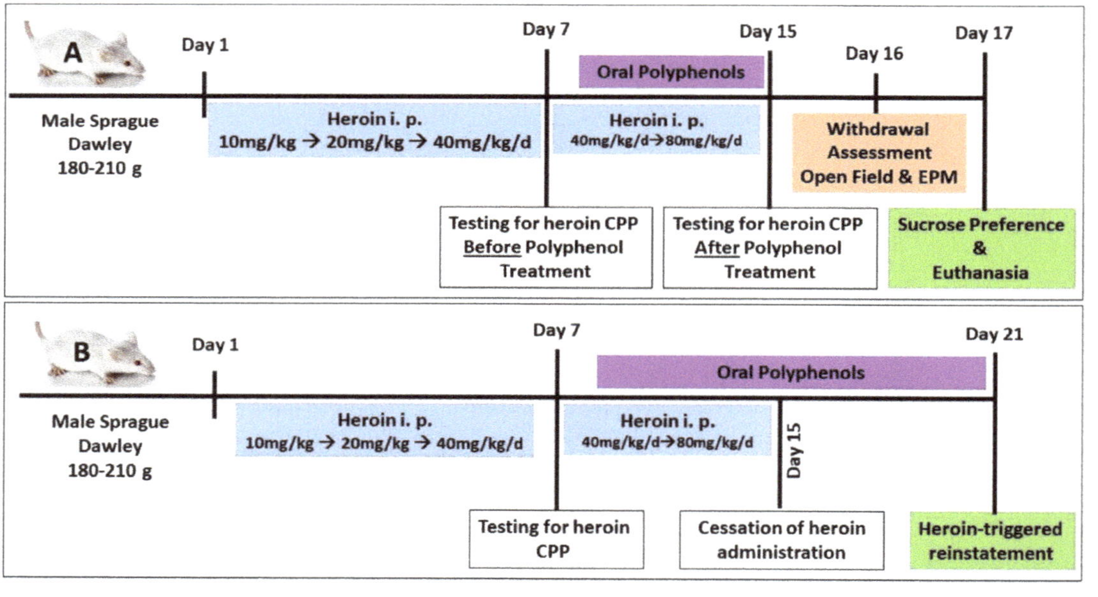

Figure 1. Study design: (**A**) naloxone-precipitated opiate withdrawal cohort; (**B**) heroin-triggered reinstatement cohort.

(B) Heroin-triggered-reinstatement (second cohort)

After establishing heroin CPP as described above, the rats were kept in normal housing conditions for 6 days, without heroin exposure. On the day 7, the rats were challenged in the CPP drug-linked chamber by a single heroin dose administration (10 mg/kg, i.p.), whereas polyphenols were administered orally from day 8 to day 21, and the last dose was administered 30 min before heroin challenge. Reinstatement was assessed as the time spent in the drug-linked chamber (Figure 1B). At the end of the experiment, the rats were euthanized, and brain tissue was collected and stored for further studies.

2.5. Striatal TNF-α and IL-6 Assay

The micro-ELISA plate was pre-coated with an antibody specific to Rat TNF-α or IL-6 (Elabscience®, Houston, TX, USA). After adding samples/standards, a biotinylated detection antibody specific for Rat TNF-α or IL-6 and Avidin-Horseradish Peroxidase (HRP) conjugate were added successively to each well and incubated. The optical density of TNF-α or IL-6 conjugated with the biotinylated detection antibody was measured spectrophotometrically at a wavelength of 450 nm using a plate reader (BMG Labtech, FLUOstar Omega, Ortenberg, Germany). The OD value is proportional to the concentration.

2.6. Statistical Analysis

Results are expressed as the mean ± SEM. Graphpad Prism software was used to perform statistical analysis, employing one way ANOVA, followed by Dunnett's multiple comparisons test. Statistical significance was considered at $p < 0.05$.

3. Results

3.1. Behavioral Effects

Resveratrol and quercetin attenuated the acquisition of heroin conditioned place preference

Daily heroin administration for 14 days successfully produced heroin CPP. Compared to the vehicle, heroin-administered rats spent significantly more time in the heroin-paired chamber ($p < 0.0001$) Concomitant administration of resveratrol and quercetin prevented the acquisition of heroin CPP (Figure 2).

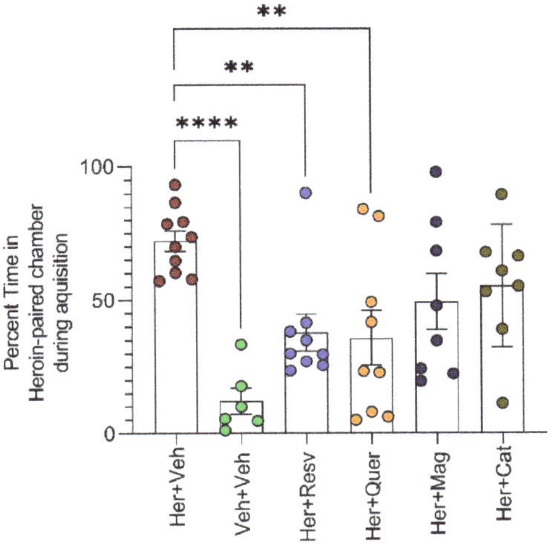

Figure 2. Heroin CPP acquisition is blocked by resveratrol and quercetin.

Daily heroin administration for 14 days established heroin dependence [F(5, 44) = 6.198, $p = 0.0002$]. Heroin administered animals (compared to the vehicle) spent significantly more time in the heroin-paired chamber [mean difference = 60.06, 95% CI = 29.19 to 90.92, $p < 0.0001$]. Resveratrol and quercetin administration concomitant with heroin successfully prevented heroin preference and reduced the percentage of time spent in the heroin-paired chamber compared to the administration of heroin alone. [Her + Veh vs. Her + Resv mean difference = 34.32, 95% CI = 6.856 to 61.78, $p = 0.009$, and Her + Veh vs. Her + Quer mean difference = 36.41, 95% CI = 8.948 to 63.87, $p = 0.005$]. Neither magnolol nor β-catechin administration reduced heroin preference, according to one way ANOVA, followed by Dunnett's multiple comparisons test, against Her + Veh control group, n = 6–10 animals per group.

Magnolol, quercetin, and β-catechin block naloxone-precipitated heroin withdrawal, prevent rapid weight loss during withdrawal, and reduce withdrawal-associated anxiety-like behavior

Opiate withdrawal manifestations measured 24 h post last heroin administration showed significantly higher scores in both spontaneous and naloxone-induced withdrawal ($p < 0.0001$ for both, Figure 3A) associated with significant weight loss (≈3% of body weight in 24 h, Figure 3B). Magnolol blocked naloxone-precipitated heroin withdrawal ($p < 0.0001$ Figure 3A), prevented weight loss (1.6% of body weight compared to heroin control animals, $p = 0.3$ vs. Her + Veh, Figure 3B), and increased the time spent in the EPM open arm in a non-significantly different manner from the heroin control animals ($p = 0.9$, Figure 3D) (** $p < 0.01$, *** $p < 0.001$, **** $p < 0.0001$).

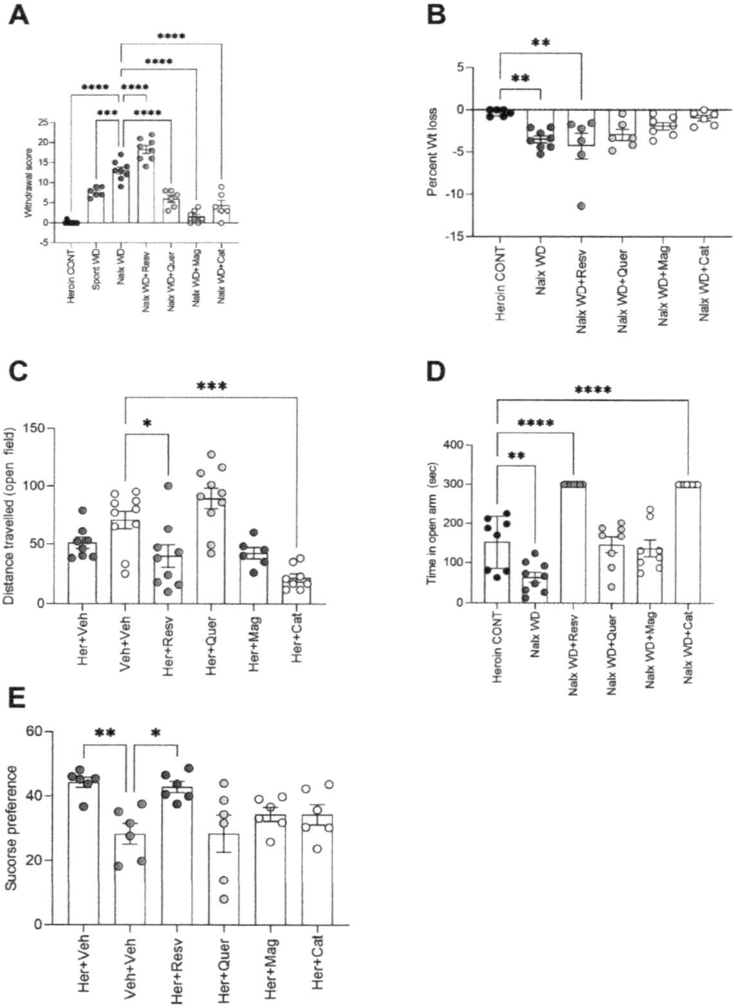

Figure 3. (**A**) Magnolol attenuates naloxoneprecipitated heroin withdrawal: magnolol blocked naloxone-precipitated heroin withdrawal. One-way ANOVA $F(6, 42) = 70.51$, $p < 0.0001$ [mean difference between heroin control and naloxone+ magnolol = −1.446, 95% CI = −4.412 to 1.519, $p = 0.6$]. Resv, Quer, and Cat all are associated with significantly higher withdrawal scores compared

to those of the control animals ($p < 0.0001$ for Resv and Quer and $p = 0.003$ for Cat). Resv specifically caused more opiate withdrawal, even more than spontaneous and naloxone-precipitated withdrawals [mean difference in withdrawal scores between Resv and heroin control = −18.3, between spontaneous WD and heroin control = −7.5, and between naloxone-precipitated WD and heroin control = −12.9, according to one way ANOVA, followed by Dunnett's multiple comparisons test against Her + Veh control group, n = 6–10 animals per group (** $p < 0.01$, *** $p < 0.001$, **** $p < 0.0001$). (**B**) Magnolol, quercetin, and β-catechin prevented the rapid wight loss associated with naloxone-precipitated heroin withdrawal: significant weight loss during naloxone-precipitated opiate withdrawal according to one-way ANOVA $F(5, 33) = 4.605$, $p = 0.002$ [mean difference in % body weight between heroin control and naloxone-precipitated withdrawal = 3.180, 95% CI = 0.6483 to 5.712, $p = 0.009$]. Mag [mean difference in % body weight between heroin control and naloxone-precipitated withdrawal + magnolol = 1.629, 95% CI = −0.9795 to 4.237, $p = 0.3$], Quer [mean difference in % body weight between heroin control and naloxone-precipitated withdrawal+ quercetin = 2.648, 95% CI = −0.05820 to 5.355, $p = 0.056$], and Cat [mean difference in % body weight between heroin control and naloxone-precipitated withdrawal + catechin = 0.6467, 95% CI = −2.060 to 3.353, $p = 0.9$] prevented weight loss, while Resv was associated with significant weight loss [mean difference in % body weight between heroin control and naloxone-precipitated withdrawal+ resveratrol = 3.973, 95% CI = 1.267 to 6.680, $p = 0.002$] by one way ANOVA followed by Dunnett's multiple comparisons test against Her + Veh control group, n = 6–10 animals per group (** $p < 0.01$, *** $p < 0.001$, **** $p < 0.0001$). (**C**) Heroin administration did not cause reduction in voluntary locomotor activity as measured in an open field: One-way ANOVA $F(5, 45)= 11.16$, $p < 0.0001$ [Veh + Veh vs. Her + Veh mean difference = 19.70, 95% CI = −7.256 to 46.66, $p = 0.2$]. However, both Resv [Veh + Veh vs. Her + Resv mean difference = 30.81, 95% CI = 4.700 to 56.92, $p = 0.01$] and cat [Veh + Veh vs. Her + Cat mean difference = 48.83, 95% CI = 21.87 to 75.78, $p = 0.0001$] caused significant reduction in distance traveled by one way ANOVA followed by Dunnett's multiple comparisons test against Her + Veh control group, n = 6–10 animals per group (* $p < 0.1$, ** $p < 0.01$, *** $p < 0.001$, **** $p < 0.0001$). (**D**) Magnolol and quercetin reduce anxiety-like behavior during naloxone-precipitated withdrawal as measured by time spent in open arm of elevated plus maze: Naloxone-precipitated withdrawal is associated with significant reduction in time spent in open arm compared to heroin controls by one-way ANOVA $F(5, 42) = 34.75$, $p < 0.0001$, [mean difference in time between heroin control and naloxone-precipitated withdrawal = 87.86, 95% CI = 29.41 to 146.3, $p = 0.001$]. Mag [mean difference in time between heroin control and naloxone-precipitated withdrawal + Mag = 14.88, 95% CI = −45.27 to 75.02, $p = 0.9$] and Quer [mean difference in time between heroin control and naloxone-precipitated withdrawal + Quer −6.125, 95% CI = −54.02 to 66.27, $p = 0.9$] increased the time in open arm to be non-significantly different from heroin control animals. However, Resv [mean difference in time between heroin control and naloxone-precipitated withdrawal + Resv = −148.3, 95% CI = −210.5 to −85.99, $p < 0.0001$] and Cat [mean difference in time between heroin control and naloxone-precipitated withdrawal + Cat = −148.3, 95% CI = −208.4 to −88.10, $p < 0.0001$] were associated with a significant increase in open arm time compared to controls by one way ANOVA followed by Dunnett's multiple comparisons test against Her + Veh control group, n = 6–8 animals per group. (** $p < 0.01$, *** $p < 0.001$, **** $p < 0.0001$). (**E**) Quercetin, magnolol, and β-catechin attenuate heroin-induced sucrose preference: Heroin administration was associated with significant increase in sucrose preference test by one-way ANOVA [$F(5, 30) = 4.300$, $p = 0.004$, heroin vs. vehicle mean difference = 15.84, 95% CI = 28.18 to 3.495, $p = 0.008$]. Resv did not reduce the increase in heroin-induced sucrose preference Resv vs. Vehicle mean difference = 14.44, 95% CI = 26.78 to 2.092, $p = 0.017$]. Quer, Mag and Cat, on the other hand attenuated heroin-induced increase in sucrose preference by one way ANOVA followed by Dunnett's multiple comparisons test against Her + Veh control group, n = 8–10 animals per group (* $p < 0.1$, ** $p < 0.01$, *** $p < 0.001$, **** $p < 0.0001$).

Quercetin and β-catechin are associated with significantly lower withdrawal scores compared to naloxone-precipitated withdrawal [mean difference in withdrawal scores between naloxone vs naloxone + quercetin = 7.00, 95% CI = 3.906 to 10.09 $p < 0.0001$ and naloxone vs naloxone + β-catechin = 8.667, 95% CI = 5.572 to 11.76 $p < 0.0001$, Figure 3A].

In addition, both quercetin and β-catechin prevented naloxone precipitated withdrawal-induced weight loss [mean difference in % body weight between heroin control and naloxone+ quercetin = 2.648, 95% CI = −0.05820 to 5.355, $p > 0.05$, and between heroin control and naloxone + B-catechin = 0.6467, 95% CI = −2.060 to 3.353, $p > 0.05$, Figure 3B]. β-catechin treatment was associated with a significant reduction in distance traveled in the open field ($p < 0.0001$, Figure 3C). Quercetin, magnolol, and β-catechin all attenuated sucrose preference associated with opiate withdrawal ($p > 0.05$, Figure 3E).

3.2. Resveratrol Worsens Naloxone-Induced Heroin Withdrawal

On the other hand, resveratrol showed a higher withdrawal score compared to naloxone [mean difference between naloxone vs naloxone + resveratrol = −5.2, 95% CI = −8.115 to −2.385 $p < 0.0001$, Figure 3A], and was associated with significant weight loss [mean difference in % body weight between heroin control and naloxone +resveratrol = 3.973, 95% CI = 1.267 to 6.680, $p < 0.01$, Figure 3B]. Resveratrol inhibited the distance traveled in the open field ($p < 0.01$ Figure 3C) but did not affect sucrose preference associated with heroin ($p < 0.05$, Figure 3E).

Resveratrol, quercetin, and magnolol prevent heroin-triggered reinstatement.

A single heroin dose (10 mg/kg) triggered reinstatement, as evidenced by the significant increase in the percentage of time spent in the heroin-paired chamber compared to the vehicle ($p = 0.0001$). Animals subjected to heroin administration along with resveratrol, quercetin, and magnolol spent significantly less time in the heroin-paired chamber compared to the heroin+ vehicle group ($p < 0.01$ each, Figure 4).

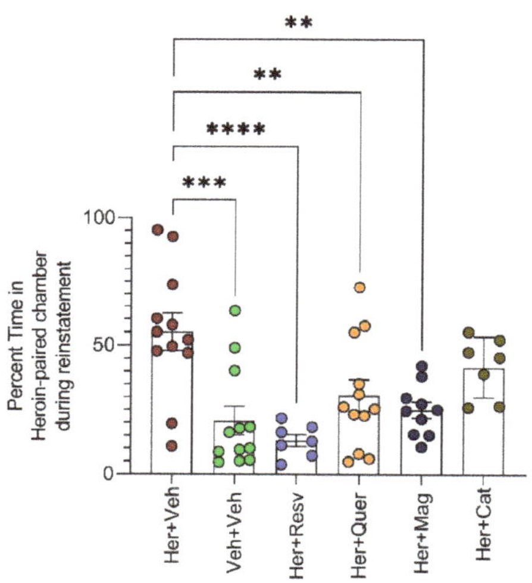

Figure 4. Resveratrol, quercetin, and magnolol prevent heroin-triggered reinstatement.

A single heroin dose triggered reinstatement, as evidenced by the percentage of time spent in the heroin-paired chamber, as assessed by one-way ANOVA $F(5, 54) = 6.983$, $p < 0.0001$, [mean difference in percentage of time spent in heroin-paired chamber between heroin and vehicle groups = 34.40, 95% CI = 14.99 to 53.82, $p = 0.0001$]. Resveratrol [mean difference in percentage of time in heroin-paired chamber between heroin and heroin + resveratrol groups = 42.06, 95% CI = 19.45 to 64.68, $p < 0.0001$], quercetin [mean difference in percentage of time in heroin-paired chamber between heroin and heroin + quercetin groups = 24.36, 95% CI = 4.947 to 43.78, $p = 0.008$], and magnolol [mean difference in percentage of time in heroin-paired chamber between heroin and heroin + magnolol groups = 30.04, 95% CI = 9.682 to 50.41, $p = 0.001$] all prevented heroin-triggered reinstatement. However, β-catechin did not [mean difference in percentage of time in heroin-paired chamber between heroin and heroin + β-catechin groups = 13.40, 95% CI = −9.222 to 36.01, $p = 0.4$], according to one way ANOVA followed by Dunnett's multiple comparisons test against Her + Veh control group, n = 7–11 animals per group (** $p < 0.01$, *** $p < 0.001$, **** $p < 0.0001$).

3.3. Molecular Effects

Resveratrol and magnolol attenuated the heroin-induced increase in striatal TNF-α concentration ($p < 0.001$, $p < 0.01$ respectively), while resveratrol and β-catechin attenuated the heroin-induced increase in striatal IL-6 concentration ($p < 0.01$, $p < 0.001$ respectively). On the other hand, quercetin accentuated IL-6 concentration ($p < 0.0001$) (Figure 5).

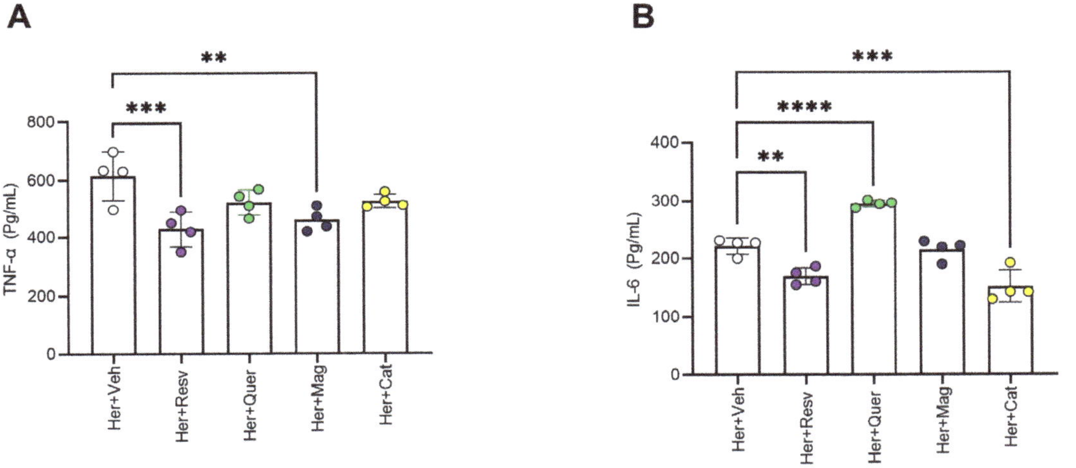

Figure 5. The effect of polyphenols on striatal TNF α and IL-6 and concentration. (**A**) Resveratrol and magnolol attenuate heroin-induced increase in striatal TNF-α concentration: Significant differences in striatal TNF-α concentrations were evident by one-way ANOVA [F(4, 15) = 6.885, $p < 0.01$]. Dunnett's multiple comparisons test against Her + Veh control group (n = 4) showed significant effect of resveratrol [mean difference in striatal TNF-α concentration between heroin and heroin+ resveratrol groups = 187.4, 95% CI = 81.40 to 293.4, $p < 0.001$], and magnolol [mean difference in striatal TNF-α concentration between heroin and heroin+ magnolol groups = 155.9, 95% CI = 49.84 to 261.9, $p < 0.01$] on TNF-α concentration. (**B**) Resveratrol and β-catechin attenuate, while quercetin accentuates, heroin-induced increase in striatal IL-6 concentration: Significant differences in striatal IL-6 concentrations were evident by one-way ANOVA [F(4, 15) = 42.74, $p < 0.0001$]. Dunnett's multiple comparisons test against Her + Veh control group (n = 4) showed significant effect of resveratrol [mean difference in striatal IL-6 concentration between heroin and heroin+ resveratrol groups = 51.11, 95% CI = 18.24 to 83.98, $p < 0.01$], quercetin [mean difference in striatal IL-6 concentration between heroin and heroin + quercetin groups = −74.66, 95% CI = −107.5 to −41.78, $p < 0.0001$], and β-catechin [mean difference in striatal IL-6 concentration between heroin and heroin+ β-catoctin groups = 68.69, 95% CI = 35.82 to 101.6, $p < 0.001$] on IL-6 concentration (** $p < 0.01$, *** $p < 0.001$, **** $p < 0.0001$).

4. Discussion

The results of this study show that different polyphenols target specific behavioral domains of heroin addiction in a CPP model and modulate the increase in striatal inflammatory cytokines TNF-α and IL-6 observed during naloxone-precipitated heroin withdrawal.

4.1. Quercetin Abolished Heroin Dependence Acquisition and Inhibited Reinstatement Attributed to Anti-Inflammatory Effects

Specifically, quercetin blocked the acquisition of heroin CPP, reduced withdrawal manifestations and heroin-triggered reinstatement, prevented heroin-induced sucrose preference, and accentuated the heroin-induced increase in striatal IL-6 concentration. Interestingly, quercetin prevented certain aspects of naloxone-precipitated heroin withdrawal, such as rapid wight loss and anxiety-like behavior.

Our results are in accordance with those of Singh et al., who reported that repeated administration of quercetin (25 and 50 mg/kg) for 9 days suppressed naloxone-precipitated morphine (10 mg/kg) withdrawal [31]. On the other hand, quercetin inhibited nicotine-triggered CPP reinstatement [42], alleviated METH-induced anxiety-like behavior in mice, attenuated the activation of astrocytes, and reduced the levels of IL-1beta and TNF-α, but not IL-6 [43]. In addition, quercetin (10, 30 and 100 mg/kg i.p.) pretreatment 30 min before ethanol administration in a CPP paradigm attenuated the acquisition and reinstatement and accelerated the extinction of ethanol-CPP [32]. Additionally, quercetin reversed morphine tolerance, attenuated morphine withdrawal expression in mice [44], and prevented ethanol-induced withdrawal somatic manifestations [36]. Taken together, it seems that quercetin is effective in blocking the acquisition and preventing the reinstatement of certain substances. Further studies are needed to investigate these specific aspects of quercetin before proposing proof of concept pilot studies in human heroin users.

4.2. Resveratrol Blocked Heroin Acquisition and Drug-Induced Reinstatement Effectively, but Accentuated Withdrawal Manifestations

The current results show that resveratrol, like quercetin, blocked the acquisition of heroin CPP and heroin-triggered reinstatement. However, it was associated with significantly higher withdrawal scores compared to heroin control, but unlike quercetin, resveratrol attenuated the heroin-induced increase in striatal TNF-α and IL-6 concentrations. Our results are the first, to the best of our knowledge, to report the effect of resveratrol on heroin addiction. Few studies have examined the effects of resveratrol on the attenuating behavioral manifestations of other substance. Yunusoglu et al. examined the effect of resveratrol on alcohol-induced conditioned place preference (CPP) in mice. Pretreatment with resveratrol (25, 50, and 75 mg/kg, i.p.) 30 min prior to ethanol administration impaired acquisition, and reinstatement of alcohol induced CPP and facilitated the extinction of alcohol CPP [30]. Moreover, pretreatment with resveratrol (10 or 100 mg/kg i.p.) remarkably attenuated methamphetamine (METH)-induced memory impairment in mice and reversed METH-induced oxidative damage and apoptosis [45]. As such, the current literature, including our results, suggest that resveratrol is also effective in blocking acquisition and preventing reinstalment of certain substances, but again, we observed a worsening of naloxone-precipitated heroin withdrawal. Calleri et al. [46] showed the antagonistic activity of resveratrol on PPARα and PPARγ. Further studies are needed to investigate whether the efficacy of resveratrol in blocking the acquisition of heroin CPP, heroin-triggered reinstatement, or its side effects in accentuating naloxone-induced heroin withdrawal is mediated through PPARα or PPARγ.

4.3. β-Catechin Blocked Reinstatement, but Not Acquisition of Heroin Dependence and Reduced the Withdrawal Manifestations

In the case of β-catechin, our results show that it failed to block the acquisition of heroin CPP or to attenuate naloxone-precipitated heroin withdrawal, but it prevented heroin-triggered reinstatement. Shutto et al. reported that resveratrol (40 mg/kg s.c.) enhanced the acute effect of cocaine on locomotor activity [47]. The authors speculated that this effect could be due to resveratrol enhancing dopamine neurotransmission through the inhibition of MAO-A and MAO-B. Further studies are needed to investigate these underlying mechanisms and to determine if indeed certain polyphenols inhibit MAO-A and MAO-B or activate $GABA_{(A)}$ receptors (as reported for quercetin), then the efficacy of these compounds in the treatment of depression and anxiety should be examined.

4.4. Magnolol Blocked Heroin-Induced Reinstatement, but Did Not Affect Acquisition and Withdrawal

Unlike quercetin and resveratrol, magnolol failed to block the acquisition of heroin CPP, but it successfully prevented heroin-triggered reinstatement. In addition, magnolol attenuated naloxone-precipitated heroin withdrawal, prevented rapid wight loss and anxiety-like behavior associated with naloxone-precipitated heroin withdrawal, and attenuate heroin-induced increase in striatal TNF-α concentration. More studies focused on this unique property of magnolol in attenuating heroin withdrawal manifestations are required.

4.5. Could Polyphenols Act through Dopaminergic Mechanisms?

Shutto et al. reported that resveratrol (40 mg/kg s.c.), enhanced the acute effect of cocaine on locomotor activity [47]. The authors speculated that this effect could be due to resveratrol enhancing dopamine neurotransmission through the inhibition of MAO-A and MAO-B. Further studies are needed to investigate these underlying mechanisms and if indeed certain polyphenols inhibit MAO-A and MAO-B or activate $GABA_{(A)}$ receptors (as reported for quercetin), then the efficacy of these compounds in the treatment of depression and anxiety should be examined.

PPAR-γ agonists can block rewarding properties of drugs through stimulating the mesolimbic dopaminergic neurotransmission [48,49]. The examined polyphenols are known to modulate PPAR-γ; thus, the acquisition blocking action of quercetin and resveratrol may be linked to dopamine transmission modulation. Quercetin was reported to increase the dopaminergic neuron density in the striatum of experimental PD in mice [50,51], and resveratrol exerted anti-depressant effects through modulating dopamine and serotonin, as shown in a previous report [52].

4.6. Could the Current Results Be Related to an Anti-Inflammatory Mechanism?

At the molecular level, we examined two neuroinflammatory markers in the striatum, IL-6 and TNF-α, following naloxone-induced heroin withdrawal. Our results show that magnolol, which successfully blocked withdrawal manifestations, attenuated the heroin-induced increase in striatal TNF-α, while quercetin was also associated with an increase in striatal IL-6 concentration. Magnolol attenuates the increase in pro-inflammatory cytokines such as IL-1β, IL-6 and TNF-α [53–59]. In addition, magnolol reduces glutamate-induced cytotoxicity in neuronal cell cultures [60], restores blood–brain barrier integrity, and reduces ischemia-associated brain edema [54], suggesting a neuroprotective property for magnolol against post ischemic stroke [61]. Several studies have shown its efficacy in reversing depressive-like behaviors in animal models using the sucrose preference test, the forced swim test [53,62], olfactory bulbectomy [63], and chronic unpredictable mild stress [64].

On the other hand, resveratrol attenuated striatal TNF-α, and β-catechin attenuated striatal IL-6 concentrations. Both resveratrol and β-catechin were of limited value in attenuating withdrawal manifestations. These results highlight the complexity of the neuroimmunological changes that take place during heroin use and the effects of different polyphenols on immune markers.

4.7. The Role of Immunomodulatory Mechanisms

Heroin and other exogenous opiates exert neuromodulatory effects through both immune suppression and activation, depending on the stage of drug use [65]. Acute morphine administration and morphine withdrawal both cause immune suppression [66]. Significant reduction in the response of T-lymphocytes to phytohemagglutinin challenge during acute withdrawal in heroin addicts has been reported [67]. Chronic heroin self-administration in rats produced a significant increase in lipopolysaccharide (LPS)-induced tumor necrosis factor-alpha (TNF-α) [68]. A similar increase in TNF-α, along with a marked elevation in total and activated B cells and IL-8 was reported in human heroin users with HIV and hepatitis C (n = 19) compared to controls (n = 19) [69]. This immune activation, with the rapid rise in cytokines, modulates the mesolimbic dopaminergic reward network, facilitating drug dependence and also contributing to the development of the acute withdrawal state [70,71].

The effects of individual polyphenols on the immune system are also complex. For example, resveratrol modulates the immune response, with both anti-inflammatory [72,73] and immune-enhancing effects [74,75], possibly in a dose-dependent manner [76]. One study reported that resveratrol interfered with the synthesis and gene expression of proinflammatory cytokines [77] through the suppression of the nuclear factor (NF)-kappaB signaling pathway. NF-kappaB plays a significant role in augmenting the inflammatory response through the release of free radicals [78]. In addition, resveratrol inhibited the production of TNF-α and IL-12 by peritoneal macrophages and blocked the activation of the transcription factor NF-kappaB, without affecting basal NF-kappaB activity [79]. On the other hand, resveratrol enhanced the immunity recovery of immunosuppressive mice through activating the NF-kappaB pathway and upregulating the expression of serum IL-2 and TNF-α in a dose-dependent manner [38]. In healthy volunteers ($n = 10$), resveratrol showed a significant increase in TNF-α levels 24 h after treatment compared to the baseline [74]. As such, resveratrol seems to exert different effects based on the underlying immune status.

Quercetin also exhibits immunomodulatory activities through inhibiting the secretion of inflammatory cytokines and improving immune function [reviewed in [80–84]]. Quercetin significantly inhibited the production of IL-6, and TNF-α in poly IC-induced RAW 264.7 mouse macrophages [85], reduced TNF-α and IL-8 mRNA expressions in a dose-dependent manner in zebrafish [86], and improved immune function via the NF-kappaB signaling pathway triggered by TNF-α in one-day-old healthy Arbor Acre broilers. [87].

β-Catechin is a natural immune enhancer present in several plants such as green tea leaves, black grapes, and cherries [88–90]. One study reported that catechin inhibited the gene expression of pro-inflammatory cytokines IL-1β and IL-6, and enhanced the gene expression of anti-inflammatory cytokines IL-4 and IL-10 [91]. At a behavioral level, catechin (25, 50, and 100 mg/kg administered orally for 11 to 25 days) was associated with significant improvement in behavioral manifestations of sociability, stereotypy, anxiety, depression, novelty, repetitive, and perseverative behaviors in rodents [92]. Another study showed that β-catechin increased life-span in a senescence accelerated mouse model of aging [93].

The role of PPAR-γ activation cannot be neglected. PPAR-γ expression functions as a vital regulator in NF-κappaB-mediated inflammation [94]. It was postulated that PPAR-γ activation by agonists such as leriglitazone reduce oxidative stress and boost biogenesis and mitochondrial functionality associated with the NF-kappaB inflammatory mechanisms [95], resulting in anti-inflammatory and anti-oxidative stress regulation [96].

Clinical data relative to the use of these compounds in drug use and dependence is not present. However, quercetin, resveratrol, and catechin were investigated clinically in many other disorders such as diabetes, cancer, arthritis, or neurodegenerative diseases, while magnolol was investigated in dental and periodontal studies (clinicaltrials.gov; review [97–101].

5. Limitations

The results of this study should be viewed in light of its limitations. First, we examined only adult male rats and could not comment on the efficacy of tested compounds in adolescents or female rats. Second, we did not examine different doses or the optimal therapeutic window for the efficacy of polyphenols for different stages of drug use. Third, we investigated only IL-6 and TNF-α following naloxone-precipitated withdrawal and did not examine other neuroimmune markers during acquisition or reinstatement. Future studies are needed to expand on the current findings and examine other neuroinflammatory markers at each stage of heroin addiction. Despite these limitations, the results of this study lend more evidence to the potential therapeutic benefits of PPAR agonists in reversing certain behavioral manifestations of heroin use and highlight the immunomodulatory function of these compounds, with some concerns for worsening behavioral manifestations of heroin withdrawal. Further preclinical research is needed to gain more insight into the utility of natural polyphenols in treating heroin use disorder in human subjects.

6. Conclusions

Despite these limitations, our current results show that four different polyphenols, with known modulatory effects at the PPAR-γ, are effective in attenuating different aspects of heroin addiction. Quercetin and resveratrol could be effective in blocking heroin relapse, while quercetin and magnolol may be utilized in reducing the severity of heroin withdrawal. Resveratrol use during early heroin abstinence might aggravate withdrawal manifestations. B-catechin was of limited value in opioid withdrawal, but it blocked reinstatement and relapse to heroin. Proof of concept pilot clinical trials are needed to test the potential efficacy of these compounds in treating patients with heroin use disorder.

Author Contributions: S.E. and O.A.A. have full access to all the data in the study and take responsibility for the integrity of the data and the accuracy of the data analysis. Concept and design, S.E. and O.A.A.; acquisition, analysis, or interpretation of data, S.E., R.E., M.E.-S. and N.M.; drafting of the manuscript, S.E., R.E. and O.A.A.; critical revision of the manuscript for important intellectual content, all authors; administrative, technical, or material support, S.E., R.E., M.E.-S., N.M. and O.A.A.; supervision, O.A.A. All authors have read and agreed to the published version of the manuscript.

Funding: This study was supported by the in-house project #12060142 funded by the National Research Center, Cairo, Egypt.

Institutional Review Board Statement: The animal study protocol was approved by the Institutional Ethics Committee) of the Egyptian National Research Centre (protocol #19-220 date of approval 20 November 2019).

Informed Consent Statement: Not applicable.

Data Availability Statement: Data is contained within the article.

Conflicts of Interest: The authors declare no conflict of interest.

References

1. Degenhardt, L.; Whiteford, H.A.; Ferrari, A.J.; Baxter, A.J.; Charlson, F.J.; Hall, W.D.; Freedman, G.; Burstein, R.; Johns, N.; Engell, R.E.; et al. Global burden of disease attributable to illicit drug use and dependence: Findings from the Global Burden of Disease Study 2010. *Lancet* **2013**, *382*, 1564–1574. [CrossRef]
2. Lipari, R.N.; Hughes, A. Trends in Heroin Use in the United States: 2002 to 2013. In *The CBHSQ Report*; Substance Abuse and Mental Health Services Administration: Rockville, MD, USA, 2013; pp. 1–11.
3. Martins, S.S.; Sarvet, A.; Santaella-Tenorio, J.; Saha, T.; Grant, B.F.; Hasin, D.S. Changes in US Lifetime Heroin Use and Heroin Use Disorder: Prevalence From the 2001–2002 to 2012–2013 National Epidemiologic Survey on Alcohol and Related Conditions. *JAMA Psychiatry* **2017**, *74*, 445–455. [CrossRef]
4. Jones, C.M. Heroin use and heroin use risk behaviors among nonmedical users of prescription opioid pain relievers—United States, 2002–2004 and 2008–2010. *Drug Alcohol Depend.* **2013**, *132*, 95–100. [CrossRef]
5. Azab, S.M.S.; Tawfik, H.; Hayes, B.D. Intoxication related to substances use in patients presenting to Ain Shams University Poisoning Treatment Center, Cairo, Egypt (2015–2019). *Drug Alcohol Rev.* **2022**, *41*, 1109–1118. [CrossRef]
6. Connery, H.S. Medication-assisted treatment of opioid use disorder: Review of the evidence and future directions. *Harv. Rev. Psychiatry* **2015**, *23*, 63–75. [CrossRef]
7. Kosten, T.R.; George, T.P. The neurobiology of opioid dependence: Implications for treatment. *Sci. Pract. Perspect* **2002**, *1*, 13–20. [CrossRef]
8. Koehl, J.L.; Zimmerman, D.E.; Bridgeman, P.J. Medications for management of opioid use disorder. *Am. J. Health Syst. Pharm.* **2019**, *76*, 1097–1103. [CrossRef]
9. Eastwood, B.; Strang, J.; Marsden, J. Effectiveness of treatment for opioid use disorder: A national, five-year, prospective, observational study in England. *Drug Alcohol Depend.* **2017**, *176*, 139–147. [CrossRef]
10. Woods, J.S.; Joseph, H. From Narcotic to Normalizer: The Misperception of Methadone Treatment and the Persistence of Prejudice and Bias. *Subst. Use Misuse* **2018**, *53*, 323–329. [CrossRef]
11. Polydorou, S.; Ross, S.; Coleman, P.; Duncan, L.; Roxas, N.; Thomas, A.; Mendoza, S.; Hansen, H. Integrating Buprenorphine Into an Opioid Treatment Program: Tailoring Care for Patients with Opioid Use Disorders. *Psychiatr. Serv.* **2017**, *68*, 295–298. [CrossRef]

12. Johnson, B.; Richert, T. Diversion of methadone and buprenorphine by patients in opioid substitution treatment in Sweden: Prevalence estimates and risk factors. *Int. J. Drug Policy* **2015**, *26*, 183–190. [CrossRef]
13. Hofford, R.S.; Russo, S.J.; Kiraly, D.D. Neuroimmune mechanisms of psychostimulant and opioid use disorders. *Eur. J. Neurosci.* **2019**, *50*, 2562–2573. [CrossRef]
14. Roy, S.; Barke, R.A.; Loh, H.H. MU-opioid receptor-knockout mice: Role of mu-opioid receptor in morphine mediated immune functions. *Brain Res. Mol. Brain Res.* **1998**, *61*, 190–194. [CrossRef]
15. Merighi, S.; Gessi, S.; Varani, K.; Fazzi, D.; Stefanelli, A.; Borea, P.A. Morphine mediates a proinflammatory phenotype via mu-opioid receptor-PKCvarepsilon-Akt-ERK1/2 signaling pathway in activated microglial cells. *Biochem. Pharmacol.* **2013**, *86*, 487–496. [CrossRef]
16. Morcuende, A.; Navarrete, F.; Nieto, E.; Manzanares, J.; Femenia, T. Inflammatory Biomarkers in Addictive Disorders. *Biomolecules* **2021**, *11*, 1824. [CrossRef]
17. Chan, Y.Y.; Yang, S.N.; Lin, J.C.; Chang, J.L.; Lin, J.G.; Lo, W.Y. Inflammatory response in heroin addicts undergoing methadone maintenance treatment. *Psychiatry Res.* **2015**, *226*, 230–234. [CrossRef]
18. Bastami, S.; Norling, C.; Trinks, C.; Holmlund, B.; Walz, T.M.; Ahlner, J.; Uppugunduri, S. Inhibitory effect of opiates on LPS mediated release of TNF and IL-8. *Acta Oncol.* **2013**, *52*, 1022–1033. [CrossRef]
19. Belkowski, S.M.; Alicea, C.; Eisenstein, T.K.; Adler, M.W.; Rogers, T.J. Inhibition of interleukin-1 and tumor necrosis factor-alpha synthesis following treatment of macrophages with the kappa opioid agonist U50, 488H. *J. Pharmacol. Exp. Ther.* **1995**, *273*, 1491–1496.
20. Wang, W.; Lin, Q.; Lin, R.; Zhang, J.; Ren, F.; Zhang, J.; Ji, M.; Li, Y. PPARalpha agonist fenofibrate attenuates TNF-alpha-induced CD40 expression in 3T3-L1 adipocytes via the SIRT1-dependent signaling pathway. *Exp. Cell Res.* **2013**, *319*, 1523–1533. [CrossRef]
21. Grabacka, M.; Pierzchalska, M.; Plonka, P.M.; Pierzchalski, P. The Role of PPAR Alpha in the Modulation of Innate Immunity. *Int. J. Mol. Sci.* **2021**, *22*, 10545. [CrossRef] [PubMed]
22. Gehm, B.D.; McAndrews, J.M.; Chien, P.Y.; Jameson, J.L. Resveratrol, a polyphenolic compound found in grapes and wine, is an agonist for the estrogen receptor. *Proc. Natl. Acad. Sci. USA* **1997**, *94*, 14138–14143. [CrossRef] [PubMed]
23. Erlund, I.; Freese, R.; Marniemi, J.; Hakala, P.; Alfthan, G. Bioavailability of quercetin from berries and the diet. *Nutr. Cancer* **2006**, *54*, 13–17. [CrossRef]
24. Johnson, R.; Bryant, S.; Huntley, A.L. Green tea and green tea catechin extracts: An overview of the clinical evidence. *Maturitas* **2012**, *73*, 280–287. [CrossRef] [PubMed]
25. Sarrica, A.; Kirika, N.; Romeo, M.; Salmona, M.; Diomede, L. Safety and Toxicology of Magnolol and Honokiol. *Planta Med.* **2018**, *84*, 1151–1164. [CrossRef]
26. Bhullar, K.S.; Rupasinghe, H.P. Polyphenols: Multipotent therapeutic agents in neurodegenerative diseases. *Oxid. Med. Cell Longev.* **2013**, *2013*, 891748. [CrossRef] [PubMed]
27. Lee, K. Transactivation of peroxisome proliferator-activated receptor alpha by green tea extracts. *J. Vet. Sci.* **2004**, *5*, 325–330. [CrossRef]
28. Marinovic, M.P.; Sousa-Filho, C.P.B.; Batista, F.A.H.; Avelino, T.M.; Cogliati, B.; Figueira, A.C.M.; Otton, R.; Rodrigues, A.C. Green tea extract increases adiponectin and PPAR alpha levels to improve hepatic steatosis. *J. Nutr. Biochem.* **2022**, *103*, 108957. [CrossRef]
29. Rana, A.; Samtiya, M.; Dhewa, T.; Mishra, V.; Aluko, R.E. Health benefits of polyphenols: A concise review. *J. Food Biochem.* **2022**, *46*, e14264. [CrossRef] [PubMed]
30. Yunusoglu, O. Resveratrol impairs acquisition, reinstatement and precipitates extinction of alcohol-induced place preference in mice. *Neurol. Res.* **2021**, *43*, 985–994. [CrossRef]
31. Singh, A.; Naidu, P.S.; Kulkarni, S.K. Quercetin, a bioflavonoid, reverses development of tolerance and dependence to morphine. *Drug Dev. Res.* **2002**, *57*, 167–172. [CrossRef]
32. Yunusoglu, O. Evaluation of the effects of quercetin on the rewarding property of ethanol in mice. *Neurosci. Lett.* **2022**, *768*, 136383. [CrossRef]
33. Capasso, A.; Piacente, S.; Pizza, C.; Sorrentino, L. Flavonoids reduce morphine withdrawal in-vitro. *J. Pharm. Pharmacol.* **1998**, *50*, 561–564. [CrossRef]
34. Abulseoud, O.A.; Miller, J.D.; Wu, J.; Choi, D.S.; Holschneider, D.P. Ceftriaxone upregulates the glutamate transporter in medial prefrontal cortex and blocks reinstatement of methamphetamine seeking in a condition place preference paradigm. *Brain Res.* **2012**, *1456*, 14–21. [CrossRef]
35. Aguilar, M.A.; Rodriguez-Arias, M.; Minarro, J. Neurobiological mechanisms of the reinstatement of drug-conditioned place preference. *Brain Res. Rev.* **2009**, *59*, 253–277. [CrossRef]
36. Joshi, D.; Naidu, P.S.; Singh, A.; Kulkarni, S.K. Protective effect of quercetin on alcohol abstinence-induced anxiety and convulsions. *J. Med. Food* **2005**, *8*, 392–396. [CrossRef] [PubMed]
37. Ganeshpurkar, A.; Saluja, A.K. Protective effect of catechin on humoral and cell mediated immunity in rat model. *Int. Immunopharmacol.* **2018**, *54*, 261–266. [CrossRef] [PubMed]
38. Lai, X.; Pei, Q.; Song, X.; Zhou, X.; Yin, Z.; Jia, R.; Zou, Y.; Li, L.; Yue, G.; Liang, X.; et al. The enhancement of immune function and activation of NF-kappaB by resveratrol-treatment in immunosuppressive mice. *Int. Immunopharmacol.* **2016**, *33*, 42–47. [CrossRef]

39. Bai, Y.; Song, L.; Dai, G.; Xu, M.; Zhu, L.; Zhang, W.; Jing, W.; Ju, W. Antidepressant effects of magnolol in a mouse model of depression induced by chronic corticosterone injection. *Steroids* **2018**, *135*, 73–78. [CrossRef] [PubMed]
40. Popik, P.; Skolnick, P. The NMDA antagonist memantine blocks the expression and maintenance of morphine dependence. *Pharmacol. Biochem. Behav.* **1996**, *53*, 791–797. [CrossRef]
41. Pum, M.E.; Huston, J.P.; Muller, C.P. The role of cortical serotonin in anxiety and locomotor activity in Wistar rats. *Behav. Neurosci.* **2009**, *123*, 449–454. [CrossRef] [PubMed]
42. Rahmadi, M.; Suasana, D.; Lailis, S.R.; Ratri, D.M.N.; Ardianto, C. The effects of quercetin on nicotine-induced reward effects in mice. *J. Basic Clin. Physiol. Pharmacol.* **2021**, *32*, 327–333. [CrossRef]
43. Chen, F.; Sun, J.; Chen, C.; Zhang, Y.; Zou, L.; Zhang, Z.; Chen, M.; Wu, H.; Tian, W.; Liu, Y.; et al. Quercetin Mitigates Methamphetamine-Induced Anxiety-Like Behavior through Ameliorating Mitochondrial Dysfunction and Neuroinflammation. *Front. Mol. Neurosci.* **2022**, *15*, 829886. [CrossRef] [PubMed]
44. Naidu, P.S.; Singh, A.; Joshi, D.; Kulkarni, S.K. Possible mechanisms of action in quercetin reversal of morphine tolerance and dependence. *Addict. Biol.* **2003**, *8*, 327–336. [CrossRef]
45. Zeng, Q.; Xiong, Q.; Zhou, M.; Tian, X.; Yue, K.; Li, Y.; Ru, Q. Resveratrol attenuates methamphetamine-induced memory impairment via inhibition of oxidative stress and apoptosis in mice. *J. Food Biochem.* **2021**, *45*, e13622. [CrossRef] [PubMed]
46. Calleri, E.; Pochetti, G.; Dossou, K.S.S.; Laghezza, A.; Montanari, R.; Capelli, D.; Prada, E.; Loiodice, F.; Massolini, G.; Bernier, M.; et al. Resveratrol and its metabolites bind to PPARs. *Chembiochem* **2014**, *15*, 1154–1160. [CrossRef]
47. Shuto, T.; Kuroiwa, M.; Koga, Y.; Kawahara, Y.; Sotogaku, N.; Toyomasu, K.; Nishi, A. Acute effects of resveratrol to enhance cocaine-induced dopamine neurotransmission in the striatum. *Neurosci. Lett.* **2013**, *542*, 107–112. [CrossRef] [PubMed]
48. Panlilio, L.V.; Justinova, Z.; Goldberg, S.R. Inhibition of FAAH and activation of PPAR: New approaches to the treatment of cognitive dysfunction and drug addiction. *Pharmacol. Ther.* **2013**, *138*, 84–102. [CrossRef]
49. Quiroga, C.; Barberena, J.J.; Alcaraz-Silva, J.; Machado, S.; Imperatori, C.; Yadollahpour, A.; Budde, H.; Yamamoto, T.; Telles-Correia, D.; Murillo-Rodríguez, E. The role of peroxisome proliferator-activated receptor in addiction: A novel drug target. *Curr. Top. Med. Chem.* **2021**, *21*, 964–975. [CrossRef]
50. Singh, S.; Jamwal, S.; Kumar, P. Neuroprotective potential of Quercetin in combination with piperine against 1-methyl-4-phenyl-1,2,3,6-tetrahydropyridine-induced neurotoxicity. *Neural Regen. Res.* **2017**, *12*, 1137–1144.
51. Jain, J.; Hasan, W.; Biswas, P.; Yadav, R.S.; Jat, D. Neuroprotective effect of quercetin against rotenone-induced neuroinflammation and alterations in mice behavior. *J. Biochem. Mol. Toxicol.* **2022**, *36*, e23165. [CrossRef]
52. Yu, Y.; Wang, R.; Chen, C.; Du, X.; Ruan, L.; Sun, J.; Li, J.; Zhang, L.; O'Donnell, J.M.; Pan, J. Antidepressant-like effect of trans-resveratrol in chronic stress model: Behavioral and neurochemical evidences. *J. Psychiatr. Res.* **2013**, *47*, 315–322. [CrossRef] [PubMed]
53. Cheng, J.; Dong, S.; Yi, L.; Geng, D.; Liu, Q. Magnolol abrogates chronic mild stress-induced depressive-like behaviors by inhibiting neuroinflammation and oxidative stress in the prefrontal cortex of mice. *Int. Immunopharmacol.* **2018**, *59*, 61–67. [CrossRef] [PubMed]
54. Kou, D.Q.; Jiang, Y.L.; Qin, J.H.; Huang, Y.H. Magnolol attenuates the inflammation and apoptosis through the activation of SIRT1 in experimental stroke rats. *Pharmacol. Rep.* **2017**, *69*, 642–647. [CrossRef]
55. Tse, A.K.; Wan, C.K.; Zhu, G.Y.; Shen, X.L.; Cheung, H.Y.; Yang, M.; Fong, W.F. Magnolol suppresses NF-kappaB activation and NF-kappaB regulated gene expression through inhibition of IkappaB kinase activation. *Mol. Immunol.* **2007**, *44*, 2647–2658. [CrossRef] [PubMed]
56. Chen, T.; Shou, L.; Guo, X.; Wei, M.; Zheng, H.; Tao, T. Magnolol attenuates the locomotor impairment, cognitive deficit, and neuroinflammation in Alzheimer's disease mice with brain insulin resistance via up-regulating miR-200c. *Bioengineered* **2022**, *13*, 531–543. [CrossRef] [PubMed]
57. Chen, H.; Fu, W.; Chen, H.; You, S.; Liu, X.; Yang, Y.; Wei, Y.; Huang, J.; Rui, W. Magnolol attenuates the inflammation and enhances phagocytosis through the activation of MAPK, NF-kappaB signal pathways in vitro and in vivo. *Mol. Immunol.* **2019**, *105*, 96–106. [CrossRef]
58. Xian, Y.F.; Qu, C.; Liu, Y.; Ip, S.P.; Yuan, Q.J.; Yang, W.; Lin, Z.X. Magnolol Ameliorates Behavioral Impairments and Neuropathology in a Transgenic Mouse Model of Alzheimer's Disease. *Oxid. Med. Cell Longev.* **2020**, *2020*, 5920476. [CrossRef]
59. Huang, S.Y.; Tai, S.H.; Chang, C.C.; Tu, Y.F.; Chang, C.H.; Lee, E.J. Magnolol protects against ischemic-reperfusion brain damage following oxygen-glucose deprivation and transient focal cerebral ischemia. *Int. J. Mol. Med.* **2018**, *41*, 2252–2262. [CrossRef]
60. Lee, W.T.; Lin, M.H.; Lee, E.J.; Hung, Y.C.; Tai, S.H.; Chen, H.Y.; Chen, T.Y.; Wu, T.S. Magnolol reduces glutamate-induced neuronal excitotoxicity and protects against permanent focal cerebral ischemia up to 4 hours. *PLoS ONE* **2012**, *7*, e39952. [CrossRef]
61. Zhou, F.; Jiang, Z.; Yang, B.; Hu, Z. Magnolol exhibits anti-inflammatory and neuroprotective effects in a rat model of intracerebral haemorrhage. *Brain Behav. Immun.* **2019**, *77*, 161–167. [CrossRef]
62. Li, L.F.; Yang, J.; Ma, S.P.; Qu, R. Magnolol treatment reversed the glial pathology in an unpredictable chronic mild stress-induced rat model of depression. *Eur. J. Pharmacol.* **2013**, *711*, 42–49. [CrossRef]
63. Matsui, N.; Akae, H.; Hirashima, N.; Kido, Y.; Tanabe, S.; Koseki, M.; Fukuyama, Y.; Akagi, M. Magnolol Enhances Hippocampal Neurogenesis and Exerts Antidepressant-Like Effects in Olfactory Bulbectomized Mice. *Phytother. Res.* **2016**, *30*, 1856–1861. [CrossRef] [PubMed]

64. Tao, W.; Hu, Y.; Chen, Z.; Dai, Y.; Hu, Y.; Qi, M. Magnolol attenuates depressive-like behaviors by polarizing microglia towards the M2 phenotype through the regulation of Nrf2/HO-1/NLRP3 signaling pathway. *Phytomedicine* **2021**, *91*, 153692. [CrossRef] [PubMed]
65. Liang, X.; Liu, R.; Chen, C.; Ji, F.; Li, T. Opioid System Modulates the Immune Function: A Review. *Transl. Perioper. Pain Med.* **2016**, *1*, 5–13. [PubMed]
66. Eisenstein, T.K.; Rahim, R.T.; Feng, P.; Thingalaya, N.K.; Meissler, J.J. Effects of opioid tolerance and withdrawal on the immune system. *J. Neuroimmune Pharmacol.* **2006**, *1*, 237–249. [CrossRef] [PubMed]
67. Govitrapong, P.; Suttitum, T.; Kotchabhakdi, N.; Uneklabh, T. Alterations of immune functions in heroin addicts and heroin withdrawal subjects. *J. Pharmacol. Exp. Ther.* **1998**, *286*, 883–889.
68. Weber, R.J.; Gomez-Flores, R.; Smith, J.E.; Martin, T.J. Immune, neuroendocrine, and somatic alterations in animal models of human heroin abuse. *J. Neuroimmunol.* **2004**, *147*, 134–137. [CrossRef]
69. Piepenbrink, M.S.; Samuel, M.; Zheng, B.; Carter, B.; Fucile, C.; Bunce, C.; Kiebala, M.; Khan, A.A.; Thakar, J.; Maggirwar, S.B.; et al. Humoral Dysregulation Associated with Increased Systemic Inflammation among Injection Heroin Users. *PLoS ONE* **2016**, *11*, e0158641. [CrossRef]
70. Sheng, W.S.; Hu, S.; Gekker, G.; Zhu, S.; Peterson, P.K.; Chao, C.C. Immunomodulatory role of opioids in the central nervous system. *Arch. Immunol. Ther. Exp.* **1997**, *45*, 359–366.
71. Coller, J.K.; Hutchinson, M.R. Implications of central immune signaling caused by drugs of abuse: Mechanisms, mediators and new therapeutic approaches for prediction and treatment of drug dependence. *Pharmacol. Ther.* **2012**, *134*, 219–245. [CrossRef]
72. Alesci, A.; Nicosia, N.; Fumia, A.; Giorgianni, F.; Santini, A.; Cicero, N. Resveratrol and Immune Cells: A Link to Improve Human Health. *Molecules* **2022**, *27*, 424. [CrossRef]
73. Shakibaei, M.; Harikumar, K.B.; Aggarwal, B.B. Resveratrol addiction: To die or not to die. *Mol. Nutr. Food Res.* **2009**, *53*, 115–128. [CrossRef]
74. Gualdoni, G.A.; Kovarik, J.J.; Hofer, J.; Dose, F.; Pignitter, M.; Doberer, D.; Steinberger, P.; Somoza, V.; Wolzt, M.; Zlabinger, G.J. Resveratrol enhances TNF-alpha production in human monocytes upon bacterial stimulation. *Biochim. Biophys. Acta* **2014**, *1840*, 95–105. [CrossRef] [PubMed]
75. Feng, Y.H.; Zhou, W.L.; Wu, Q.L.; Li, X.Y.; Zhao, W.M.; Zou, J.P. Low dose of resveratrol enhanced immune response of mice. *Acta Pharmacol. Sin.* **2002**, *23*, 893–897. [PubMed]
76. Falchetti, R.; Fuggetta, M.P.; Lanzilli, G.; Tricarico, M.; Ravagnan, G. Effects of resveratrol on human immune cell function. *Life Sci.* **2001**, *70*, 81–96. [CrossRef]
77. Malaguarnera, L. Influence of Resveratrol on the Immune Response. *Nutrients* **2019**, *11*, 946. [CrossRef]
78. Xu, L.; Botchway, B.O.A.; Zhang, S.; Zhou, J.; Liu, X. Inhibition of NF-kappaB Signaling Pathway by Resveratrol Improves Spinal Cord Injury. *Front. Neurosci.* **2018**, *12*, 690. [CrossRef]
79. Gao, X.; Xu, Y.X.; Janakiraman, N.; Chapman, R.A.; Gautam, S.C. Immunomodulatory activity of resveratrol: Suppression of lymphocyte proliferation, development of cell-mediated cytotoxicity, and cytokine production. *Biochem. Pharmacol.* **2001**, *62*, 1299–1308. [CrossRef] [PubMed]
80. Li, Y.; Yao, J.; Han, C.; Yang, J.; Chaudhry, M.T.; Wang, S.; Liu, H.; Yin, Y. Quercetin, Inflammation and Immunity. *Nutrients* **2016**, *8*, 167. [CrossRef]
81. Gangwar, V.; Garg, A.; Lomore, K.; Korla, K.; Bhat, S.S.; Rao, R.P.; Rafiq, M.; Kumawath, R.; Uddagiri, B.V.; Kareenhalli, V.V. Immunomodulatory Effects of a Concoction of Natural Bioactive Compounds-Mechanistic Insights. *Biomedicines* **2021**, *9*, 1522. [CrossRef]
82. Zhang, L.; Ma, J.; Yang, F.; Li, S.; Ma, W.; Chang, X.; Yang, L. Neuroprotective Effects of Quercetin on Ischemic Stroke: A Literature Review. *Front. Pharmacol.* **2022**, *13*, 854249. [CrossRef]
83. Azeem, M.; Hanif, M.; Mahmood, K.; Ameer, N.; Chughtai, F.R.S.; Abid, U. An insight into anticancer, antioxidant, antimicrobial, antidiabetic and anti-inflammatory effects of quercetin: A review. *Polym. Bull.* **2022**, *80*, 241–262. [CrossRef] [PubMed]
84. Mlcek, J.; Jurikova, T.; Skrovankova, S.; Sochor, J. Quercetin and Its Anti-Allergic Immune Response. *Molecules* **2016**, *21*, 623. [CrossRef] [PubMed]
85. Kim, Y.J.; Park, W. Anti-Inflammatory Effect of Quercetin on RAW 264.7 Mouse Macrophages Induced with Polyinosinic-Polycytidylic Acid. *Molecules* **2016**, *21*, 450. [CrossRef] [PubMed]
86. Wang, J.; Zhang, C.; Zhang, J.; Xie, J.; Yang, L.; Xing, Y.; Li, Z. The effects of quercetin on immunity, antioxidant indices, and disease resistance in zebrafish (*Danio rerio*). *Fish Physiol Biochem.* **2020**, *46*, 759–770. [CrossRef]
87. Yang, J.X.; Maria, T.C.; Zhou, B.; Xiao, F.L.; Wang, M.; Mao, Y.J.; Li, Y. Quercetin improves immune function in Arbor Acre broilers through activation of NF-kappaB signaling pathway. *Poult. Sci.* **2020**, *99*, 2305. [CrossRef]
88. Li, J.; Du, L.; He, J.N.; Chu, K.O.; Guo, C.L.; Wong, M.O.M.; Pang, C.P.; Chu, W.K. Anti-inflammatory Effects of GTE in Eye Diseases. *Front. Nutr.* **2021**, *8*, 753955. [CrossRef]
89. Kim, J.M.; Heo, H.J. The roles of catechins in regulation of systemic inflammation. *Food Sci. Biotechnol.* **2022**, *31*, 957–970. [CrossRef]
90. Baranwal, A.; Aggarwal, P.; Rai, A.; Kumar, N. Pharmacological Actions and Underlying Mechanisms of Catechin: A Review. *Mini Rev. Med. Chem.* **2022**, *22*, 821–833. [CrossRef]

91. Cheng, A.W.; Tan, X.; Sun, J.Y.; Gu, C.M.; Liu, C.; Guo, X. Catechin attenuates TNF-alpha induced inflammatory response via AMPK-SIRT1 pathway in 3T3-L1 adipocytes. *PLoS ONE* **2019**, *14*, e0217090. [CrossRef]
92. Mehta, R.; Bhandari, R.; Kuhad, A. Effects of catechin on a rodent model of autism spectrum disorder: Implications for the role of nitric oxide in neuroinflammatory pathway. *Psychopharmacology* **2021**, *238*, 3249–3271. [CrossRef]
93. Kumari, M.V.; Yoneda, T.; Hiramatsu, M. Effect of "beta CATECHIN" on the life span of senescence accelerated mice (SAM-P8 strain). *Biochem. Mol. Biol. Int.* **1997**, *41*, 1005–1011. [PubMed]
94. Wang, J.; Guo, X.; Chen, C.; Sun, S.; Liu, G.; Liu, M.; Hao, M.; Che, H. Gender differences in food allergy depend on the PPAR gamma/NF-kappaB in the intestines of mice. *Life Sci.* **2021**, *278*, 119606. [CrossRef] [PubMed]
95. Santambrogio, P.; Cozzi, A.; Di Meo, I.; Cavestro, C.; Vergara, C.; Rodriguez-Pascau, L.; Martinell, M.; Pizcueta, P.; Tiranti, V.; Levi, S. PPAR Gamma Agonist Leriglitazone Recovers Alterations Due to Pank2-Deficiency in hiPS-Derived Astrocytes. *Pharmaceutics* **2023**, *15*, 202. [CrossRef] [PubMed]
96. Yao, H.; Zhao, J.; Song, X. Protective effects of fraxin on cerebral ischemia-reperfusion injury by mediating neuroinflammation and oxidative stress through PPAR-γ/NF-KB pathway. *Brain Res. Bull.* **2022**, *187*, 49–62. [CrossRef] [PubMed]
97. Leyva-Soto, A.; Chavez-Santoscoy, R.A.; Porras, O.; Hidalgo-Ledesma, M.; Serrano-Medina, A.; Ramírez-Rodríguez, A.A.; Castillo-Martinez, N.A. Epicatechin and quercetin exhibit in vitro antioxidant effect, improve biochemical parameters related to metabolic syndrome, and decrease cellular genotoxicity in humans. *Food Res. Int.* **2021**, *142*, 110101. [CrossRef]
98. Chhabra, G.; Singh, C.K.; Ndiaye, M.A.; Fedorowicz, S.; Molot, A.; Ahmad, N. Prostate cancer chemoprevention by natural agents: Clinical evidence and potential implications. *Cancer Lett.* **2018**, *422*, 9–18. [CrossRef] [PubMed]
99. Moussa, C.; Hebron, M.; Huang, X.; Ahn, J.; Rissman, R.A.; Aisen, P.S.; Turner, R.S. Resveratrol regulates neuro-inflammation and induces adaptive immunity in Alzheimer's disease. *J. Neuroinflamm.* **2017**, *14*, 1. [CrossRef] [PubMed]
100. Delpino, F.M.; Figueiredo, L.M. Resveratrol supplementation and type 2 diabetes: A systematic review and meta-analysis. *Crit. Rev. Food Sci. Nutr.* **2022**, *62*, 4465–4480. [CrossRef]
101. Henning, S.M.; Wang, P.; Lee, R.P.; Trang, A.; Husari, G.; Yang, J.; Aronson, W.J. Prospective randomized trial evaluating blood and prostate tissue concentrations of green tea polyphenols and quercetin in men with prostate cancer. *Food Funct.* **2020**, *11*, 4114–4122. [CrossRef]

toxics

MDPI

Article

Effects of *Rosmarinus officinalis* L. Extract on Neurobehavioral and Neurobiological Changes in Male Rats with Pentylenetetrazol-Induced Epilepsy

Jawaher Alrashdi [1], Gadah Albasher [1,*], Mohammed M. Alanazi [2], Wedad Saeed Al-Qahtani [3], Abdulkareem A. Alanezi [4] and Fawaz Alasmari [2]

1 Department of Zoology, College of Science, King Saud University, Riyadh 11451, Saudi Arabia
2 Department of Pharmacology and Toxicology, College of Pharmacy, King Saud University, Riyadh 11451, Saudi Arabia; ffalasmari@ksu.edu.sa (F.A.)
3 Department of Forensic Sciences, College of Criminal Justice, Naif Arab University for Security Sciences, Riyadh 11452, Saudi Arabia
4 Department of Pharmaceutics, College of Pharmacy, University of Hafr Al Batin, Hafr Al Batin 31991, Saudi Arabia
* Correspondence: galbeshr@ksu.edu.sa

Abstract: This study investigated the effect of *Rosmarinus officinalis* L. (RO) extract on neurobehavioral and neurobiological changes in male rats with pentylenetetrazol (PTZ)-induced epilepsy. Rats were assigned into five groups: (1) control rats, (2) RO-treated rats, (3) PTZ-treated rats, (4) PTZ + RO-treated rats, and (5) PTZ + valproic acid (VA)-treated rats. The PTZ-treated rats required a significantly longer time and distance to find the platform in the Morris water maze test than the control and RO-treated rats. Additionally, PTZ-treated rats showed a decrease in tendency to cross over the platform compared to PTZ group. PTZ + RO-treated rats showed decreased swimming time and distance to find the platform compared to PTZ group. PTZ + RO-treated rats showed a significant decrease in seizure score, a reduced number of myoclonic jerks, and an increased onset of the first myoclonic jerk compared to PTZ group. PTZ reduced the time required to enter the dark room in the passive avoidance learning test, which was reversed by RO treatment. Biochemical results revealed that PTZ-treated rats had higher levels of oxidative stress markers. RO significantly increased the antioxidant markers levels and maintained normal rat brain histology. This study revealed that RO can shield the brain and neural tissues from PTZ.

Keywords: epilepsy; pentylenetetrazol; *Rosmarinus officinalis* L.; neurobehavioral changes; neurobiological changes

Citation: Alrashdi, J.; Albasher, G.; Alanazi, M.M.; Al-Qahtani, W.S.; Alanezi, A.A.; Alasmari, F. Effects of *Rosmarinus officinalis* L. Extract on Neurobehavioral and Neurobiological Changes in Male Rats with Pentylenetetrazol-Induced Epilepsy. *Toxics* **2023**, *11*, 826. https://doi.org/10.3390/toxics11100826

Academic Editor: Remco H.S. Westerink

Received: 14 September 2023
Revised: 24 September 2023
Accepted: 25 September 2023
Published: 30 September 2023

1. Introduction

According to the World Health Organization (WHO), herbal medicine delivers primary healthcare to >80% of the population in low-income countries and 60% of the global population. Phytochemicals and their chemical analogs have given rise to many clinically useful drugs for treating acute and chronic diseases [1], and additional research is being conducted on therapeutic substances from medicinal plants.

Rosmarinus officinalis L. (RO) is a member of the Lamiaceae family [2]. RO has been widely used in traditional medicine and cooking, especially to change and enhance flavors. It is also a highly regarded medicinal plant for cold, rheumatism, and muscle and joint pain [3]. This plant exhibits many pharmacological properties, such as antibacterial, antidiabetic, anti-inflammatory, anticancer, and antioxidant effects [4]. RO is a medicinal plant and a highly abundant source of many bioactive substances found naturally in RO's extracts, including carnozole, carnosic acid, and triterpenes, which are also important components of nutritional supplements, complementary and alternative medicine, and antioxidants [5].

Antioxidant components decrease lipid peroxidation and interact with the detrimental free radical chain reactions [6].

Neurological disorders are characterized by impairment of the central nervous system. Older adults were found to develop neurological conditions, including dementia, Alzheimer's disease, migraine, Parkinson's disease, stroke, and epilepsy worldwide. Several therapeutic techniques can treat the symptoms in the early stages of these conditions; however, individuals become increasingly incapacitated as time passes and with potential drug side effects [7]. Therefore, natural products are promoted globally to maximize safety.

Epilepsy is a serious, chronic neurological disorder that affects individuals of all ages. It is strongly associated with significant morbidity, mortality, and reduced quality of life [8]. It is one of the most prevalent chronic neurological disorders worldwide and is characterized by recurrent, unpredictable, and typically unprovoked seizures. According to the WHO, at least 50 million people worldwide are affected by epilepsy, and approximately 0.5–1% of people worldwide have epilepsy [8]. Several studies have revealed bidirectional relationships between epilepsy and depression, suicidal ideation, and attention deficit disorders. Two population-based studies revealed that patients with epilepsy were four to seven times more likely to have a depression before the onset of their condition than controls [9].

Previous studies have shown the effectiveness of RO extract in induced seizures [10,11]. A study also revealed that RO acid counteracts the effects of hypoxia-ischemia and improves movement, cognition, and spatial memory [12]. RO extract has also been shown to enhance long-term memory in scopolamine-administered rats [13]. Another study revealed the neuroprotective benefits of RO in combination with fluoxetine (FLX) [14].

This study investigated the effect of *Rosmarinus officinalis* L. (RO) extract on neurobehavioral changes in male rats with pentylenetetrazol (PTZ)-induced epilepsy. RO can modulate seizure scores and memory functioning through molecular pathways. We hypothesized that RO can normalize oxidative stress and neuroinflammatory markers in rats that have developed epilepsy. Apoptotic proteins can be also modulated by RO treatments in epileptic models.

2. Materials and Methods

2.1. Animals

Forty male Wistar rats (150–170 g, 7–9 weeks old) were provided by the Animal House of the College of Science at King Saud University, Riyadh, Saudi Arabia. During the adaptation and experimental periods, all animals were housed in plastic cages under ambient controlled conditions (25 ± 1 °C, 50 ± 10% humidity, and 12 h light/dark cycle). All study protocols were approved by the Institutional Ethics Committee (IACUC) (KSU_SE_21_13).

2.2. Induction of Epilepsy

This was performed as previously described [15] using repetitive intraperitoneal (i.p.) injections of PTZ (50 mg/kg) dissolved in normal saline (Cat. No. P6500, Sigma Aldric, Cambridge, UK). Accordingly, PTZ was injected every 2 days until the treated rats showed a total seizure score of 5 (development of kindling), 24 days post PTZ injection.

2.3. Preparation of RO Aqueous Extract

The RO leaves were bought in Riyadh, Saudi Arabia, from licensed local marketplaces. After that, distilled water was used to wash the leaves. Samples were freeze dried at −80 °C, pulverized into powder, and then frozen at −80 °C. A quantity of 500 g of powder was soaked in 1 L of methanol in a clean bottle for 24 h while being constantly stirred at about 27 °C. This phase was repeated three times. After that, the extract was run through sterile filter paper. Under low pressure, the extract had its solvent removed. After concentration, all extracts were kept at −20 °C until use [16].

2.4. Experimental Design

After a 1-week adaptation period, all rats were divided into five groups of eight rats each, as follows:

- Control group: Normal saline was administered.
- RO-treated group: RO extract (100 mg/kg) was administered.
- PTZ-treated group: Rats were treated with pentylenetetrazol (PTZ) (30 mg/kg).
- PTZ + RO-treated group: Rats were treated with RO (100 mg/kg) for 30 min before treatment with PTZ (30 mg/kg).
- PTZ + valproic acid (VA)-treated group (a positive control group): Rats were treated with VA (300 mg/kg) 30 min before treatment with PTZ (30 mg/kg).

All drugs were administered via oral gavage needle. The PTZ-treated rats achieved a total seizure score of 5/5 by the end of day 24; therefore, the experiments were terminated on this day.

2.5. Dose Selection

The dose of RO was based on a study by Naderali et al. (2018), which showed the protective potential of RO against kainic acid (KA)-induced hippocampal damage and memory deficits by suppressing neurodegeneration [17]. The dose and route of valproic acid (VA) administration were based on previous studies that showed potent anti-elliptic and neuroprotective effects in PTZ-treated rats [18]. Lower doses of 100 mg/kg also attenuated PTZ-induced catamenial epilepsy (caused by the estrous cycle) in female rats [19].

2.6. Scoring of Epileptic Seizures

Immediately after each PTZ treatment, each rat was placed in an individual transparent Plexiglass cage and monitored for 30 min to evaluate the seizure score using a previously established five-scale racing scoring system [15,20,21]. The following scores were considered: (A) Score 0 (no response), (B) Score 1: facial movements with saccades of ears and whiskers, (C) Score 2: myoclonic jerks without rearing, (D) Score 3: unilateral or bilateral limb clonus, (E) Score 4: rearing and forelimb clonic seizures, (F) Score 5: Generalized tonic-clonic seizures with falling.

2.7. Morris Water Maze (MWM) Test

After scoring the seizures, the spatial memory of all rats was tested using the MWM test as described previously [22]. The test was performed in a large swimming pool with a 1.7 m diameter and 60 cm depth. The pool was hypothetically divided into four directional quadrants (S, E, W, and N), and a hidden platform was placed in the NW quadrant 2 cm below the water surface. On the training day, all the rats were released from one quadrant and could locate and step over the visible platform. The next day (1–4), the platform was submerged in water (2 cm), and milk was added for camouflage. The test was repeated for the next 4 days (test days). The escape time (latency) to find the platform and swimming distance (m) were calculated as intelligence and intact memory indicators. An additional probe trial was conducted on day 5, when the platform was removed, and the test procedure was repeated for each rat to calculate the number of times it passed over the rescue platform.

2.8. Passive Avoidance Learning (PAL) Test

The emotional memory of all rats was tested using the dark room PAL test, as previously described [22]. The test apparatus contained a large illuminated room and a small dark room with an electrical grid floor. The rooms were separated by a door that could be opened or closed. The test consisted of training and test procedures. During training, each rat was placed in an illuminated area with an open door and was given three trials, each lasting 3 min, to explore the machine. In the fourth trial, the animal was allowed to re-explore, and once it entered the dark room, the door was closed, and the rat was exposed to a foot electrical stimulation (50 Hz, 1.5 mA for 2 s), after which the door was opened. All the animals were returned to their cages. Two hours later, each rat was returned to the apparatus and placed in a light room with an open door. The time required by the animals to enter the dark room was also recorded.

2.9. Tissue Collection and Processing

After behavioral assessment, all rats were anesthetized with a ketamine/xylazine mixture (90/10 mg/mg). Each rat's skull was opened, and the brains were removed and placed on ice. The brains were divided into two equal parts. One portion was placed in 10% buffered formalin and processed for histological evaluation. The other halves were cut into smaller pieces and preserved at −80 °C for further procedures.

2.10. Biochemical Analysis in the Brain Homogenates

Part of the brain tissue was homogenized in phosphate-buffered saline (pH 7.4). The homogenates were centrifuged at 1000× *g* to collect the supernatants. These supernatants were stored at −20 °C and used later to measure the levels of several markers. Malondialdehyde (MDA), an indicator of lipid peroxides, was measured using an assay kit (cat no. MBS268427, MyBioSource, San Diego, CA, USA). Total levels of superoxide dismutase (SOD), tumor necrosis factor-alpha (TNF-α), glutathione peroxidase (GPX), interleukine-6 (IL-6), and glutathione (GSH) were measured using ELISA-based kits (Cat. No. RTFI00215, Cat. No. RTFI01177, Cat. No. RTEB0206, Cat. No. RTEB0061, and Cat. No. RTEB1811, all supplied by Assay Genie, London, UK). Total levels of cytochrome c, Bcl2, Bax, and caspase-3 in the homogenates were measured using ELISA (Cat. No. MBS9304546, Cat. No. MBS2881713, Cat. No. MBS935667, and Cat. No. MBS018987 MyBiosources, San Diego, CA, USA, respectively). All measurements were performed for eight samples/groups, according to the manufacturer's instructions [23–25].

2.11. Histological Evaluations

Brain sections were preserved in 10% buffered formalin for 24 h. Tissues were rehydrated in ascending ethanol concentrations (70–100%). All the slides were cleared with xylene and embedded in paraffin. Afterward, all tissues were sectioned at 3–5 µm using a rotatory microtome. The sections were placed on glass slides and stained with hematoxylin and eosin (H&E). The mounting medium and coverslips were used to cover each section. All slides were examined under a light microscope (Nikon Eclipse E200, Tokyo, Japan) and photographed at 200× [26,27].

2.12. Statistical Analysis

One-way analysis of variance (ANOVA) was used to evaluate all data using GraphPad Prism software. Utilizing the Kolmogorov−Smirnov test, normality was evaluated. Tukey's post hoc test was used to compare various groups. At $p < 0.05$, data were deemed statistically different [28].

3. Results

3.1. Scoring of Epileptic Seizures

3.1.1. Assessment of Seizure Scores

Kindling (seizure score 5/5) was achieved in model rats after repetitive administration of PTZ every 2 days for 24 days (Figure 1). No seizure episodes or jerks were observed in the control or RO-treated rats throughout the 24 days of the study (Figure 1A,B). The seizure score was increased in PTZ-treated rats and reached a score of 5 by the end of day 24 (Figure 1A,B). RO and VA could attenuate this effect.

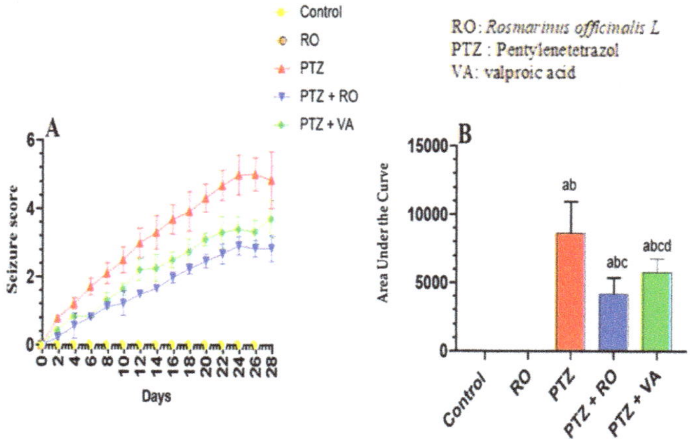

Figure 1. Assessment of (**A**) seizure scores during 24 days and (**B**) area under the curve of episode score in all groups of rats. Data are presented as means ± SD of 8 rats per group. Statistical significance was set at $p < 0.05$. a: significantly different when compared to control rats; b: significantly different when compared to RO-treated rats; c: significantly different when compared to PTZ-treated rats; and d: significantly different when compared to PTZ + RO-treated rats. VA, valproic acid; SD, standard deviation; RO, *Rosmarinus officinalis* L.; PTZ, pentylenetetrazol.

3.1.2. Total Number of Myoclonic Jerks

No myoclonic jerks were observed in the control and RO-treated rats throughout the experiment; in contrast, the number of myoclonic jerks increased significantly in the PTZ-treated rats compared to the control and RO-treated rats. The average number of myoclonic tremors was 59.7 ± 7.1 in the PTZ-treated rats. Compared with the PTZ-treated rats, the number of myoclonic jerks were decreased in the PTZ + VA-treated rats, and a more significant reduction was observed in the PTZ + RO-treated rats compared with the PTZ- and PTZ + VA-treated rats (Figure 2).

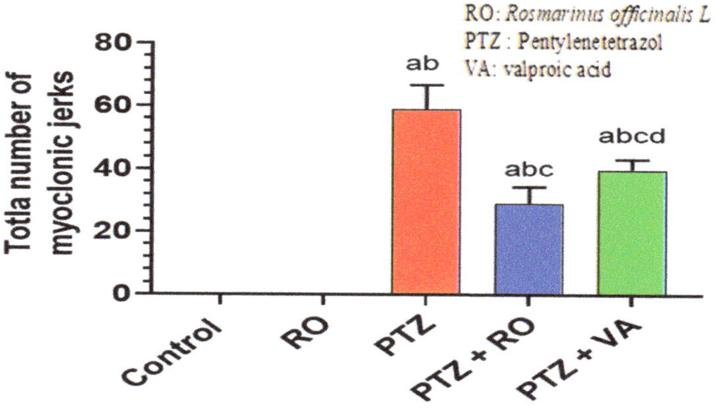

Figure 2. Total number of myoclonic jerks during the 24 days of all treatments in all groups of rats. Data are presented as means ± SD of 8 rats per group. Statistical significance was set at $p < 0.05$. a: significantly different when compared to control rats; b: significantly different when compared to RO-treated rats; c: significantly different when compared to PTZ-treated rats; and d: significantly different when compared to PTZ + RO-treated rats. VA, valproic acid; SD, standard deviation; RO, *Rosmarinus officinalis* L.; PTZ, pentylenetetrazol.

3.1.3. Onset of the First Myoclonic Jerk

The average onset of the first myoclonic jerk was observed after 44.8 s. Daily epilepsy shift score and AUC were reduced. The first jerk onset was significantly increased in PTZ + RO- and PTZ + VA-treated rats; however, PTZ + RO-treated rats showed a significantly longer duration of first jerk onset than PTZ + VA-treated rats. PTZ-treated rats had a lower first jerk onset time than those treated with PTZ + RO or PTZ + VA groups (Figure 3).

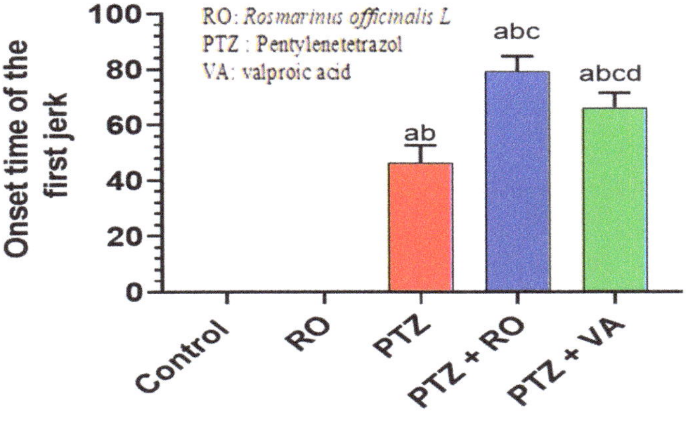

Figure 3. The onset of the first myoclonic jerk in all groups of rats. Data are presented as means ± SD of 8 rats per group. Statistical significance was set at $p < 0.05$. a: significantly different when compared to control rats; b: significantly different when compared to RO-treated rats; c: significantly different when compared to PTZ-treated rats; and d: significantly different when compared to PTZ + RO-treated rats. VA, valproic acid; SD, standard deviation; RO, *Rosmarinus officinalis* L.; PTZ, pentylenetetrazol.

3.1.4. Behavioral Tests

Time Required to Find the Hidden Platform during the MWM Test

A significant and progressive decline in swimming time to find the hidden platform was observed over the 4 days of the MWM test in the control and RO-treated groups. No significant variations in the time to find the hidden platform were observed between the control and RO-treated rats; however, the time to find the hidden platform, as measured on days 1–4, was significantly higher in PTZ-treated rats than in the control or RO-treated rats. Compared to rats treated with PTZ, the time to find the hidden platform was decreased in rats treated with PTZ + RO and PTZ + VA, and the time to find the hidden platform was less in PTZ + RO-treated rats than in PTZ + VA-treated rats (Figure 4A,B).

Total Swimming Distance to Find the Hidden Platform during the MWM Test

Over the 4 days of the MWM test, the studied groups of rats showed significant and progressive differences in the distance required to discover the hidden platform. The distance to find the hidden platform did not differ noticeably between the RO-treated and control rats. The distance to find the hidden platform was significantly greater in the PTZ-treated rats than in the RO-treated or control rats. Compared to PTZ-treated rats, the distance to find the hidden platform decreased in the PTZ + RO- and PTZ + VA-treated rats and was reduced in the PTZ + RO-treated rats compared to PTZ + VA group (Figure 5A,B).

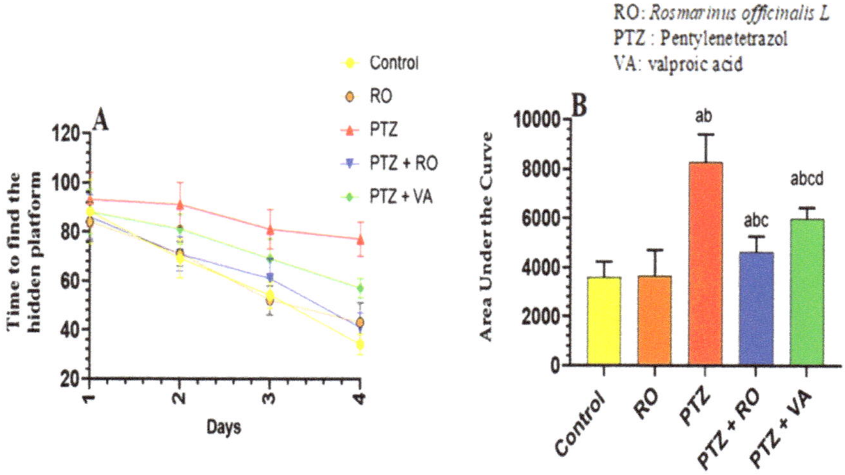

Figure 4. Time required to find the hidden platform (**A**) daily readings and (**B**) area under the curve in the MWM test in rats of all groups. Data are presented as means ± SD of 8 rats per group. Statistical significance was set at $p < 0.05$. a: significantly different when compared to control rats; b: significantly different when compared to RO-treated rats; c: significantly different when compared to PTZ-treated rats; and d: significantly different when compared to PTZ + RO-treated rats. VA, valproic acid; SD, standard deviation; RO, *Rosmarinus officinalis* L.; PTZ, pentylenetetrazol; MWM, Morris water maze.

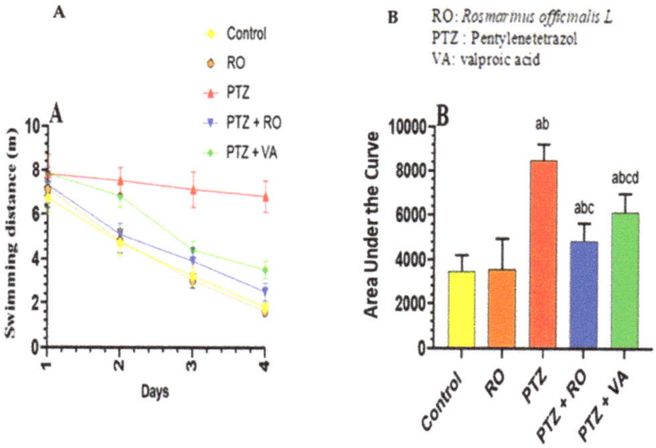

Figure 5. The total swimming distance to find the hidden platform (**A**) daily readings and (**B**) area under the curve in the MWM test in rats of all groups. Data are presented as means ± SD of 8 rats per group. Statistical significance was set at $p < 0.05$. a: significantly different when compared to control rats; b: significantly different when compared to RO-treated rats; c: significantly different when compared to PTZ-treated rats; and d: significantly different when compared to PTZ+RO-treated rats. VA, valproic acid; SD, standard deviation; RO, *Rosmarinus officinalis* L.; PTZ, pentylenetetrazol; MWM, Morris water maze.

Number of Crossing Times over the Removed Platform during the Probe Trial of the MWM Test

There were no significant differences in the number of crossings over the removed platform in the probe trial between the control and RO-treated rats (Figure 6). Compared to the control and RO-treated rats, PTZ-treated rats significantly decreased the number of crossings over the removed platform. Compared with PTZ-treated rats, PTZ + RO- and PTZ+VA-treated rats exhibited an increase in crossings over the platform, and the number of crossings was increased higher in PTZ + RO-treated rats compared to PTZ-VA group (Figure 6).

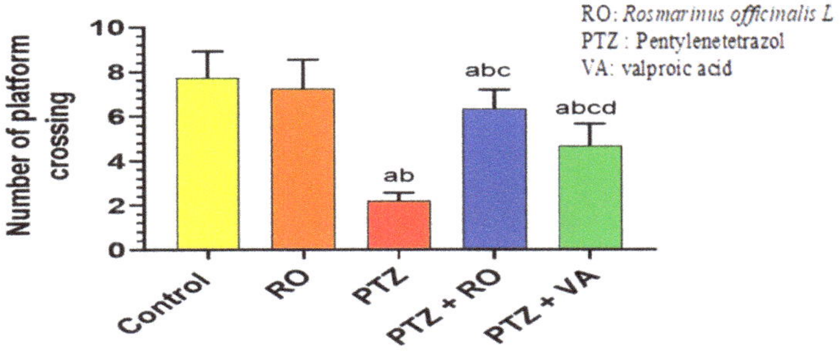

Figure 6. The number of crossing times over the removed platform during the probe trial of the MWM test in rats of all groups. Data are presented as means ± SD of 8 rats per group. Statistical significance was set at $p < 0.05$. a: significantly different when compared to control rats; b: significantly different when compared to RO-treated rats; c: significantly different when compared to PTZ-treated rats; and d: significantly different when compared to PTZ + RO-treated rats. VA, valproic acid; SD, standard deviation; RO, *Rosmarinus officinalis* L.; PTZ, pentylenetetrazol; MWM, Morris water maze.

PAL Test

The time required to enter the dark area in the PAL test was not significantly different between the control and RO-treated rats. In addition, compared with control or RO-treated rats, PTZ-treated rats required less time to enter the dark area during the PAL test. Moreover, PTZ + RO- and PTZ + VA-treated rats showed an increased time required to enter the dark room compared to PTZ-treated rats; however, PTZ + VA-treated rats showed a decrease in time required to enter the dark room compared to PTZ + RO-treated rats (Figure 7).

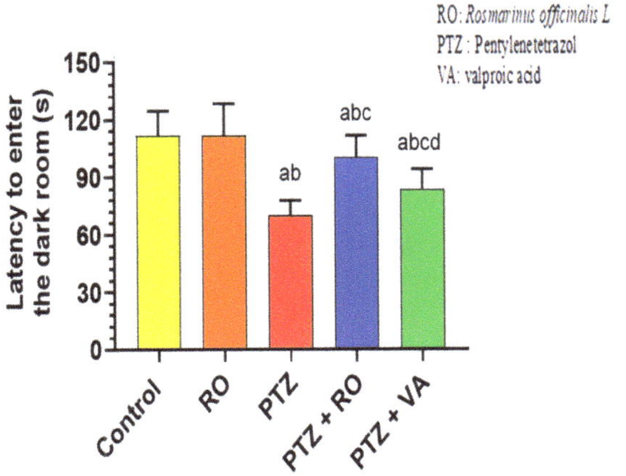

Figure 7. The time each rat spent entering the darkroom during the PAL test in rats of all groups. Data are presented as means ± SD of 8 rats per group. Statistical significance was set at $p < 0.05$. a: significantly different when compared to control rats; b: significantly different when compared to RO-treated rats; c: significantly different when compared to PTZ-treated rats; and d: significantly different when compared to PTZ + RO-treated rats. VA, valproic acid; SD, standard deviation; RO, *Rosmarinus officinalis* L.; PTZ, pentylenetetrazol; PAL, passive avoidance learning.

3.1.5. Biochemical Analysis

Levels of Antioxidant Markers

MDA levels were significantly reduced in the brains of RO-treated rats compared to the control rats. The brain homogenates of PTZ-treated rats showed a significant increase in the MDA levels compared to control and RO groups, and MDA levels were significantly reduced in the PTZ + RO and PTZ + VA groups compared to PTZ group. However, the MDA levels were significantly lower in PTZ + RO-treated rats than in the PTZ + VA-treated rats (Figure 8A). SOD levels were significantly increased in the brains of RO-treated rats compared to those of the control rats. Brain homogenates of PTZ-treated rats showed a significant decrease in total SOD compared to those of the control and RO-treated rats. SOD levels were significantly higher in the PTZ + RO- and PTZ + VA-treated rats than in the PTZ-treated rats. However, SOD levels was significantly higher in the PTZ + RO-treated rats than in the PTZ + VA-treated rats. Furthermore, no significant differences were observed in the levels of these markers when PTZ + RO-treated rats were compared to control rats (Figure 8B). GSH levels in RO-treated rats increased significantly compared to in the control rats; in contrast, GSH levels were decreased significantly in the PTZ-treated rats compared to the control and RO-treated rats.

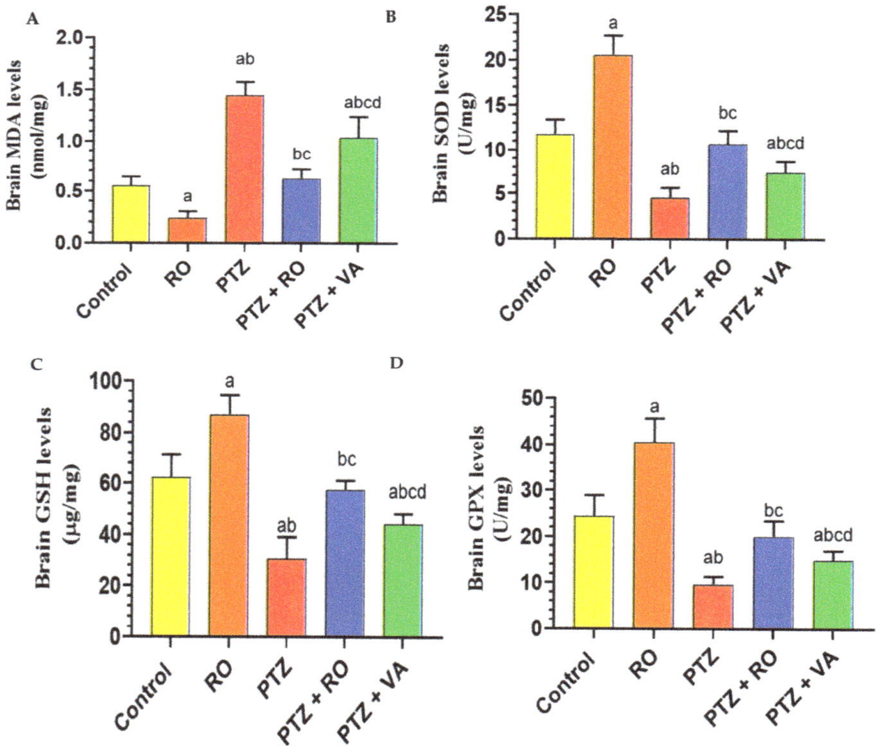

Figure 8. Levels of (**A**) MDA, (**B**) total SOD, (**C**) total GSH, and (**D**) total GPX in the brains of all groups of rats. Data are presented as means ± SD of 8 rats per group. Statistical significance was set at $p < 0.05$. a: significantly different when compared to control rats; b: significantly different when compared to RO-treated rats; c: significantly different when compared to PTZ-treated rats; and d: significantly different when compared to PTZ+RO-treated rats. VA, valproic acid; SD, standard deviation; RO, *Rosmarinus officinalis* L.; PTZ, pentylenetetrazol; MDA, malondialdehyde; SOD, superoxide dismutase; GSH, glutathione; GPX, glutathione peroxidase.

Additionally, GSH levels were increased significantly in the PTZ + RO- and PTZ + VA-treated rats compared to in the PTZ-treated rats (Figure 8C). GPX levels were significantly increased in the RO-treated rats compared to in the control rats; however, there was a significant decrease in GPX levels in the PTZ-treated rats compared to the control and RO-treated rats. Additionally, there was an increase in GPX levels in the PTZ + RO- and PTZ + VA-treated rats compared to the PTZ-treated rats (Figure 8D).

Levels of Inflammatory Markers

The brain homogenates of the PTZ-treated rats showed a significant increase in the levels of TNF-α as compared to the control and RO-treated rats. In contrast, the levels of TNF-α were significantly reduced in the brains of the PTZ + RO- and PTZ + VA-treated rats compared to the PTZ-treated rats. However, the brain levels of TNF-α were significantly lower in the PTZ + RO-treated rats than in the PTZ + VA-treated rats. In contrast, the levels of TNF-α remained significantly higher in the brains of the PTZ + RO and PTZ + VA -treated rats than in the control rats (Figure 9A). The RO-treated rats and the control rats did not show any significant variation in the brain levels of IL-6. When compared to the control and RO-treated rats, the brain homogenates of model rats that had received PTZ treatment revealed significantly higher levels of IL-6; in contrast, the brain IL-6 levels in the PTZ + RO- and PTZ + VA-treated rats were both lower than those in the PTZ-treated rats. Although brain IL-6 was significantly lower in the PTZ + RO-treated rats than in the PTZ + VA-treated rats, it was also significantly higher in the PTZ + RO-treated rats than in the control rats (Figure 9B).

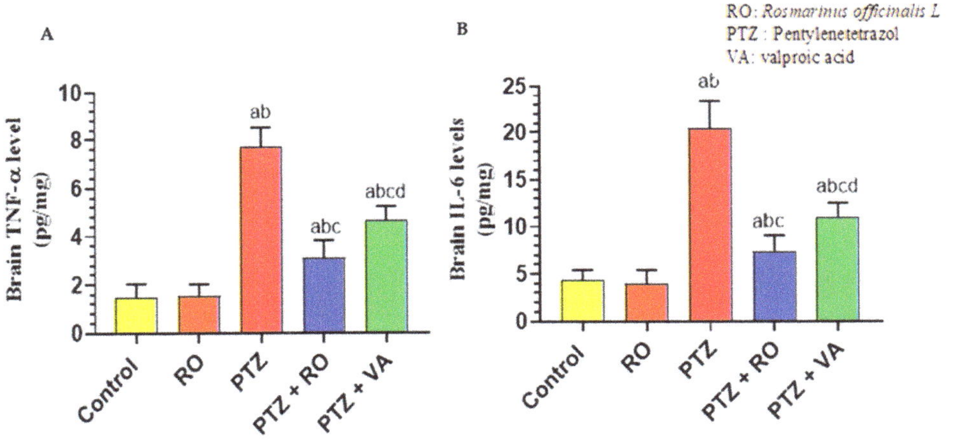

Figure 9. Levels of (**A**) TNF-α and (**B**) IL-6 in the brains of all groups of rats. Data are presented as means ± SD of 8 rats per group. Statistical significance was set at $p < 0.05$. a: significantly different when compared to control rats; b: significantly different when compared to RO-treated rats; c: significantly different when compared to PTZ-treated rats; and d: significantly different when compared to PTZ + RO-treated rats. VA, valproic acid; SD, standard deviation; RO, *Rosmarinus officinalis* L.; PTZ, pentylenetetrazol; TNF- α, tumor necrosis factor; IL-6, interleukin-6.

Levels of Apoptotic Proteins in the Brain

The brain levels of Bax protein were not significantly different between the PTZ-treated, RO-treated, and control rats. Bax protein levels were significantly higher in the brains of the PTZ-treated rats than in the control and RO-treated rats (Figure 10A). Brain homogenates of the PTZ + RO- and PTZ + VA-treated rats showed a significant decrease in Bax protein levels compared with the PTZ-treated rats. Bax protein levels were significantly higher in the PTZ + VA-treated rats than in the PTZ + RO-treated rats (Figure 10A). As shown in Figure 10B, brain levels of caspase-3 were slightly lower in the RO-treated rats than in the control rats. Caspase-3 levels were significantly higher in the brains of the PTZ-treated rats than in the control and RO-treated rats (Figure 10B). Brain homogenates of the PTZ + RO- and PTZ + VA-treated rats showed a significant decrease in caspase-3 levels compared with the PTZ-treated rats. Levels of caspase-3 were higher in the PTZ + VA-treated rats than in the PTZ + RO-treated rats. Compared with the control rats, RO-treated rats had lower brain cytochrome c levels. The PTZ-treated rats had significantly higher cytochrome c levels in their brains than the RO-treated rats, and brain homogenates from the PTZ + RO- and PTZ + VA-treated rats revealed a substantial drop in cytochrome c compared to PTZ group. The PTZ + VA-treated rats had higher cytochrome c levels than the PTZ + RO-treated rats (Figure 1C). There was no statistically significant difference in the brain levels of Bcl2 between the RO-treated and control rats, and the brains of the PTZ-treated rats showed considerably reduced Bcl2 levels compared to controls (Figure 10D).

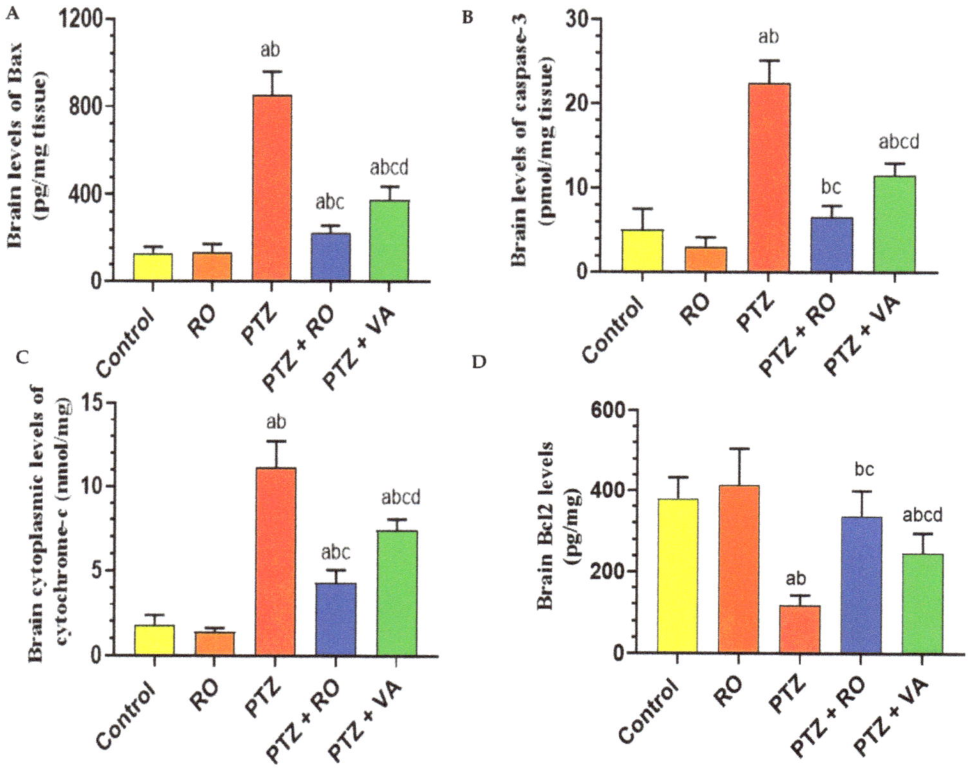

Figure 10. Levels of (**A**) Bax protein, (**B**) caspase-3, (**C**) cytochrome c, and (**D**) Bcl2 in the brains of all groups of rats. Data are presented as means ± SD of 8 rats per group. Statistical significance was set at $p < 0.05$. a: significantly different when compared to control rats; b: significantly different when compared to RO-treated rats; c: significantly different when compared to PTZ-treated rats; and d: significantly different when compared to PTZ + RO-treated rats. VA, valproic acid; SD, standard deviation; RO, *Rosmarinus officinalis* L.; PTZ, pentylenetetrazol.

The brain homogenates of PTZ + RO- and PTZ + VA-treated rats showed a significant increase in Bcl2 levels compared to PTZ-treated rats; in contrast, Bcl2 levels were significantly lower in the brains of the PTZ + VA-treated rats than in the PTZ + RO-treated rats (Figure 10D).

3.1.6. Histological Analysis of Dental Gyrus of the Hippocampi

As shown in Figure 11A,B, the dental gyrus area of the hippocampi of the control and RO-treated rats showed normal features, with the presence of all three layers (glandular, polymorphic, and molecular layers) and normally sized blood vessels. The molecular layer showed 4–6 intact layers of cells, and the polymorphic and molecular layers showed many baskets and glial cells. According to Figure 11C,D, the PTZ-treated hippocampi showed an obvious reduction in the number of cell layers forming the glandular layer; in contrast, the remaining cells showed swelling. In addition, this group of rats showed reduced basket and glial cells and dilation of their blood vessels.

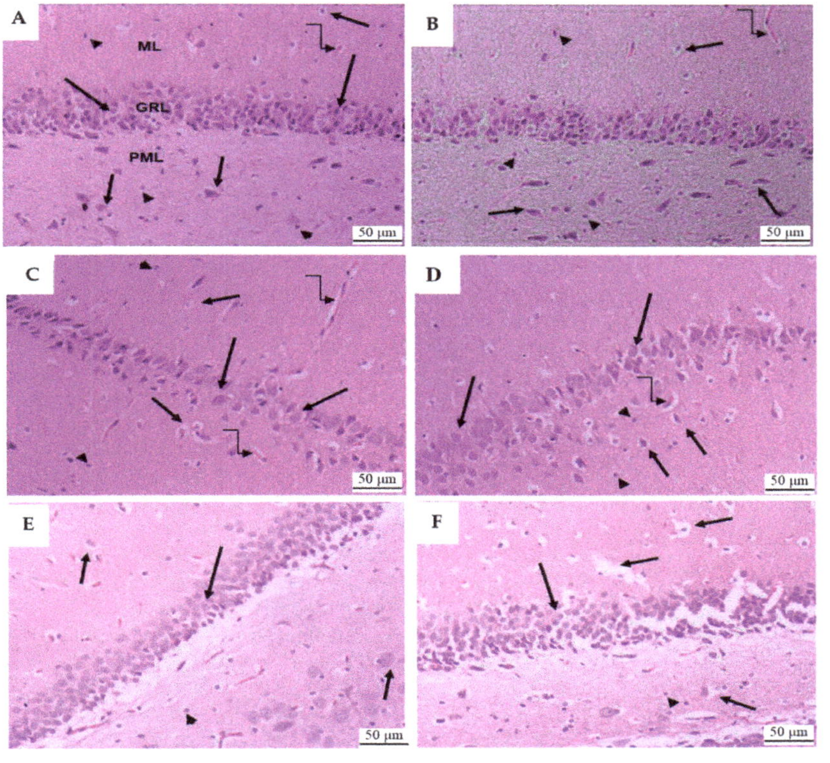

Figure 11. The histological section of the dental gyrus of the hippocampi of all groups of rats. (**A**) and (**B**) Taken from control rats, showing a normal structure composed of three intact layers, namely, the glandular layer (GRL) lying between the polymorphic (PML) and molecular layers (ML). The GRL comprises 4–6 cell layers with rounded pale vesicular nuclei (long arrow). The basket cells (short arrow) and glial cells (arrowhead) were also observed, and the PLM and ML layers have normally sized blood vessels (curved arrow). (**C**,**D**) Taken from PTZ-treated rats and showing an obvious reduction in the number of cell layers forming the GRL with swelling cells (long arrow). A lower number of basket and glial cells (shot arrow and arrowhead, respectively) were also observed in this group of rats with dilated blood vessels (curved arrow). (**E**,**F**) Taken from the PTZ + RO- and PTZ + VA-treated rats, respectively, and showing significant improvement in the structure of their hippocampi, with an obvious increase in the number of cell layers forming the GRL, as well as an increase in the number of basket (short arrow) and glial cells (arrowhead) of the PLM and ML. Both groups also showed normally sized blood vessels. However, some degeneration in the GRL and basket cells (short arrow) is still seen in the PTZ + VA-treated rats.

The hippocampi of the PTZ + RO- and PTZ + VA-treated rats showed an obvious improvement in their dental gyrus structures. They had normally sized blood vessels and an almost normal count of basket and glial cells in the molecular and polymorphic layers (Figure 11E,F). However, some degeneration in the glandular layer and basket cells was observed in the dental gyrus of the PTZ + VA-treated rats (Figure 11F).

4. Discussion

Herbs have a long history of being effective in treating diseases, with few side effects or toxicity [29]. Although many antiepileptic medications have been developed, most do not improve the cognitive impairment caused by refractory epilepsy. These medications also have numerous adverse effects, such as psychosis, heightened irritability, and aggressive behavior [30,31]. Different RO extraction and purification methods and numerous antioxidant assays have shown that RO is rich in phenolic compounds, such as carnosic acid and carnosol. RO oil's antioxidant, antimicrobial, and cognitive properties and its extracts have also been investigated. Therefore, they provide a number of naturally derived antioxidants recognized by the food industry [32]. Numerous studies have highlighted the neuropharmacological benefits of RO extracts. RO has significant antibacterial, anti-inflammatory, antioxidant, anti-tumor, anti-pain, and neuroprotective effects. It also has significant clinical benefits for mood, memory, learning, pain, anxiety, and sleep [33].

In a previous study [10], treating animals with different doses of RO extract delayed the onset of picrotoxin-induced seizures, and the most effective dose was 50 mg/kg. These results are consistent with those of the present study, in which treatment of rats with RO extract (100 mg/kg) delayed the onset of seizures caused by PTZ (30 mg/kg). These results agree with Boroushaki, 2002. It was found that the most effective dose of the plant extract (12 mL/kg) reduced the onset of seizures, their duration, and the number of deaths after 24 h in PTZ-induced seizures, and the results showed that all concentrations of the plant extract delayed seizure onset [11]. In a previous study [34] consistent with our results, it was found that the latency to myoclonic jerks and generalized seizures were increased in the PTZ model treated with rosmarinic acid (30 mg/kg); moreover, the latency to myoclonic jerks induced by pilocarpine was delayed following rosmarinic acid treatment. This study also found that the appearance of spontaneous recurrent seizures was not abolished with rosmarinic acid in a chronic epilepsy model. However, rosmarinic acid could increase the time of immobility using forced swim assay, and increase the time spent at the center and crossing numbers using the open field test.

The effect of the aerial portions of RO in rats undergoing the MWM test and the effects of the essential oil on intact memory and scopolamine-induced learning were examined. Rats were administered an intraperitoneal injection of the oil 30 min before training for 5 days in a row, and the latency time to discover the platform was reduced when the oil was administered at levels of 125–250 mg/kg. Rats with learning deficits due to hyoscine (0.5 mg/kg) were also tested to determine the effect of the RO oil. The memory-impairment effects of hyoscine were reduced by RO oil [35]. This is consistent with the results of this study, which showed that RO improved learning and memory. A previous study revealed that RO significantly improved movement problems, cognition, and spatial memory [12]. In addition, these results are consistent with Rad et al., 2021, who found that RO extract (100 mg/kg) improved spatial memory retrieval using the MWM test. There was also a significant difference in the time each group spent in the target quadrant during the experiment [36].

The results of this study showed that RO extract (100 mg/kg) significantly improved long-term memory during the PAL test, which is consistent with a previous study [37] where it was found that doses (50 and 100 mg/kg) of the natural oil of RO resulted in significant improvements in long-term memory. These results also agree with Ozarowsk et al., 2013, who found that RO extracts enhanced long-term memory in rats administered scopolamine [13]. This study showed that treatment with RO extract in a rat model of epilepsy led to a decrease in lipid peroxide (MDA) levels, which agrees with the findings of recent studies [38–40]. Treatment with RO extract significantly increased antioxidant enzymes, such as SOD, GSH, and GPX; the results of this study agreed with prior studies [41,42]. Administration of RO increased the levels of all these enzymes and a glutathione-reducing antioxidant.

Other studies have confirmed the results of this study, which showed that RO in separate dosage forms (RO essential oil, aqueous extract, and crude plant powder) enhanced the activity of SOD and the total antioxidant capacity [43]. In a recent study, doses of RO increased the levels of several enzymes, including SOD [44]. Increased blood flow and vascular permeability, and the buildup of fluid, leukocytes, and inflammatory mediators, including cytokines, are characteristics of inflammation in its acute phase, an immediate response to tissue injury. IL-1 and TNF-α, two cytokines that are extremely strong inflammatory agents, are important in mediating acute inflammatory reactions [45]. Our work suggests that RO showed the ability to restore the neuroinflammatory cytokines levels in animal models of epilepsy. The production and expression of extracellular SOD were stimulated by pro-inflammatory cytokines [46]. Mitochondrial SOD in fibroblasts was stimulated following exposure to inflammatory cytokines such as TNF-α [46]. Here, we found that the PTZ group had higher brain antioxidant parameters, including SOD, which suggest a compensatory feedback mechanism for increased brain inflammatory cytokine levels [47]. Through the suppression of immune cell infiltration and immunological responses, extracellular SOD may produce blocking effects on IL-23-induced-inflammatory conditions such as psoriasis [48]. In response to lipopolysaccharide, SOD 3 was released from intracellular stores of macrophages producing biological responses [49]. Lactate has been considered a useful seizure-like activity biomarker, as demonstrated by an increase in its levels in the extracellular environment of the human hippocampus during seizures; rosmarinic acid (30 mg/kg/body volume) was able to reduce pilocarpine-stimulated lactate release in the brains of epileptic male rats [50]. Gap junctions that allow direct transfer of ions and molecules between neighboring cells, and gap junctions between astrocytes, play an important role in the development of brain diseases such as epilepsy [51]. Astrocytic coupling entirely disappears in hippocampal sclerosis, according to functional data from human patients and animal models, while the gap-junction-forming proteins connexin43 and connex30 are still present. Additionally, astrocyte dissociation was found to be a contributing factor in the onset of temporal lobe epilepsy [52]. In a recent study [53], human SH-SY5Y neuronal-like cells and A-172 glial-like cells were used in a hypoxic environment in vitro, examining the effects mediated by spray-dried RO extract (SDROE) on cell viability, apoptosis, and Cx43-based intercellular communication. It was discovered that SDROE had a protective effect on cells damaged by glucose deficiency (OGD), promoting metabolic turnover and cell survival while reducing Cx43-based cell coupling.

Treatment with PTZ led to decreased glial cells and expansion of blood vessels; in contrast, these effects were mitigated by treatment with RO. These results agree with those of a recent study conducted by Atef et al. (2021), where it was found that treatment with RO extract can reduce the toxic effects of monosodium glutamate (MSG) [14]. In a rat brain tissue study, the MSG group showed glial cell shrinkage and significant vasodilation, which was attenuated by RO extract and/or FLX treatment. Treatment with RO extract and FLX reduced the toxic effects of MSG in the rat hippocampus. Our results agree with those of Rad et al. (2021), who studied the beneficial effects of rosemary extract and adipose-tissue-derived stem cells on memory and hippocampal neurogenesis in Parkinson's rat models [36]. The hippocampi of rats administered RO extract or water contained significantly more neurons than the treatment groups.

5. Conclusions

This study investigated the preventive and therapeutic efficacies of RO in promoting brain health in rats with epilepsy. Based on these proven properties in many studies, RO is a promising candidate for attenuating epilepsy. Therefore, RO can protect the brain via its antioxidant (GSH, SOD, and GPX), anti-apoptotic, and anti-inflammatory properties, as depicted following the oxidative stress, an effect associated with improved neurobehavioral parameter. Future studies are warranted to link the positive behavioral and neuroinflammation effects of RO to the gene and protein levels of the involved pathways and biomarkers. Future studies can investigate and compare RO treatment durations on the neurobehavioral changes in epilepsy models.

Author Contributions: Conceptualization, G.A. and F.A.; methodology, J.A., G.A., M.M.A., W.S.A.-Q., A.A.A. and F.A.; software, J.A., G.A. and F.A.; validation G.A. and F.A.; formal analysis, J.A., G.A., M.M.A., W.S.A.-Q., A.A.A. and F.A.; investigation, G.A. and F.A.; resources, G.A. and F.A.; data curation, J.A., G.A., M.M.A., W.S.A.-Q., A.A.A. and F.A.; writing—original draft preparation, J.A., G.A., F.A. and A.A.A.; writing—review and editing, J.A., G.A., M.M.A., W.S.A.-Q., A.A.A. and F.A.; visualization, G.A. and F.A.; supervision, G.A.; project administration, G.A.; funding acquisition, F.A. All authors have read and agreed to the published version of the manuscript.

Funding: This work was supported by the researchers supporting project number (RSP2023R235), King Saud University, Riyadh, Saudi Arabia.

Institutional Review Board Statement: All study protocols were approved by the Institutional Ethics Committee (IACUC) (KSU_SE_21_13).

Informed Consent Statement: Not applicable.

Data Availability Statement: The data presented in this study are available within the article.

Acknowledgments: This work was supported by the researchers supporting project number (RSP2023R235), King Saud University, Riyadh, Saudi Arabia.

Conflicts of Interest: The authors declare no conflict of interest.

References

1. Khan, M.S.A.; Ahmad, I. Herbal medicine: Current trends and future prospects. In *New Look to Phytomedicine*; Elsevier: Amsterdam, The Netherlands, 2019; pp. 3–13.
2. Rotblatt, M. *Herbal medicine: Expanded commission E monographs*; American College of Physicians: Philadelphia, PA, USA, 2000.
3. Zhang, Y.; Adelakun, T.A.; Qu, L.; Li, X.; Li, J.; Han, L.; Wang, T. New terpenoid glycosides obtained from *Rosmarinus officinalis* L. aerial parts. *Fitoterapia* **2014**, *99*, 78–85. [CrossRef]
4. Andrade, J.M.; Faustino, C.; Garcia, C.; Ladeiras, D.; Reis, C.P.; Rijo, P. *Rosmarinus officinalis* L.: An update review of its phytochemistry and biological activity. *Future Sci. OA* **2018**, *4*, FSO283. [CrossRef]
5. Borrás-Linares, I.; Stojanović, Z.; Quirantes-Piné, R.; Arráez-Román, D.; Švarc-Gajić, J.; Fernández-Gutiérrez, A.; Segura-Carretero, A. *Rosmarinus officinalis* leaves as a natural source of bioactive compounds. *Int. J. Mol. Sci.* **2014**, *15*, 20585–20606. [CrossRef]
6. Doolaege, E.H.; Vossen, E.; Raes, K.; De Meulenaer, B.; Verhe, R.; Paelinck, H.; De Smet, S. Effect of rosemary extract dose on lipid oxidation, colour stability and antioxidant concentrations, in reduced nitrite liver pates. *Meat Sci.* **2012**, *90*, 925–931. [CrossRef]
7. Alvi, S.S.; Ahmad, P.; Ishrat, M.; Iqbal, D.; Khan, M.S. Secondary metabolites from rosemary (*Rosmarinus officinalis* L.): Structure, biochemistry and therapeutic implications against neurodegenerative diseases. In *Natural Bio-Active Compounds: Volume 2: Chemistry, Pharmacology and Health Care Practices*; Springer: Singapore, 2019; pp. 1–24.
8. Jennum, P.; Christensen, J.; Ibsen, R.; Kjellberg, J. Long-term socioeconomic consequences and health care costs of childhood and adolescent-onset epilepsy. *Epilepsia* **2016**, *57*, 1078–1085. [CrossRef]
9. Kanner, A.M. Epilepsy, suicidal behaviour, and depression: Do they share common pathogenic mechanisms? *Lancet Neurol.* **2006**, *5*, 107–108. [CrossRef]
10. Heidari, M.; Assadipour, A.; Rashid, F.P. Effect of *Rosmarinus officinalis* L. Extract on the seizure induced by picrotoxin in mice. *Pak. J. Biol. Sci.* **2005**, *8*, 1807–1811.
11. Boroushaki, M.; Baharloo, A.; Malek, F. A comparative study on the anticonvulsive effects of the aqueous extract of the *Rosmarinus officinalis* plant with phenobarbital in pentylentetrazol-induced seizures in mice. *Koomesh* **2002**, *3*, 53–58.
12. Li, M.; Cui, M.-M.; Kenechukwu, N.A.; Gu, Y.-W.; Chen, Y.-L.; Zhong, S.-J.; Gao, Y.-T.; Cao, X.-Y.; Wang, L.; Liu, F.-M. Rosmarinic acid ameliorates hypoxia/ischemia induced cognitive deficits and promotes remyelination. *Neural Regen. Res.* **2020**, *15*, 894. [CrossRef]
13. Ozarowski, M.; Mikolajczak, P.L.; Bogacz, A.; Gryszczynska, A.; Kujawska, M.; Jodynis-Liebert, J.; Piasecka, A.; Napieczynska, H.; Szulc, M.; Kujawski, R. *Rosmarinus officinalis* L. leaf extract improves memory impairment and affects acetylcholinesterase and butyrylcholinesterase activities in rat brain. *Fitoterapia* **2013**, *91*, 261–271. [CrossRef]
14. Atef, R.M.; Fattah, I.O.A.; Mahmoud, O.M.; Abdel-Rahman, G.M.; Salem, N.A. Protective effects of Rosemary extract and/or Fluoxetine on Monosodium Glutamate-induced hippocampal neurotoxicity in rat. *Rom. J. Morphol. Embryol.* **2021**, *62*, 169. [CrossRef] [PubMed]

15. Hassanzadeh, P.; Arbabi, E.; Atyabi, F.; Dinarvand, R. Ferulic acid exhibits antiepileptogenic effect and prevents oxidative stress and cognitive impairment in the kindling model of epilepsy. *Life Sci.* **2017**, *179*, 9–14. [CrossRef] [PubMed]
16. Hasan, T.N.; Ahmed, S.N.; Aalam, S.M.; Kumar, C.; Shafi, G. Evaluation of cichorium extract for the growth supporting property in rat hepatocyte primary culture. *Asian J. Plant Sci.* **2007**, *6*, 431–434. [CrossRef]
17. Naderali, E.; Nikbakht, F.; Ofogh, S.N.; Rasoolijazi, H. The role of rosemary extract in degeneration of hippocampal neurons induced by kainic acid in the rat: A behavioral and histochemical approach. *J. Integr. Neurosci.* **2018**, *17*, 31–43. [CrossRef]
18. Uma Devi, P.; Kolappa Pillai, K.; Vohora, D. Modulation of pentylenetetrazole-induced seizures and oxidative stress parameters by sodium valproate in the absence and presence of N-acetylcysteine. *Fundam. Clin. Pharmacol.* **2006**, *20*, 247–253. [CrossRef]
19. Kaboutari, J.; Zendehdel, M.; Habibian, S.; Azimi, M.; Shaker, M.; Karimi, B. The antiepileptic effect of sodium valproate during different phases of the estrous cycle in PTZ-induced seizures in rats. *J. Physiol. Biochem.* **2012**, *68*, 155–161. [CrossRef]
20. Kilinc, E.; Ankarali, S.; Ayhan, D.; Ankarali, H.; Torun, I.E.; Cetinkaya, A. Protective effects of long-term probiotic mixture supplementation against pentylenetetrazole-induced seizures, inflammation and oxidative stress in rats. *J. Nutr. Biochem.* **2021**, *98*, 108830. [CrossRef]
21. Kola, P.K.; Akula, A.; NissankaraRao, L.S.; Danduga, R.C.S.R. Protective effect of naringin on pentylenetetrazole (PTZ)-induced kindling; possible mechanisms of antikindling, memory improvement, and neuroprotection. *Epilepsy Behav.* **2017**, *75*, 114–126. [CrossRef]
22. Alshammari, G.M.; Al-Qahtani, W.H.; Alshuniaber, M.A.; Yagoub, A.E.A.; Al-Khalifah, A.S.; Al-Harbi, L.N.; Alhussain, M.H.; AlSedairy, S.A.; Yahya, M.A. Quercetin improves the impairment in memory function and attenuates hippocampal damage in cadmium chloride-intoxicated male rats by suppressing acetylcholinesterase and concomitant activation of SIRT1 signaling. *J. Funct. Foods* **2021**, *86*, 104675. [CrossRef]
23. Attia, G.M.; Elmansy, R.A.; Elsaed, W.M. Neuroprotective effect of nilotinib on pentylenetetrazol-induced epilepsy in adult rat hippocampus: Involvement of oxidative stress, autophagy, inflammation, and apoptosis. *Folia Neuropathol.* **2019**, *57*, 146–160. [CrossRef]
24. Koguchi, Y.; Kawakami, K.; Kon, S.; Segawa, T.; Maeda, M.; Uede, T.; Saito, A. Penicillium marneffei causes osteopontin-mediated production of interleukin-12 by peripheral blood mononuclear cells. *Infect. Immun.* **2002**, *70*, 1042–1048. [CrossRef] [PubMed]
25. Yoshida, N.; Ikemoto, S.; Narita, K.; Sugimura, K.; Wada, S.; Yasumoto, R.; Kishimoto, T.; Nakatani, T. Interleukin-6, tumour necrosis factor α and interleukin-1β in patients with renal cell carcinoma. *Br. J. Cancer* **2002**, *86*, 1396–1400. [CrossRef] [PubMed]
26. Zhang, M.; Lv, X.-Y.; Li, J.; Xu, Z.-G.; Chen, L. The characterization of high-fat diet and multiple low-dose streptozotocin induced type 2 diabetes rat model. *J. Diabetes Res.* **2008**, *2008*, 704045. [CrossRef] [PubMed]
27. Suzuki, Y.; Imada, T.; Yamaguchi, I.; Yoshitake, H.; Sanada, H.; Kashiwagi, T.; Takaba, K. Effects of prolonged water washing of tissue samples fixed in formalin on histological staining. *Biotech. Histochem.* **2012**, *87*, 241–248. [CrossRef] [PubMed]
28. Kim, T.K. Understanding one-way ANOVA using conceptual figures. *Korean J. Anesthesiol.* **2017**, *70*, 22–26. [CrossRef]
29. Liu, H.; Bai, Y.; Huang, C.; Wang, Y.; Ji, Y.; Du, Y.; Xu, L.; Yu, D.-G.; Bligh, S.W.A. Recent progress of electrospun herbal medicine nanofibers. *Biomolecules* **2023**, *13*, 184. [CrossRef] [PubMed]
30. Hixson, J.D. When and how to stop antiepileptic drugs. In *Epilepsy*; Wiley: Hoboken, NJ, USA, 2014; pp. 118–121.
31. Wang, X.; Huang, S.; Liu, Y.; Li, D.; Dang, Y.; Yang, L. Effects of ketogenic diet on cognitive function in pentylenetetrazol-kindled rats. *Epilepsy Res.* **2021**, *170*, 106534. [CrossRef]
32. Pappachan, F.; Suku, A.; Mohanan, S. *Rosmarinus officinalis. In Herbs, Spices and Their Roles in Nutraceuticals and Functional Foods*; Elsevier: Amsterdam, The Netherlands, 2023; pp. 149–170.
33. Rahbardar, M.G.; Hosseinzadeh, H. Therapeutic effects of rosemary (*Rosmarinus officinalis* L.) and its active constituents on nervous system disorders. *Iran. J. Basic. Med. Sci.* **2020**, *23*, 1100.
34. Grigoletto, J.; de Oliveira, C.V.; Grauncke, A.C.B.; de Souza, T.L.; Souto, N.S.; de Freitas, M.L.; Furian, A.F.; Santos, A.R.S.; Oliveira, M.S. Rosmarinic acid is anticonvulsant against seizures induced by pentylenetetrazol and pilocarpine in mice. *Epilepsy Behav.* **2016**, *62*, 27–34. [CrossRef]
35. Hosseinzadeh, H.; Karimi, G.; Nobakht, N. Effects of *Rosmarinus officinalis* L. aerial parts essential oil on intact memory and scopolamine-induced learning deficits in rats performing the Morris water maze task. *J. Med. Plants* **2004**, *3*, 51–57.
36. Rad, S.N.H.; Kashani, M.H.G.; Abrari, K. Pre-treatment by rosemary extract or cell transplantation improves memory deficits of parkinson's disease: When tradition meets the future. *Braz. Arch. Biol. Technol.* **2021**, *64*, e21180392. [CrossRef]
37. Al-Tawarah, N.M.; Al-Dmour, R.H.; Abu Hajleh, M.N.; Khleifat, K.M.; Alqaraleh, M.; Al-Saraireh, Y.M.; Jaradat, A.Q.; Al-Dujaili, E.A. *Rosmarinus officinalis* and Mentha piperita Oils Supplementation Enhances Memory in a Rat Model of Scopolamine-Induced Alzheimer's Disease-like Condition. *Nutrients* **2023**, *15*, 1547. [CrossRef]
38. Motlagh, M.K.; Sharafi, M.; Zhandi, M.; Mohammadi-Sangcheshmeh, A.; Shakeri, M.; Soleimani, M.; Zeinoaldini, S. Antioxidant effect of rosemary (*Rosmarinus officinalis* L.) extract in soybean lecithin-based semen extender following freeze–thawing process of ram sperm. *Cryobiology* **2014**, *69*, 217–222. [CrossRef] [PubMed]
39. Asl, J.F.; Goudarzi, M.; Shoghi, H. The radio-protective effect of rosmarinic acid against mobile phone and Wi-Fi radiation-induced oxidative stress in the brains of rats. *Pharmacol. Rep.* **2020**, *72*, 857–866. [CrossRef] [PubMed]
40. Cetin, I.; Yesilbag, D.; SS, C.; Belenli, D. Effects of supplementation with rosemary (*Rosmarinus officinalis* L.) volatile oil on growth performance, meat MDA level and selected plasma antioxidant parameters in quail diets. *Kafkas Üniversitesi Vet. Fakültesi Derg.* **2017**, *23*, 283–288.

41. Khalil, O.A.; Ramadan, K.S.; Danial, E.N.; Alnahdi, H.S.; Ayaz, N.O. Antidiabetic activity of *Rosmarinus officinalis* and its relationship with the antioxidant property. *Afr. J. Pharm. Pharmacol.* **2012**, *6*, 1031–1036.
42. Rašković, A.; Milanović, I.; Pavlović, N.; Ćebović, T.; Vukmirović, S.; Mikov, M. Antioxidant activity of rosemary (*Rosmarinus officinalis* L.) essential oil and its hepatoprotective potential. *BMC Complement. Altern. Med.* **2014**, *14*, 225. [CrossRef]
43. Wu, Y.-n.; Huang, J.; Zuo, A.-l.; Yao, L. Research on the Effects of Rosemary (*Rosmarinus officinalis* L.) on the Blood Lipids and Anti-lipid Peroxidation in Rats. *J. Essent. Oil Res.* **2011**, *23*, 26–34. [CrossRef]
44. Karataş, T.; Korkmaz, F.; Karataş, A.; Yildirim, S. Effects of Rosemary (Rosmarinus officinalis) extract on growth, blood biochemistry, immunity, antioxidant, digestive enzymes and liver histopathology of rainbow trout, Oncorhynchus mykiss. *Aquac. Nutr.* **2020**, *26*, 1533–1541. [CrossRef]
45. Mengoni, E.S.; Vichera, G.; Rigano, L.A.; Rodriguez-Puebla, M.L.; Galliano, S.R.; Cafferata, E.E.; Pivetta, O.H.; Moreno, S.; Vojnov, A.A. Suppression of COX-2, IL-1β and TNF-α expression and leukocyte infiltration in inflamed skin by bioactive compounds from *Rosmarinus officinalis* L. *Fitoterapia* **2011**, *82*, 414–421. [CrossRef]
46. Marklund, S. Regulation by cytokines of extracellular superoxide dismutase and other superoxide dismutase isoenzymes in fibroblasts. *J. Biol. Chem.* **1992**, *267*, 6696–6701. [CrossRef] [PubMed]
47. Bowler, R.P.; Nicks, M.; Tran, K.; Tanner, G.; Chang, L.-Y.; Young, S.K.; Worthen, G.S. Extracellular superoxide dismutase attenuates lipopolysaccharide-induced neutrophilic inflammation. *Am. J. Respir. Cell Mol. Biol.* **2004**, *31*, 432–439. [CrossRef] [PubMed]
48. Lee, Y.S.; Cheon, I.-S.; Kim, B.-H.; Kwon, M.-J.; Lee, H.-W.; Kim, T.-Y. Loss of extracellular superoxide dismutase induces severe IL-23-mediated skin inflammation in mice. *J. Investig. Dermatol.* **2013**, *133*, 732–741. [CrossRef] [PubMed]
49. Hu, L.; Zachariae, E.D.; Larsen, U.G.; Vilhardt, F.; Petersen, S.V. The dynamic uptake and release of SOD3 from intracellular stores in macrophages modulates the inflammatory response. *Redox Biol.* **2019**, *26*, 101268. [CrossRef] [PubMed]
50. Neuberger, B.; Mello, F.K.; Mallmann, M.P.; da Costa Sobral, K.G.; Fighera, M.R.; Royes, L.F.F.; Furian, A.F.; Sampaio, T.B.; Oliveira, M.S. Beneficial Effects of Rosmarinic Acid In Vitro and In Vivo Models of Epileptiform Activity Induced by Pilocarpine. *Brain Sci.* **2023**, *13*, 289. [CrossRef]
51. Goodenough, D.A.; Paul, D.L. Gap junctions. *Cold Spring Harb. Perspect. Biol.* **2009**, *1*, a002576. [CrossRef] [PubMed]
52. Bedner, P.; Steinhäuser, C. Role of Impaired Astrocyte Gap Junction Coupling in Epileptogenesis. *Cells* **2023**, *12*, 1669. [CrossRef]
53. Zappalà, A.; Vicario, N.; Calabrese, G.; Turnaturi, R.; Pasquinucci, L.; Montenegro, L.; Spadaro, A.; Parenti, R.; Parenti, C. Neuroprotective effects of *Rosmarinus officinalis* L. extract in oxygen glucose deprivation (OGD)-injured human neural-like cells. *Nat. Prod. Res.* **2021**, *35*, 669–675. [CrossRef]

toxics MDPI

Article

Mechanistic Assessment of Anise Seeds and Clove Buds against the Neurotoxicity Caused by Metronidazole in Rats: Possible Role of Antioxidants, Neurotransmitters, and Cytokines

Amira M. El-Moslemany [1], Mai Hussein Abd-Elfatah [1], Nawal A. Tahoon [2], Rasha M. Bahnasy [1], Badriyah S. Alotaibi [3,*], Heba I. Ghamry [4] and Mustafa Shukry [5,*]

[1] Nutrition and Food Science Department, Faculty of Home Economics, Al-Azhar University, Tanta 31732, Egypt; amiraelmoslemany@azhar.edu.eg (A.M.E.-M.); rashaomar@azhar.edu.eg (R.M.B.)
[2] Department of Home Economics, Faculty of Specific Education, Banha University, Banha 13511, Egypt; nawal.tahon@fsed.bu.edu.eg
[3] Department of Pharmaceutical Sciences, College of Pharmacy, Princess Nourah Bint Abdulrahman University, P.O. Box 84428, Riyadh 11671, Saudi Arabia
[4] Nutrition and Food Sciences, Department of Home Economics, Faculty of Home Economics, King Khalid University, P.O. Box 960, Abha 61421, Saudi Arabia; hgmry@kku.edu.sa
[5] Physiology Department, Faculty of Veterinary Medicine, Kafrelsheikh University, Kafrelsheikh 33516, Egypt
* Correspondence: badriyah.salotaibi@gmail.com (B.S.A.); mostafa.ataa@vet.kfs.edu.eg (M.S.)

Abstract: Long-term use of the nitroimidazole-derived antibiotic metronidazole has been associated with neuronal damage due to its ability to cross the blood–brain barrier. Polyphenol-rich plants, such as anise seeds and clove buds, are suggested to have neuroprotective effects. However, their intracellular protective pathway against metronidazole-induced neurotoxicity remains unexplored. This study aims to evaluate the potential neuroprotective benefits of anise seeds and clove buds and elucidate the proposed metronidazole-induced neurotoxicity mechanism. This study divided rats into six groups, each containing six rats. In Group I, the control group, rats were administered saline orally. Group II rats received 200 mg/kg of metronidazole orally. Group III rats received 250 mg/kg b.w. of anise seed extract and metronidazole. Group IV rats received 500 mg/kg b.w. of anise seed extract (administered orally) and metronidazole. Group V rats received 250 mg/kg b.w. of clove bud extract (administered orally) and metronidazole. Group VI rats were administered 500 mg/kg b.w. of clove bud extract and metronidazole daily for 30 consecutive days. The study evaluated the phenolic compounds of anise seeds and clove buds. Moreover, it assessed the inflammatory and antioxidant indicators and neurotransmitter activity in brain tissues. A histological examination of the brain tissues was conducted to identify neuronal degeneration, brain antioxidants, and apoptotic mRNA expression. The study found that metronidazole treatment significantly altered antioxidant levels, inflammatory mediators, and structural changes in brain tissue. Metronidazole also induced apoptosis in brain tissue and escalated the levels of inflammatory cytokines. Oral administration of metronidazole resulted in a decrease in GABA, dopamine, and serotonin and an increase in ACHE in brain tissue. Conversely, oral administration of anise and clove extracts mitigated the harmful effects of metronidazole. The neurotoxic effects of metronidazole appear to stem from its ability to reduce antioxidants in brain tissue and increase nitric oxide production and apoptosis. The study concludes that neuronal damage caused by metronidazole is significantly mitigated by treatment with anise and clove extracts.

Keywords: anise seeds; clove buds; metronidazole; brain injury; antioxidant markers; neurotransmitters

Citation: El-Moslemany, A.M.; Abd-Elfatah, M.H.; Tahoon, N.A.; Bahnasy, R.M.; Alotaibi, B.S.; Ghamry, H.I.; Shukry, M. Mechanistic Assessment of Anise Seeds and Clove Buds against the Neurotoxicity Caused by Metronidazole in Rats: Possible Role of Antioxidants, Neurotransmitters, and Cytokines. *Toxics* **2023**, *11*, 724. https://doi.org/10.3390/toxics11090724

Academic Editor: Youssef Sari

Received: 24 July 2023
Revised: 18 August 2023
Accepted: 21 August 2023
Published: 24 August 2023

1. Introduction

The global use of prescribed and non-prescribed antibiotics is rising, provoking significant safety concerns. Neurological side effects associated with antibiotics are likely underdiagnosed and necessitate increased awareness and attention [1]. Metronidazole (MET) is a widely used antimicrobial drug due to its effectiveness against various bacteria and yeasts [2]. Metronidazole (MET) is generally considered a safe medication and is known to penetrate the brain rapidly. However, some potential side effects include mild abdominal pain, headaches, nausea, and a metallic taste in the mouth [3]. However, chronic exposure to metronidazole (MET) in humans at doses exceeding 2 g per day can lead to potential side effects such as peripheral neuropathy, seizures, cerebellar ataxia, and optic neuropathy [4]. Clinical symptoms associated with metronidazole-induced neurotoxicity, including ataxia, dysarthria, and altered mental status, are typical neurological manifestations [5]. For thousands of years, people across the globe have relied on traditional medicinal plants for various purposes, from maintaining overall health to treating illnesses [6]. Herbal medicine continues to be widely used as it has historically been the primary means of disease treatment. However, only a fraction of the plant kingdom has been explored for its phytochemicals. Plant extracts may contain bioactive compounds stimulating or inhibiting specific biological processes [7]. *Anise, or Pimpinella anisum* L. *(Family: Apiaceae)*, is a plant native to the Middle East and cultivated across the Mediterranean region. The fruit comprises oil, fatty acids, coumarins, flavonoids, glycosides, proteins, and carbohydrates. The seeds and essential oil have many uses, including antioxidants, antispasmodics, antimicrobials, digestive stimulants, and galactagogues. Traditional Iranian medical practitioners have even utilized aniseed to treat epilepsy and convulsions [7]. Research has demonstrated that *Pimpinella anisum* L. can mitigate lead-induced neurotoxicity [8]. *Pimpinella anisum* L. has been shown to improve memory in mice by reducing oxidative stress in the brain [9]. Clove, or Syzygium (S.) aromaticum, is a dried flower bud from the Myrtaceae family, traditionally cultivated exclusively in the Maluku Islands of Indonesia [10].

Numerous studies have demonstrated that fragrant herbs, including cinnamon, oregano, clove, thyme, and mint, possess antibacterial, antiviral, anticarcinogenic, and antifungal properties. However, clove has garnered significant attention among these spices due to its potent antimicrobial and antioxidant effects [11]. Various chemical constituents with antioxidant properties contribute to clove's crucial role in preventing degenerative diseases [12]. Clove essential oil (CEO) can benefit multiple ailments, including burns, wounds, dental pain, tooth infections, and toothaches. It is also incorporated into soaps and perfumes and cleans histology slides. Historically, cloves have alleviated symptoms such as indigestion, abdominal pain, diarrhea, and motion sickness. In tropical Asia, cloves treat scabies, cholera, malaria, and tuberculosis. In the United States, clove has been utilized to combat food-borne infections caused by bacteria, protozoa, and even viruses and worms [12].

Furthermore, eugenol, a component of clove, has found extensive use in dentistry due to its ability to reach the bloodstream via the dental pulp tissue [13]. Sesquiterpenes, compounds isolated from clove, have been reported to possess anticarcinogenic activity [10]. In another study, Syzygium aromaticum (clove) had promising antioxidant and neuroprotective properties [14]. Due to its high concentration of bioactive compounds rich in antioxidants (phenolics, flavonoids, and tannins), the extract of Syzygium aromaticum (clove) has demonstrated significant biological benefits. It has shown potential in alleviating brain injury induced by CeCl3 and oxidative stress [15].

Clove oil (CO) is derived from the lilac plant and is primarily composed of eugenol. It has been documented that CO possesses remarkable antimicrobial and antioxidant properties [16]. Clove oil has received FDA approval and is permitted for food and medicine applications. The volatile oil extracted from clove buds exhibits significant antioxidant activity, which can be attributed to its high content of phenolic compounds, particularly eugenol and eugenol acetate, as reported by [17,18]. Clove oil can potentially be utilized in the food and pharmaceutical industries to reduce or prevent oxidation. Doing so can help hinder the formation of harmful oxidation by-products, preserve the nutritional value of products, and extend their shelf life. This application of clove oil can be beneficial in maintaining the quality and stability of both food and pharmaceutical products [19]. The antimicrobial properties of clove oil and eugenol make them suitable for inhibiting the growth of these bacteria, which are associated with oral health issues. The use of clove oil and eugenol as natural alternatives for controlling these bacterial infections (cariogenic and periodontopathogenic) could have promising implications in oral care [20]. Eugenol has the potential to restrain the assembly of essential enzymes within bacteria and cause damage to the cell wall of bacteria. These actions contribute to eugenol's antimicrobial and hydrophobic nature and its potential as a natural antibacterial agent [21,22].

Therefore, this study aims to delineate the potential protective mechanisms of anise and/or clove essential oil that may alleviate the toxic effects induced by metronidazole. The research explores various cellular, molecular, and biochemical signaling pathways that regulate oxidative stress, apoptosis, inflammation, fibrosis, and anti-apoptotic indicators.

2. Materials and Methods

2.1. Plant Material and Animals

Anise seeds and clove buds were procured from the National Research Center located in El-Dokki, Giza, Egypt. Thirty-six male albino rats, weighing between 140–160 g, were obtained from Helwan Farm, an animal colony associated with the Vaccine and Immunity Organization in Cairo, Egypt. All animal specimens underwent thorough health checks. Before the experiment, they were allowed a week to acclimate to the laboratory environment. During this period, the rats were maintained in a calm atmosphere with natural airflow and a 12 h light–dark cycle. Food and water were provided ad libitum. The Animal Care and Use Committee of the Faculty of Veterinary Medicine at the University of Kafr El-Sheikh in Egypt approved the guidelines for the care and use of these animals.

2.2. Chemicals and Kits

Casein, vitamins, minerals, cellulose, choline chloride, DL-methionine, and other necessary chemicals were provided by the El-Gomhoreya Company, a provider of medications, chemicals, and medical appliances based in Cairo, Egypt. Gama Trade Company, Cairo, Egypt, supplied the kits for biochemical determinations. Both corn starch and corn oil were freshly purchased from the market in Tanta City, Al-Gharbia Governorate, Egypt. Metronidazole, under the trade name of Flagyl® (250 mg tablets), was produced by Rhone-Poulenc, U.K., and supplied by the Alexandria Pharmaceutical Co., Alexandria, Egypt. The dosing and selection of metronidazole dose were based on previous research findings. According to Ogbonye et al. [23], administering metronidazole at high dosages to rats can harm the brain. It can lead to cell distortion or displacement in the cerebellum and cause damage to the cells of the pituitary gland. Metronidazole at this dose caused neurotoxicity, as shown by Oda [24].

2.3. Extract Preparation

The plant seeds were washed with distilled water and dried in the shade before being ground into a fine powder. This powdered substance (50 g) was added to 250 milliliters of ethanol in a beaker equipped with a magnetic stirrer, and the mixture was heated to 60 °C for 15 min. The mixture was then allowed to steep in a dark glass bottle for 24 h to ensure complete extraction. After this period, the ethanol was evaporated from the supernatant of the extracted liquid, and the resulting pure extract was filtered through a 0.2 μm membrane.

2.4. Phenolic Compounds Analysis

High-performance Liquid Chromatography (HPLC) was employed to separate the polyphenolic compounds of the seed extract, as per the outlined procedure. This method was used to identify the phenolic and flavonoid compounds present in the sample [25]. Once dissolved in the mobile phase, the standard phenolic acid was introduced into the

HPLC system. The concentration of phenolic compounds was calculated based on the retention time and the peak area.

2.5. Study Model

The rats were divided into six groups, with six rats per group. In Group I, which served as the control group, the rats were administered saline orally. In Group II, designated as the metronidazole group, the rats were given orally 200 mg/kg of metronidazole dissolved in saline [23]. In Group III, the rats were given 250 mg/kg b.w. of anise seed extract and metronidazole (200 mg/kg). In Group IV, the rats were administered 500 mg/kg b.w. of anise seed extract orally, in addition to MET (200 mg/kg). In Group V, the rats were given orally 250 mg/kg b.w. of clove bud extract, along with MET. Finally, in Group VI, the rats were gavaged 500 mg/kg b.w. of clove bud extract, combined with MET, daily for 30 consecutive days. Feed intake and growth parameters were recorded every week.

2.6. Sampling and Biochemical Investigation

At the end of the study period, the rats were fasted overnight before being euthanized. Blood samples were collected from each rat and then centrifuged for 10 min at 3000 rpm to separate the serum. The serum was carefully transferred into dry, clean Eppendorf tubes and stored at −20 degrees Celsius for subsequent analysis, following the methodology proposed by Schermer [26]. Each rat's brain was carefully dissected, cleaned from the adhering matter with a saline solution (0.9%), dried with filter paper, and weighed. The brain was divided into four sections: one was stored at −80 °C for later use in isolating total RNA and performing molecular analyses; another was fixed in 10% formalin for histopathological examination; a third was used fresh in the comet assay; and a fourth was homogenized for use in determining the brain's antioxidant status.

2.7. Preparation of Brain Homogenates and Biochemical Analysis

The cerebrum and cerebellum were separated and washed with an ice-cold saline solution. Brain tissue was homogenized in a 1:10 (*w*/*v*) solution of ice-cold KCL buffer (1.15%, pH 7.2). The homogenate was centrifuged at 10,000× *g* for ten minutes at 4 °C, producing a post-mitochondrial supernatant (PMS). This PMS was utilized for the measurement of gamma-aminobutyric acid (GABA), following the specified methodology of Lasley et al. [27]; the level of acetylcholinesterase (AChE) was evaluated following the prescribed method by Carageorgiou et al. [28]. The levels of dopamine (DA) and serotonin (ST) were analyzed following the established protocol by Sasa and Blank [29]. The supernatant was also used to measure the concentrations of malondialdehyde (MDA), a marker for lipid peroxidation, and nitric oxide (NO), following the specified procedures by Uchiyama and Mihara [30] and Giustarini et al. [31], respectively, using ELISA plate reader at 540 nm. The enzymatic activity of Superoxide Dismutase (SOD) was determined using the method described by Marklund and Marklund [32]. Catalase (CAT) activity was measured spectrophotometrically at 240 nm by estimating the rate of H_2O_2 degradation, following the method described by Oberley et al. [33]. Reduced glutathione (GSH) activity was measured as previously described [34].

2.8. Histopathological Examination

The brain samples (cerebrum and cerebellum) were fixed in 10% formalin. They were then rinsed with tap water and placed in a dehydration bath containing successive dilutions of graded alcohol (methyl, ethyl). At 56 °C, the samples were cleaned with xylene and embedded in liquid paraffin. Using a light microscope, sections of 4 µm thickness were cut, deparaffinized, and stained with hematoxylin and eosin for histological evaluation.

2.9. qPCR

Gene expression in the brain was evaluated using a reverse transcription-polymerase chain reaction (RT-PCR). Approximately 100 mg of brain tissue was used to isolate total

RNA using TRIzol (Invitrogen, Life Technologies, Carlsbad, CA, USA). RNA samples with a specific A260/A280 ratio greater than 1.8 were used for cDNA synthesis using a kit from Fermentas (Waltham, MA, USA). The GAPDH gene (a housekeeping gene) was amplified using the SYBR Green master mix and the primers in Table S1. The amplification data were analyzed using the $2-\Delta\Delta T$ method [35].

2.10. Statistical Analysis

Data were analyzed using the SPSS software (SPSS Inc., Chicago, IL, USA). Statistics were reported as mean ± standard deviation. The Shapiro–Wilks test was used to test for normality. Differences between groups for normally distributed variables were analyzed using ANOVA. In cases where the analysis of variance was significant, a post hoc Tukey test was used to identify statistically substantial pairings. A p-value of less than 0.05 was considered statistically significant.

3. Results

3.1. The Polyphenolic Compounds of the Extract

Phenolic compounds in anise seeds and clove buds were identified using High-performance Liquid Chromatography (HPLC), with standard compounds as a reference (see Table 1). The primary components of anise, in descending order of concentration, were chlorogenic acid (153.15 µg/mL), naringenin (133.64 µg/mL), and gallic acid (63.38 µg/mL). These were followed by taxifolin, caffeic acid, ellagic acid, and syringic acid.

Table 1. Phenolic compounds of anise and clove seed extract (µg/mL).

Compounds	Anise (1 g/15 mL)		Clove (1 g/15 mL)	
	Area	Conc. (µg/mL)	Area	Conc. (µg/mL)
Gallic acid	721.26	63.38	14,131.40	1241.87
Chlorogenic acid	2050.14	153.15	3252.87	243.00
Catechin	28.44	3.41	2029.36	243.16
Methyl gallate	0.00	0.00	1014.85	13.78
Coffeic acid	1085.13	46.38	0.00	0.00
Syringic acid	475.21	20.80	4503.74	197.08
Pyro catechol	202.00	14.33	473.29	33.58
Rutin	89.69	12.21	10.48	1.43
Ellagic acid	645.97	43.22	3844.39	257.22
Coumaric acid	48.63	0.83	1967.15	33.44
Vanillin	259.17	5.83	103.12	2.32
Ferulic acid	66.09	2.31	0.00	0.00
Naringenin	2417.75	133.64	2310.65	127.72
Taxifolin	800.27	58.17	313.15	22.76
Cinnamic acid	570.17	5.76	220.99	2.23
Kaempferol	9.14	0.36	20.42	0.80

Conversely, the primary constituents of clove seeds were gallic acid (1241.87 µg/mL), catechin (257.22 µg/mL), and ellagic acid (243.16 µg/mL), followed by chlorogenic acid, syringic acid, naringenin, and pyrocatechol. Please refer to Figure 1 and Supplementary Figures S1 and S2 for the characterization of anise and clove seed extracts.

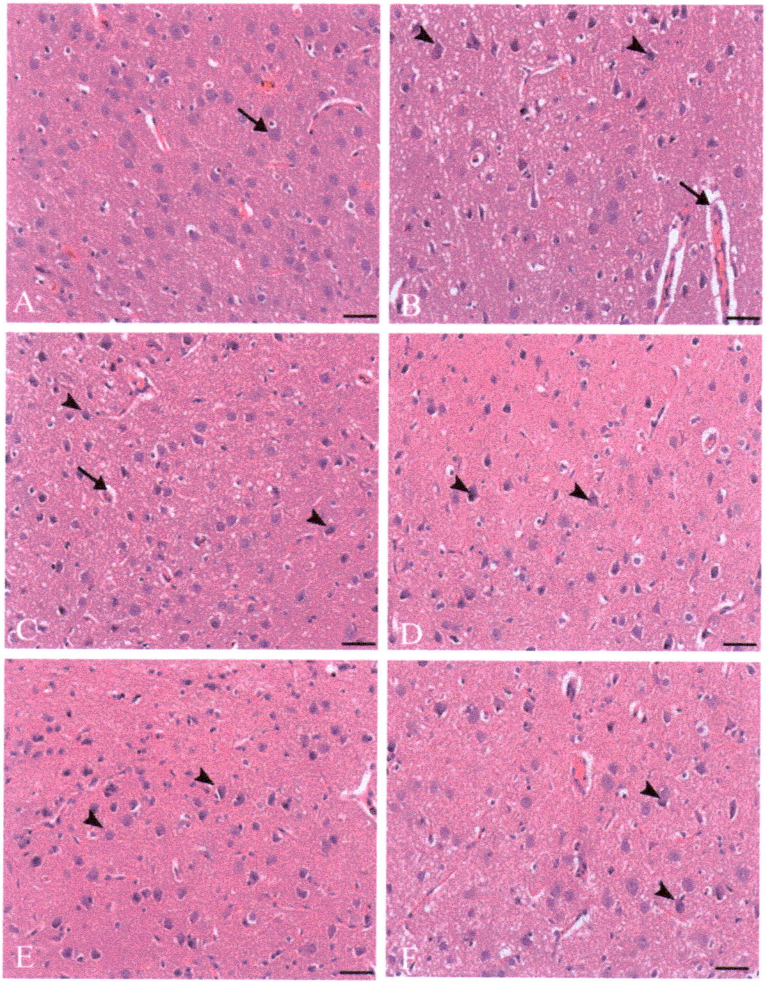

Figure 1. Photomicrographs of H&E stained the cerebral cortex of different treated groups—control group (**A**). The pyramidal layer of the cerebral cortex exhibited normal neuronal cells (indicated by an arrow). The metronidazole-treated group (**B**) showed severe ischemic neuronal injury in the same region, characterized by shrunken cytoplasm, nuclear pyknosis (indicated by arrowheads), and accompanied by pericellular and perivascular vacuolation (indicated by an arrow). The group treated with anise (250 mg/kg) in combination with metronidazole (**C**) showed a decrease in these ischemic neuronal degenerative changes (indicated by arrowheads) and a reduction in perivascular edema (indicated by an arrow). The group treated with anise (500 mg/kg) + metronidazole (**D**) showed decreased ischemic neuronal injury (arrowheads indicate the affected neurons). The groups treated with clove (250 mg/kg) combined with metronidazole (**E**) also displayed decreased ischemic neuronal injury (indicated by arrowheads). The group treated with clove (500 mg/kg) + metronidazole (**F**) exhibited minimal ischemic neuronal injury (indicated by arrowheads). All images were taken using H&E staining at a magnification of $\times 200$, bar = 100 μm.

3.2. Body and Brain Weights

Feed intake (FI), body weight gain percentage (BWG%), and feed efficiency ratio (FER) showed a significant ($p < 0.05$) decrease in MET-treated rats compared to the control group. The administration of anise seeds and clove bud extracts in combination with MET caused a significant ($p < 0.05$) increase in these parameters compared to the MET-treated rats. Additionally, the MET-treated group displayed a notably ($p < 0.05$) lower brain weight percentage than the control group. In contrast, groups treated with seed extracts showed a significant increase in brain weight compared to the MET-treated group (see Table 2).

Table 2. Protective effect of anise and clove seeds extract on FI (g/d), BWG%, FER, and brain weight % in MET-treated male rats.

Groups	FI (g/d)	BWG %	FER	Brain %
Control	23.35 ± 0.28 ^{a}	29.90 ± 2.12 ^{a}	0.19 ± 0.01 ^{a}	1.58 ± 0.13 ^{a}
Metronidazole	17.87 ± 0.12 ^{e}	8.85 ± 0.23 ^{e}	0.02 ± 0.005 ^{d}	0.76 ± 0.04 ^{c}
Anise (250 mg/kg) + Metronidazole	22.49 ± 0.12 ^{ab}	26.17 ± 0.64 ^{b}	0.04 ± 0.002 ^{c}	0.97 ± 0.01 ^{b}
Anise (500 mg/kg) + Metronidazole	20.95 ± 0.64 ^{c}	23.94 ± 0.68 ^{c}	0.05 ± 0.004 ^{c}	1.56 ± 0.98 ^{a}
Clove (250 mg/kg) + Metronidazole	19.87 ± 0.94 ^{d}	22.47 ± 1.02 ^{c}	0.07 ± 0.003 ^{b}	1.02 ± 0.003 ^{b}
Clove (500 mg/kg) + Metronidazole	21.79 ± 0.19 ^{bc}	19.49 ± 1.55 ^{d}	0.06 ± 0.005 ^{b}	0.95 ± 0.133 ^{b}

The data are presented as mean ± SD. The statistical analysis was performed using one-way ANOVA followed by the Tukey multiple range test. ^{a,b,c,d,e} means within the same column having different superscripts indicate statistically significant differences ($p < 0.05$).

3.3. Protective Effect of Anise and Clove Seeds Extract on GABA, ACHE, DA, and ST Quantity in the Brain Rats Administered MET

Table 3 summarizes the biochemical parameters of the tested groups' brain tissue. MET supplementation triggered a reduction of the brain GABA, DA, and ST and increased ACHE. Moreover, oral administration with anise and clove extract elevated GABA, DA, and ST suppression in brain tissue. Conversely, treated groups with anise and clove extract recorded a significant decrease in ACHE.

Table 3. Protective effect of anise and clove seeds extract on GABA, ACHE, DA, and ST levels in MET-treated male rats.

Groups	GABA (Pg/mL)	ACHE (Pg/mL)	DA (ng/mL)	ST (ng/mL)
control	404.02 ± 22 ^{a}	11.21 ± 21 ^{e}	2.51 ± 0.12 ^{a}	67.5 ± 6.5 ^{a}
Metronidazole	46.86 ± 4.26 ^{f}	87.12 ± 6.1 ^{a}	0.15 ± 0.005 ^{f}	5.21 ± 0.9 ^{e}
Anise (250 mg/kg) + Metronidazole	97.41 ± 12.9 ^{e}	67.35 ± 2 ^{b}	0.37 ± 0.01 ^{e}	25.31 ± 6.8 ^{d}
Anise (500 mg/kg) + Metronidazole	274.75 ± 26.55 ^{c}	20.51 ± 3.5 ^{d}	0.94 ± 0.03 ^{c}	50.41 ± 20 ^{c}
Clove (250 mg/kg) + Metronidazole	157.5 2 ± 9.5 ^{d}	42.51 ± 5.5 ^{c}	0.76 ± 0.04 ^{d}	32.21 ± 5.7 ^{d}
Clove (500 mg/kg) + Metronidazole	335.51 ± 10.5 ^{b}	19.11 ± 1.1 ^{d}	1.12 ± 0.03 ^{b}	59.14 ± 1.2 ^{b}

The data are presented as mean ± SD. The statistical analysis was performed using one-way ANOVA followed by the Tukey multiple range test. ^{a,b,c,d,e,f} means within the same column having different superscripts indicate statistically significant differences ($p < 0.05$).

3.4. Anise and Clove Extract to Improve the Altered Antioxidant Status Due to MET

Exposure to MET significantly decreased the activity of both SOD and CAT enzymes. However, anise and clove extract showed a dose-dependent reverse effect. Anise and clove extract at two doses significantly reduced the elevated levels of MDA and NO in the MET group ($p < 0.05$). Moreover, the MET group exhibited a significant decrease in GSH levels ($p < 0.001$), which was restored by the anise and clove extract (see Table 4)

Table 4. Effect of anise and clove seeds extract on MDA, NO, CAT, SOD, and GSH levels in MET-treated male rats.

Groups	MDA (nmol/gm)	NO (umol/L)	SOD (U/mg)	CAT (U/mg)	GSH (mg/gm)
Control	1.24 ± 0.51^{f}	5 ± 1^{e}	222 ± 16^{a}	9.91 ± 1.09^{a}	192.5 ± 11.5^{a}
Metronidazole	25.75 ± 2^{a}	86 ± 7^{a}	30.5 ± 2.5^{e}	0.59 ± 0.03^{d}	24.5 ± 2.5^{d}
Anise (250 mg/kg) + Metronidazole	18.6 ± 1.12^{b}	45 ± 5^{c}	48.5 ± 9.5^{d}	2.22 ± 0.26^{c}	70 ± 3^{c}
Anise (500 mg/kg) + Metronidazole	9.92 ± 0.92^{d}	19.5 ± 1.5^{d}	130 ± 9^{b}	5.52 ± 0.72^{b}	95 ± 5^{b}
Clove (250 mg/kg) + Metronidazole	12.65 ± 0.15^{c}	56.5 ± 6.5^{b}	80.5 ± 0.13^{c}	1.86 ± 0.5^{c}	73.5 ± 5.74^{c}
Clove (500 mg/kg) + Metronidazole	6.45 ± 1.39^{e}	27.5 ± 3.5^{d}	126 ± 7^{b}	5.08 ± 0.82^{b}	93 ± 3^{b}

The data are presented as mean ± SD. The statistical analysis was performed using one-way ANOVA followed by the Tukey multiple range test. ^{a,b,c,d,e,f} means within the same column having different superscripts indicate statistically significant differences ($p < 0.05$).

3.5. Histopathological Results

The metronidazole-treated group displayed severe ischemic neuronal injury in the pyramidal layer of the cerebral cortex, characterized by shrunken cytoplasm, nuclear pyknosis, and pericellular and perivascular vacuolation. However, co-treatment with anise and clove extracts and metronidazole decreased ischemic neuronal degenerative changes and reduced perivascular edema, as shown in Figure 1.

3.6. Effect of the Anise and Clove Extract on the Antioxidants and Apoptotic Gene Expression of the Brain Subjected to MET

Our results demonstrated that oral administration of MET significantly downregulated the mRNA expression of SOD, CAT, and Nrf2 compared to the other treated groups. In contrast, the groups treated with anise and clove extract showed a significant upregulation of SOD, CAT, and Nrf2 mRNA expression compared to the MET-treated group. Notably, the larger doses of anise and clove extract led to a pronounced increase in the expression of these genes. These findings are illustrated in Figure 2.

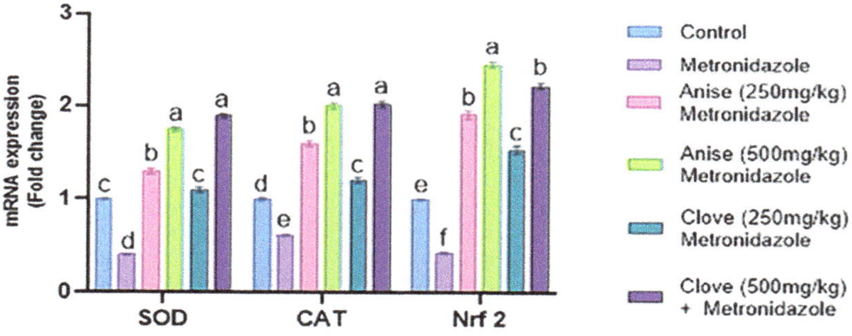

Figure 2. Effect of anise and clove seeds extract on the SOD, CAT, and Nrf2 mRNA expression in MET-treated male rats. Data are shown as mean ± SEM. The data were analyzed using one-way ANOVA, followed by Bonferroni multiple comparisons test. ^{a,b,c,d,e,f} Values with different letters are statistically different at $p < 0.05$. (N = 5).

As depicted in Figure 3, a significant upregulation in the mRNA expression of Bax, caspase 3, and P53 was observed, coupled with a substantial decrease in the mRNA expression of Bcl2 compared to the other treated groups. In contrast, the groups treated with anise and clove extract showed a notable downregulation of the mRNA expression of Bax, caspase-3, and P53 compared to the MET-treated group. This was accompanied by a pronounced increase in the mRNA expression of Bcl2, with the larger doses of anise and clove extract showing a marked improvement.

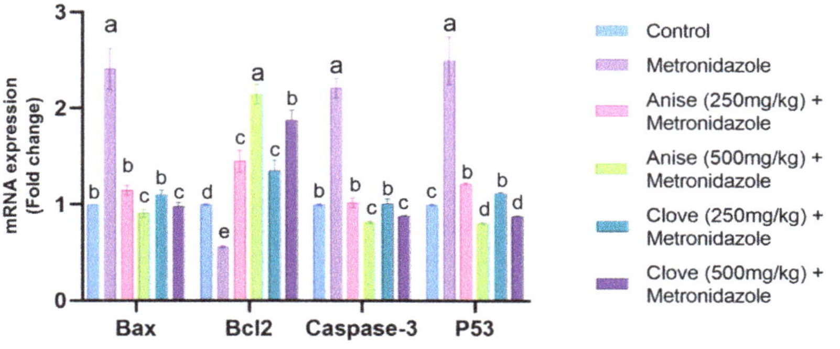

Figure 3. Effect of anise and clove seeds extract on the Bax, Bcl2, Caspase-3, and P53 mRNA expression in MET-treated male rats. Data are shown as mean ± SEM. The data were analyzed using one-way ANOVA, followed by Bonferroni multiple comparisons test. [a,b,c,d,e] Values with different letters are statistically different at $p < 0.05$. (N = 5).

4. Discussion

Long-term use of metronidazole, a nitroimidazole-derived antibiotic, has been associated with neuronal damage due to its ability to cross the blood–brain barrier. However, chronic use of metronidazole in humans at doses at or above 2 g per day can result in potential side effects, including peripheral neuropathy, seizures, cerebellar ataxia, and optic neuropathy [4]. Metronidazole can cause a range of neurological syndromes, including cerebellar syndrome, encephalopathy, seizures, optic neuropathy, autonomic neuropathy, and peripheral neuropathy [36]. Adverse effects on the central nervous system due to metronidazole toxicity have been observed in humans and various animal species, such as dogs, rats, and cats. Additionally, there have been reports of cerebellar syndrome occurring due to prolonged exposure to metronidazole [24]. A previous study has provided clear evidence that metronidazole can cross the blood–brain barrier and accumulate in specific brain regions, including the hippocampus, olfactory bulb, and cerebellum [37]. Metronidazole can lead to adverse effects on the central nervous system, resulting in a condition known as metronidazole-induced encephalopathy [5]. Later, Evans et al. [38] GABA (gamma-aminobutyric acid) has been identified as the main inhibitory neurotransmitter in the cerebellar and vestibular systems, which are affected by metronidazole intoxication.

Herbal treatments are gaining increasing recognition due to their natural origins, perceived safety, and demonstrated effectiveness against a broad spectrum of health conditions, including cardiovascular disease, diabetes, and cancer [39,40]. The seeds of *Pimpinella anisum* L., commonly known as anise, have been extensively researched for their numerous beneficial effects on human health [41]. Their medicinal properties include functioning as a diuretic, an antihypertensive, an anti-diabetic, an anti-cancer, an immunomodulator, an antibiotic, an anti-inflammatory, an analgesic, and even an anti-stress agent [42]. The natural phenolic antioxidant properties found in clove bud extract make it an effective treatment for disorders caused by oxidative stress [43]. Various compounds extracted from alcoholic and aqueous extracts of clove buds, such as tannins, ellagic acid, gallic acid, flavonoids, and their glycosides, have been attributed with numerous health benefits. These include antithrombotic, antiprotozoal, hypoglycemic, anti-inflammatory, gastro-protective, and aphrodisiac effects [44]. Anise seeds and clove bud extract, which contain highly bioactive phenolic components, have been identified as potential treatments for the neurotoxic effects of MET.

In the MET-treated group, adverse effects on brain function parameters (GABA, DA, and ST) were observed, along with increases in ACHE and the brain level of oxidative stress markers (MDA, NO). There were also decreases in the activity of brain antioxidant enzymes (SOD, CAT, and GSH). These findings agree with Tahoun [45], who demonstrated that significant weight loss and the onset of neurological symptoms were induced by daily doses of 500 mg/kg body weight of metronidazole for 60 days. These results do not align with the findings of Sohrabi [46], who demonstrated that a dose of 400 mg/kg of metronidazole had no significant impact on body weight after 60 days. Additionally, Chukwu et al. [47] showed that administering metronidazole at 200 and 400 mg/kg doses for 28 days did not affect body weight. These discrepancies could explain variations in metronidazole dosage and study duration. It could also be attributed to anorexia (lack of appetite) in the animals, as was observed and corroborated in an earlier investigation [48]. While several theories have been proposed to explain how metronidazole (MTZ) causes cerebellar toxicity, the exact mechanism remains unclear. One hypothesis suggests that MTZ and its metabolites cause reversible axonal swelling and symmetrical damage to the cerebellar nuclei by binding to neuronal RNA and inhibiting protein synthesis [49]. Another theory posits that the inhibitory neurotransmitter gamma-aminobutyric acid (GABA) receptors in the vestibular and cerebellar systems may also be influenced by MTZ [38]. In addition, MTZ may stimulate the production of harmful radicals, including semiquinone and nitro anion radicals [50].

Congestion and some displacement of Purkinje cells were also noted in experimental mice treated with 200 mg/kg/day of metronidazole (MTZ). This observation suggests that chronic exposure to escalating doses of metronidazole may result in cerebellar damage, as highlighted in a study by Agarwal et al. [51]. Metronidazole has been demonstrated to have toxicological effects on brain cells [52]. Biochemical evaluations in brain tissue indicated elevated levels of MDA and NO in the metronidazole (MET)-treated group compared to the control group, with the increase being statistically significant ($p < 0.05$). Researchers discovered that mice administered with metronidazole experienced increased MDA activity, leading to higher lipid peroxidation (LPO) levels, a marker suggestive of potential cellular damage [53].

Metronidazole significantly affects the activity of brain antioxidant enzymes such as SOD, CAT, and GSH. Additionally, it dramatically alters inflammatory mediators and induces morphological changes [3]. The gene expression data from our study supports these findings, demonstrating that oral administration of MET significantly suppressed the mRNA expression of SOD, CAT, and Nrf2 compared to other treatment groups. Neurons in the MET group exhibited severe ischemic injury, characterized by reduced cytoplasm, nuclear pyknosis, and pericellular and perivascular vacuolation. In addition, Purkinje cells selectively underwent cell death and showed signs of degeneration, including swelling, vacuolation, and clumping of protoplasm [24].

Additionally, the results indicated that treatment with anise seeds and clove buds extract improved tissue levels of brain function parameters (GABA, DA, and ST), decreased ACHE, and increased feed intake, body weight gain percentage, feed efficiency ratio, and brain weight percentage. Improvements in brain histology and a reduction in the concentration of oxidative stress markers (MDA and NO) were also observed. These results aligned with Cabuk et al. [54], which demonstrated a substantial increase in body weight following treatment with 750 mg/kg of *Pimpinella anisum* L. aqueous extract for 15 days.

The bio-active compounds in aniseed—including Anethole, Eugenol, Anisaldehyde, Estragol, and Methylchavicol—may stimulate the digestive system and contribute to this effect. Anethole, the primary compound in *Pimpinella anisum* L., has been found to inhibit the growth of pathogenic microorganisms in the digestive tract, leading to improved weight gain and feed conversion. Anise seeds contain a volatile oil that comprises 1.5–6% of their makeup, with trans-anethole making up 88% of this oil [55]. Cabuk et al. [56] suggested that the oil extracted from anise seeds could potentially modify the performance of medications targeting the central nervous system. According to our results, anise seeds effectively reduce lipid peroxidation and enhance the activity of antioxidant enzymes (SOD, CAT, and GSH). The radical scavenging phytochemicals present in these plants may be responsible for this effect. Therefore, Bekara et al. [7] demonstrated the effectiveness of polyphenols in reducing lipid peroxidation. Polyphenols are a class of bioactive compounds found in nearly all plant species. Multiple studies have shown that the polyphenols in *Pimpinella anisum* L. seeds donate an electron to free radicals and then react with them to form more stable compounds. This process restores the pro-oxidant/antioxidant balance, thereby reducing lipid peroxidation [57].

Anise and clove extract caused a considerable upregulation of SOD, CAT, and Nrf2 mRNA expression compared to the MET-treated group. A pronounced enhancement in the expression of these genes was observed at higher dosages of the anise and clove extract, corroborating our results. In addition, the essential oil of this plant may function as an effective antidepressant [58]. *Pimpinella anisum* L. is preventive against the development of cerebrovascular diseases and is therapeutic in treating neurological conditions such as epilepsy and seizures [59]. Seeds that contain anise oil have demonstrated a neuroprotective effect, likely through enhanced modulation of NMDA activities, including activating the glycine site NMDA receptor [60]. Anise oil, possibly by activating GABA A receptors, exhibits an inverse effect, reducing hyper-locomotor activity [58]. Treatment with anise seed extract alleviated the histological deterioration induced by MET. The overall cellular structure improved following the anise seed extract treatment, although vacuolization persisted [7]. In addition, [61] demonstrated the inhibition of AChE by clove extract.

Furthermore, eugenol and isoeugenol mitigated the increase in AChE activity and intracellular Ca^{2+} levels in the cerebral cortex and cerebellum regions of rat brains after exposure to acrylamide [62]. However, treatment with clove bud extract counteracted the toxicity of MET by enhancing the activity of CAT, SOD, and GSH. The results align with Gülçin et al. [63], who discovered that clove buds are rich in antioxidants. Concurrently, the physiological functions of CAT, SOD, and GPx are interconnected. SOD catalyzes the conversion of superoxide anions (O_2) into molecular oxygen and hydrogen peroxide. Subsequently, catalase and peroxidase cooperate to convert H_2O_2 into water. Reduced activity of SOD is consistent with lower levels of intracellular H_2O_2, which consequently reduces CAT and GPx activity [64]. The antioxidant status may have been restored due to the ability of clove components to scavenge free radicals induced by artesunate stress. When S. aromaticum bud extract was administered, the antioxidant status improved and lipid levels decreased [65]. Amber et al. [14] demonstrated the antioxidant and neuroprotective potential of S. aromaticum. AlC13-treated rats on a diet including clove aqueous extract showed a significant increase in their GSH content and GPx activity after 14 days of daily gavage [62]. In addition, eugenol, a component of clove, exhibits anti-oxidative and anti-amyloid beta peptide activity, as well as cholinomimetic action. Studies on mice have shown that acute treatment with an ethanolic clove extract enhances learning and memory recall [66].

The mRNA levels of Bax, caspase-3, and P53 were significantly upregulated, while the expression of Bcl2 mRNA was notably downregulated. These findings are consistent with previous studies on MET-treated rats [3]. Furthermore, in the anise-treated group, the high estrogen level—known for its anti-apoptotic action—was restored, along with the downregulation of caspase 3. This provides further evidence that MET induces apoptosis in neuronal cells. Given that several components in aniseed possess antioxidant properties, it may be reasonable to consider it a significant natural estrogen source [67]. In addition, clove extract mitigates oxidative stress and cellular apoptotic death [68,69].

5. Conclusions

This study illustrates how MET is associated with modifications in various biomarkers, including oxidative stress markers, inflammatory mediators, determinants of apoptosis, neurotransmitters, and nitric oxide signaling molecules. The present research reveals the neuroprotective qualities of anise seeds and clove buds when co-administered with MET at diverse doses. While further exploration is needed to elucidate the precise mechanism underpinning MET-induced neuronal degeneration, this work may lay the groundwork for future investigations.

Supplementary Materials: The following supporting information can be downloaded at: https://www.mdpi.com/article/10.3390/toxics11090724/s1, Figure S1: HPLC of Anise characterization; Figure S2: HPLC of Clove characterization; Table S1: Primers for gene expression by RT-PCR.

Author Contributions: A.M.E.-M. and M.H.A.-E.: supervised, designed, and performed experiments and revised the manuscript. N.A.T. and R.M.B.: performed experiments, analyzed data, and wrote and finalized the manuscript. B.S.A., H.I.G. and M.S.: performed experiments. A.M.E.-M. and M.S.: performed experiments, interpreted data, and wrote and finalized the manuscript. All authors have read and agreed to the published version of the manuscript.

Funding: The authors extend their appreciation to the Deanship of Scientific Research at King Khalid University for funding this work through a large group Research Project under grant number RGP2/435/44. We appreciate the resources the Princess Nourah bint Abdulrahman University Researchers Supporting Project number (PNURSP2023R73), Princess Nourah bint Abdulrahman University, Riyadh, Saudi Arabia.

Institutional Review Board Statement: The Animal Care and Use Committee of the Faculty of Veterinary Medicine at the University of Kafr El-Sheikh in Egypt approved the guidelines for the care and use of these animals, code No. KFS-IACUC/95/2019.

Informed Consent Statement: Not applicable.

Data Availability Statement: Not applicable.

Acknowledgments: The authors thank the Deanship of Scientific Research at King Khalid University for funding this work through a large group Research Project under grant number RGP2/435/44. This work was supported by Princess Nourah bint Abdulrahman University Researchers Supporting Project number (PNURSP2023R73), Princess Nourah bint Abdulrahman University, Riyadh, Saudi Arabia.

Conflicts of Interest: The authors declare no conflict of interest.

References

1. AbdRabou, M.A.; Alrashdi, B.M.; Alruwaili, H.K.; Elmazoudy, R.H.; Alwaili, M.A.; Othman, S.I.; Alghamdi, F.A.; Fahmy, G.H. Exploration of Maternal and Fetal Toxicity Risks for Metronidazole-Related Teratogenicity and Hepatotoxicity through an Assessment in Albino Rats. *Toxics* **2023**, *11*, 303. [CrossRef]
2. Stoian, I.-A.; Iacob, B.-C.; Dudaș, C.-L.; Barbu-Tudoran, L.; Bogdan, D.; Marian, I.O.; Bodoki, E.; Oprean, R. Biomimetic electrochemical sensor for the highly selective detection of azithromycin in biological samples. *Biosens. Bioelectron.* **2020**, *155*, 112098. [CrossRef]
3. Chaturvedi, S.; Malik, M.Y.; Rashid, M.; Singh, S.; Tiwari, V.; Gupta, P.; Shukla, S.; Singh, S.; Wahajuddin, M. Mechanistic exploration of quercetin against metronidazole induced neurotoxicity in rats: Possible role of nitric oxide isoforms and inflammatory cytokines. *Neurotoxicology* **2020**, *79*, 1–10. [CrossRef] [PubMed]
4. Vilian, A.E.; Ranjith, K.S.; Lee, S.J.; Umapathi, R.; Hwang, S.-K.; Oh, C.W.; Huh, Y.S.; Han, Y.-K. Hierarchical dense Ni–Co layered double hydroxide supported carbon nanofibers for the electrochemical determination of metronidazole in biological samples. *Electrochim. Acta* **2020**, *354*, 136723. [CrossRef]
5. Sørensen, C.G.; Karlsson, W.K.; Amin, F.M.; Lindelof, M. Metronidazole-induced encephalopathy: A systematic review. *J. Neurol.* **2020**, *267*, 1–13. [CrossRef]
6. Shahrajabian, M.H.; Sun, W.; Cheng, Q. Clinical aspects and health benefits of ginger (*Zingiber officinale*) in both traditional Chinese medicine and modern industry. *Acta Agric. Scand. Sect. B—Soil Plant Sci.* **2019**, *69*, 546–556. [CrossRef]
7. Bekara, A.; Hamadouche, N.A.; Kahloula, K.; Harouat, S.; Tabbas, D.; Aoues, A. Effect of *Pimpinella anisum* L (Aniseed) Aqueous Extract against Lead (Pb) Neurotoxicity: Neurobehavioral Study. *Int. J. Neurosci. Behav. Sci.* **2015**, *3*, 32–40. [CrossRef]
8. Bekara, A.; Aithamadouche, N.; Kahloula, K.; Sadi, N.; Aoues, A. Effect of *Pimpinella anisum* L. on Histological and Biochemical Damage in Cerebrum and Cerebellum of Young Rats Intoxicated by Lead Acetate. *Group* **2016**, *11*, 12.
9. Mushtaq, A.; Anwar, R.; Ahmad, M. Methanolic extract of *Pimpinella anisum* L. prevents dementia by reducing oxidative stress in neuronal pathways of hypermnesic mice. *Pak. J. Zool.* **2020**, *52*, 1779–1786. [CrossRef]
10. Batiha, G.E.-S.; Beshbishy, A.M.; Tayebwa, D.S.; Shaheen, H.M.; Yokoyama, N.; Igarashi, I. Inhibitory effects of *Syzygium aromaticum* and *Camellia sinensis* methanolic extracts on the growth of *Babesia* and *Theileria* parasites. *Ticks Tick-Borne Dis.* **2019**, *10*, 949–958. [CrossRef]
11. Shaaban, H.A. Essential oil as antimicrobial agents: Efficacy, stability, and safety issues for food application. In *Essential Oils-Bioactive Compounds, New Perspectives and Applications*; IntechOpen: London, UK, 2020; pp. 1–33.

12. El-Saber Batiha, G.; Magdy Beshbishy, A.; El-Mleeh, A.; Abdel-Daim, M.M.; Prasad Devkota, H. Traditional uses, bioactive chemical constituents, and pharmacological and toxicological activities of *Glycyrrhiza glabra* L. (Fabaceae). *Biomolecules* **2020**, *10*, 352. [CrossRef] [PubMed]
13. Martínez-Herrera, A.; Pozos-Guillén, A.; Ruiz-Rodríguez, S.; Garrocho-Rangel, A.; Vértiz-Hernández, A.; Escobar-García, D.M. Effect of 4-allyl-1-hydroxy-2-methoxybenzene (eugenol) on inflammatory and apoptosis processes in dental pulp fibroblasts. *Mediat. Inflamm.* **2016**, *2016*, 9371403. [CrossRef] [PubMed]
14. Amber, S.; Shah, S.A.A.; Ahmed, T.; Zahid, S. Syzygium aromaticum ethanol extract reduces $AlCl_3$-induced neurotoxicity in mice brain through regulation of amyloid precursor protein and oxidative stress gene expression. *Asian Pac. J. Trop. Med.* **2018**, *11*, 123.
15. Kadri, Y.; Nciri, R.; Bardaa, S.; Brahmi, N.; Saber, S.; Harrath, A.H.; Aldahmash, W.; Alwasel, S.; Mohany, M.; El Feki, A.; et al. Syzygium Aromaticum Alleviates Cerium Chloride-Induced Neurotoxic Effect In The Adult Mice. *Toxicol. Mech. Methods* **2019**, *29*, 26–34. [CrossRef] [PubMed]
16. El-Mesallamy, A.M.; El-Gerby, M.; Azim, M.H.A.E.; Awad, A. Antioxidant, antimicrobial activities and volatile constituents of clove flower buds oil. *J. Essent. Oil Bear. Plants* **2012**, *15*, 900–907. [CrossRef]
17. Lee, K.-G.; Shibamoto, T. Antioxidant property of aroma extract isolated from clove buds [*Syzygium aromaticum* (L.) Merr. et Perry]. *Food Chem.* **2001**, *74*, 443–448. [CrossRef]
18. Nassar, M.I.; Gaara, A.H.; El-Ghorab, A.H.; Farrag, A.; Shen, H.; Huq, E.; Mabry, T.J. Chemical constituents of clove (*Syzygium aromaticum*, Fam. Myrtaceae) and their antioxidant activity. *Rev. Latinoam. Química* **2007**, *35*, 47.
19. Gülçin, İ.; Elmastaş, M.; Aboul-Enein, H.Y. Antioxidant activity of clove oil—A powerful antioxidant source. *Arab. J. Chem.* **2012**, *5*, 489–499. [CrossRef]
20. Moon, S.-E.; Kim, H.-Y.; Cha, J.-D. Synergistic effect between clove oil and its major compounds and antibiotics against oral bacteria. *Arch. Oral Biol.* **2011**, *56*, 907–916. [CrossRef]
21. Helander, I.M.; Alakomi, H.-L.; Latva-Kala, K.; Mattila-Sandholm, T.; Pol, I.; Smid, E.J.; Gorris, L.G.; von Wright, A. Characterization of the action of selected essential oil components on Gram-negative bacteria. *J. Agric. Food Chem.* **1998**, *46*, 3590–3595. [CrossRef]
22. Burt, S. Essential oils: Their antibacterial properties and potential applications in foods—A review. *Int. J. Food Microbiol.* **2004**, *94*, 223–253. [CrossRef]
23. Ogbonye, E.; Ejimofor, O.; Ogbodo, E.; Ezeugwunne, I.; Madukwe, D.; Odumodu, I.; Agada, U.; Okezie, A.; Amah, A. The effect of metronidazole on the histology of the cerebellum and pituitary gland in female wistar rats. *IP Indian J. Neurosci.* **2020**, *6*, 67–72.
24. Oda, S.S. Histopathological and biochemical alterations of metronidazole-induced toxicity in male rats. *GV* **2012**, *9*, 303–310.
25. Tarola, A.M.; Van de Velde, F.; Salvagni, L.; Preti, R. Determination of phenolic compounds in strawberries (*Fragaria ananassa* Duch) by high performance liquid chromatography with diode array detection. *Food Anal. Methods* **2013**, *6*, 227–237. [CrossRef]
26. Schermer, S. *The Blood Morphology of Laboratory Animals*, 3rd ed.; F.A. Davis Company: Philadelphia, PA, USA, 1967; pp. 42–48.
27. Lasley, S.M.; Greenland, R.D.; Michaelson, I.A. Determination of gamma-aminobutyric and glutamic acids in rat brain by liquid chromatography with electrochemical detection. *Life Sci.* **1984**, *35*, 1921–1930. [CrossRef]
28. Carageorgiou, H.; Tzotzes, V.; Sideris, A.; Zarros, A.; Tsakiris, S. Cadmium effects on brain acetylcholinesterase activity and antioxidant status of adult rats: Modulation by zinc, calcium and L-cysteine co-administration. *Basic Clin. Pharmacol. Toxicol.* **2005**, *97*, 320–324. [CrossRef] [PubMed]
29. Sasa, S.; Blank, C.L. Determination of serotonin and dopamine in mouse brain tissue by high performance liquid chromatography with electrochemical detection. *Anal. Chem.* **1977**, *49*, 354–359. [CrossRef]
30. Fossati, P.; Prencipe, L.; Berti, G. Use of 3, 5-dichloro-2-hydroxybenzenesulfonic acid/4-aminophenazone chromogenic system in direct enzymic assay of uric acid in serum and urine. *Clin. Chem.* **1980**, *26*, 227–231. [CrossRef]
31. Giustarini, D.; Rossi, R.; Milzani, A.; Dalle-Donne, I. Nitrite and nitrate measurement by Griess reagent in human plasma: Evaluation of interferences and standardization. *Methods Enzymol.* **2008**, *440*, 361–380.
32. Marklund, S.; Marklund, G. Involvement of the superoxide anion radical in the autoxidation of pyrogallol and a convenient assay for superoxide dismutase. *Eur. J. Biochem.* **1974**, *47*, 469–474. [CrossRef] [PubMed]
33. Oberley, L.; Spitz, D.; Greenwald, R. CRC handbook of methods for oxygen radical research. In *CRC Handbook of Methods for Oxygen Radical Research*; CRC Press: Boca Raton, FL, USA, 1985; Volume 2.
34. Ellman, G.L. Tissue sulfhydryl groups. *Arch. Biochem. Biophys.* **1959**, *82*, 70–77. [CrossRef] [PubMed]
35. Livak, K.J.; Schmittgen, T.D. Analysis of relative gene expression data using real-time quantitative PCR and the $2^{-\Delta\Delta CT}$ method. *Methods* **2001**, *25*, 402–408. [CrossRef]
36. Sarna, J.R.; Brownell, A.K.W.; Furtado, S. Reversible cerebellar syndrome caused by metronidazole. *Can. Med Assoc. J.* **2009**, *181*, 611–613. [CrossRef] [PubMed]
37. Sarna, J.R.; Furtado, S.; Brownell, A.K.W. Neurologic complications of metronidazole. *Can. J. Neurol. Sci.* **2013**, *40*, 768–776. [CrossRef]
38. Evans, J.; Levesque, D.; Knowles, K.; Longshore, R.; Plummer, S. Diazepam as a treatment for metronidazole toxicosis in dogs: A retrospective study of 21 cases. *J. Vet. Intern. Med.* **2003**, *17*, 304–310. [CrossRef] [PubMed]
39. Mechchate, H.; Es-Safi, I.; Mohamed Al Kamaly, O.; Bousta, D. Insight into gentisic acid antidiabetic potential using in vitro and in silico approaches. *Molecules* **2021**, *26*, 1932. [CrossRef]

40. Es-Safi, I.; Mechchate, H.; Amaghnouje, A.; Calarco, A.; Boukhira, S.; Noman, O.M.; Mothana, R.A.; Nasr, F.A.; Bekkari, H.; Bousta, D. Defatted Hydroethanolic Extract of Ammodaucus leucotrichus Cosson and Durieu Seeds: Antidiabetic and Anti-Inflammatory Activities. *Appl. Sci.* **2020**, *10*, 9147. [CrossRef]
41. Mosavat, S.H.; Jaberi, A.R.; Sobhani, Z.; Mosaffa-Jahromi, M.; Iraji, A.; Moayedfard, A. Efficacy of Anise (*Pimpinella anisum* L.) oil for migraine headache: A pilot randomized placebo-controlled clinical trial. *J. Ethnopharmacol.* **2019**, *236*, 155–160. [CrossRef]
42. Shojaii, A.; Abdollahi Fard, M. Review of pharmacological properties and chemical constituents of *Pimpinella anisum*. *Int. Sch. Res. Not.* **2012**, *2012*, 510795. [CrossRef]
43. Yashin, A.; Yashin, Y.; Xia, X.; Nemzer, B. Antioxidant activity of spices and their impact on human health: A review. *Antioxidants* **2017**, *6*, 70. [CrossRef]
44. Johannah, N.; Renny, R.; Gopakumar, G.; Maliakel, B.; Sureshkumar, D.; Krishnakumar, I. Beyond the flavour: A de-flavoured polyphenol rich extract of clove buds (*Syzygium aromaticum* L.) as a novel dietary antioxidant ingredient. *Food Funct.* **2015**, *6*, 3373–3382.
45. Tahoun, E.A.E.-A.M. Protective Effect of Moringa Oleifera against Metronidazole-induced toxicity in male albino rats. *J. Biosci. Appl. Res.* **2017**, *3*, 137–149.
46. Sohrabi, D.; Alipour, M.; Melati, A. Effect of metronidazole on spermatogenesis, plasma gonadotrophins and testosterone in rats. *Iran. J. Pharm. Res.* **2007**, *6*, 279–283.
47. Chukwu, V.; Akudike, C.; Ezejindu, D.; Ofoego, U.; Chukwuocha, C. The protective effects of turmeric on testicular tissues, after treatment with metronidazole in adult male Wistar rats. CIBTech. *J. Pharm. Sci.* **2015**, *4*, 62–68.
48. Dusengeyezu, E.; Kadima, J.N. How do metronidazole drawbacks impact patient compliance and therapeutic outcomes in treating amoebiasis in Rwanda. *Int. J. Trop. Dis. Health* **2016**, *17*, 1–7. [CrossRef]
49. Bradley, W.G.; Karlsson, I.; Rassol, C. Metronidazole neuropathy. *Br. Med. J.* **1977**, *2*, 610. [CrossRef]
50. Hassan, M.; Awadalla, E.; Ali, R.; Fouad, S.; Abdel-Kahaar, E. Thiamine deficiency and oxidative stress induced by prolonged metronidazole therapy can explain its side effects of neurotoxicity and infertility in experimental animals: Effect of grapefruit co-therapy. *Hum. Exp. Toxicol.* **2020**, *39*, 834–847. [CrossRef]
51. Agarwal, A.; Kanekar, S.; Sabat, S.; Thamburaj, K. Metronidazole-induced cerebellar toxicity. *Neurol. Int.* **2016**, *8*, 6365. [CrossRef]
52. Kalia, V.; Saggar, K. Case report: MRI of the brain in metronidazole toxicity. *Indian J. Radiol. Imaging* **2010**, *20*, 195–197. [CrossRef]
53. Ligha, A.; Paul, C. Oxidative effect of metronidazole on the testis of Wistar rats. *Aus. J. Bas. Appl. Sci.* **2011**, *5*, 1339–1344.
54. Cabuk, M.; Alcicek, A.; Bozkurt, M.; Imre, N. Antimicrobial properties of the essential oils isolated from aromatic plants and using possibility as alternative feed additives. In Proceedings of the National Animal Nutrition Congress, Konya, Turkey, 18–20 September 2003; pp. 184–187.
55. Koch, C.; Reichling, J.; Schneele, J.; Schnitzler, P. Inhibitory effect of essential oils against herpes simplex virus type 2. *Phytomedicine* **2008**, *15*, 71–78. [PubMed]
56. Samojlik, I.; Mijatović, V.; Petković, S.; Škrbić, B.; Božin, B. The influence of essential oil of aniseed (*Pimpinella anisum*, L.) on drug effects on the central nervous system. *Fitoterapia* **2012**, *83*, 1466–1473. [CrossRef] [PubMed]
57. Andallu, B.; Rajeshwari, C.U. Chapter 20—Aniseeds (*Pimpinella anisum* L.) in Health and Disease. In *Nuts and Seeds in Health and Disease Prevention*; Preedy, V.R., Watson, R.R., Patel, V.B., Eds.; Academic Press: San Diego, CA, USA, 2011; pp. 175–181. [CrossRef]
58. Kahloula, K.; Slimani, M.; Adli, D.E.H.; Rachdi, S.; Boumediene, D. Neuro beneficial effects of *Pimpinella anisum* against lead exposure. *Int. J. Green Pharm. (IJGP)* **2013**, *7*, 18. [CrossRef]
59. Gorji, A.; Ghadiri, M.K. History of epilepsy in Medieval Iranian medicine. *Neurosci. Biobehav. Rev.* **2001**, *25*, 455–461. [CrossRef] [PubMed]
60. Karimzadeh, F.; Hosseini, M.; Mangeng, D.; Alavi, H.; Hassanzadeh, G.R.; Bayat, M.; Jafarian, M.; Kazemi, H.; Gorji, A. Anticonvulsant and neuroprotective effects of *Pimpinella anisum* in rat brain. *BMC Complement. Altern. Med.* **2012**, *12*, 76. [CrossRef]
61. Kassab, R.B.; Bauomy, A.A. The neuroprotective efficency of the aqueous extract of clove (*Syzygium aromaticum*) in aluniniumin-duced neurotoxicity. *Int. J. Pharm. Pharm. Sci.* **2014**, *6*, 503–508.
62. Prasad, S.N. Neuroprotective efficacy of eugenol and isoeugenol in acrylamide-induced neuropathy in rats: Behavioral and biochemical evidence. *Neurochem. Res.* **2013**, *38*, 330–345. [CrossRef]
63. Gülçin, İ.; Şat, İ.G.; Beydemir, Ş.; Elmastaş, M.; Küfrevioğlu, Ö.İ. Comparison of antioxidant activity of clove (*Eugenia caryophylata* Thunb) buds and lavender (*Lavandula stoechas* L.). *Food Chem.* **2004**, *87*, 393–400. [CrossRef]
64. Lu, C.-Y.; Wang, E.K.; Lee, H.-C.; Tsay, H.-J.; Wei, Y.-H. Increased expression of manganese-superoxide dismutase in fibroblasts of patients with CPEO syndrome. *Mol. Genet. Metab.* **2003**, *80*, 321–329. [CrossRef]
65. Adegbola, M.V.; Anyim, G.; Ntwasa, M.; Ayeleso, A.O.; Oyedepo, T.A. Potential Effect of *Syzygium aromaticum* (Cloves) Extract on Serum Antioxidant Status and Lipid Profiles in Wistar Rats with Artesunate Toxicity. *Appl. Sci.* **2022**, *12*, 8216. [CrossRef]
66. Dashti-R, M.H.; Morshedi, A. The effects of *Syzygium aromaticum* (clove) on learning and memory in mice. *Asian J. Tradit. Med.* **2009**, *4*, 128–133.
67. Faried, M.A.; Issa, N.M. Comparative study on the effect of the pimpinella anisum and estradiol on the hippocampus and dentate gyrus of ovariectomized rats. *Anat. Physiol. Biochem. Int. J.* **2017**, *2*. [CrossRef]

68. Moghimian, M.; Abtahi-Evari, S.-H.; Shokoohi, M.; Amiri, M.; Soltani, M. Effect of *Syzygium aromaticum* (clove) extract on seminiferous tubules and oxidative stress after testicular torsion in adult rats. *Physiol. Pharmacol.* **2017**, *21*, 343–350.
69. Soltani, M.; Moghimian, M.; Abtahi-Evari, S.-H.; Esmaeili, S.-A.; Mahdipour, R.; Shokoohi, M. The Effects of Clove Oil on The Biochemical and Histological Parameters, and Autophagy Markers in Polycystic Ovary Syndrome-Model Rats. *Int. J. Fertil. Steril.* **2023**, *17*, 187–194. [PubMed]

toxics MDPI

Article

Exploration of Tilmicosin Cardiotoxicity in Rats and the Protecting Role of the *Rhodiola rosea* Extract: Potential Roles of Cytokines, Antioxidant, Apoptotic, and Anti-Fibrotic Pathways

Salwa A. Elgendy [1], Mohamed Mohamed Soliman [2,3], Heba I. Ghamry [4], Mustafa Shukry [5,*], Lina Abdelhady Mohammed [6], Hend Elsayed Nasr [6], Badriyah S. Alotaibi [7,*], Ibrahim Jafri [8], Samy Sayed [9,10], Amira Osman [11,12] and Heba A. Elnoury [1]

1 Department of Pharmacology, Faculty of Medicine, Benha University, Benha 13511, Egypt
2 Department of Clinical Laboratory Sciences, Turabah University College, Taif University, P.O. Box 11099, Taif 21944, Saudi Arabia; mmsoliman@tu.edu.sa
3 Biochemistry Department, Faculty of Veterinary Medicine, Benha University, Toukh 13736, Egypt
4 Nutrition and Food Science, Department of Home Economics, Faculty of Home Economics, King Khalid University, P.O. Box 960, Abha 61421, Saudi Arabia; hgmry@kku.edu.sa
5 Department of Physiology, Faculty of Veterinary Medicine, Kafrelsheikh University, Kafrelsheikh 33516, Egypt
6 Department of Medical Biochemistry and Molecular Biology, Faculty of Medicine, Benha University, Benha 13511, Egypt; hend.mosalm@fmed.bu.edu.eg (H.E.N.)
7 Department of Pharmaceutical Sciences, College of Pharmacy, Princess Nourah bint Abdulrahman University, P.O. Box 84428, Riyadh 11671, Saudi Arabia
8 Department of Biotechnology, College of Science, Taif University, P.O. Box 11099, Taif 21944, Saudi Arabia
9 Department of Economic Entomology and Pesticides, Faculty of Agriculture, Cairo University, Giza 12613, Egypt; samy_mahmoud@hotmail.com
10 Department of Science and Technology, University College-Ranyah, Taif University, P.O. Box 11099, Taif 21944, Saudi Arabia
11 Department of Basic Medical and Dental Sciences, Faculty of Dentistry, Zarqa University, Zarqa 13110, Jordan; mero.osman@med.kfs.edu.eg
12 Department of Histology and Cell Biology, Faculty of Medicine, Kafrelsheikh University, Kafrelsheikh 33516, Egypt
* Correspondence: mostafa.ataa@vet.kfs.edu.eg (M.S.); bsalotaibi@pnu.edu.sa (B.S.A.)

Citation: Elgendy, S.A.; Soliman, M.M.; Ghamry, H.I.; Shukry, M.; Mohammed, L.A.; Nasr, H.E.; Alotaibi, B.S.; Jafri, I.; Sayed, S.; Osman, A.; et al. Exploration of Tilmicosin Cardiotoxicity in Rats and the Protecting Role of the *Rhodiola rosea* Extract: Potential Roles of Cytokines, Antioxidant, Apoptotic, and Anti-Fibrotic Pathways. *Toxics* **2023**, *11*, 857. https://doi.org/10.3390/toxics11100857

Academic Editor: Youssef Sari

Received: 22 August 2023
Revised: 18 September 2023
Accepted: 20 September 2023
Published: 13 October 2023

Abstract: Tilmicosin (TIL) is a common macrolide antibiotic in veterinary medicine. High doses of TIL can have adverse cardiovascular effects. This study examined the effects of *Rhodiola rosea* (RHO) that have anti-inflammatory, antioxidant, and anti-fibrotic effects on tilmicosin (TIL)-induced cardiac injury targeting anti-inflammatory, antioxidant, apoptotic, and anti-apoptotic signaling pathways with anti-fibrotic outcomes. Thirty-six male Wistar albino rats were randomly divided into groups of six rats each. Rats received saline as a negative control, CARV 1 mL orally (10 mg/kg BW), and RHO 1 mL orally at 400 mg/kg BW daily for 12 consecutive days. The TIL group once received a single subcutaneous injection (SC) dose of TIL (75 mg/kg BW) on the sixth day of the experiment to induce cardiac damage. The standard group (CARV + TIL) received CARV daily for 12 consecutive days with a single TIL SC injection 1 h after CARV administration only on the sixth day of study and continued for another six successive days on CARV. The protective group (RHO + TIL) received RHO daily for the same period as in CARV + TIL-treated rats and with the dosage mentioned before. Serum was extracted at the time of the rat's scarification at 13 days of study and examined for biochemical assessments in serum lactate dehydrogenase (LDH), cardiac troponin I (cTI), and creatine phosphokinase (CK-MB). Protein carbonyl (PC) contents, malondialdehyde (MDA), and total antioxidant capacity (TAC) in cardiac homogenate were used to measure these oxidative stress markers. Quantitative RT-PCR was used to express interferon-gamma (INF-γ), cyclooxygenase-2 (COX-2), OGG1, BAX, caspase-3, B-cell lymphoma-2 (Bcl-2), and superoxide dismutase (SOD) genes in cardiac tissues, which are correlated with inflammation, antioxidants, and apoptosis. Alpha-smooth muscle actin (α-SMA), calmodulin (CaMKII), and other genes associated with Ca^{2+} hemostasis and fibrosis were examined using IHC analysis in cardiac cells (myocardium). TIL administration significantly increased the examined cardiac markers, LDH, cTI, and CK-MB. TIL administration also increased ROS, PC, and MDA while decreasing antioxidant activities (TAC and SOD mRNA) in cardiac tissues. Serum inflammatory cytokines and genes of inflammatory markers, DNA damage (INF-γ, COX-2), and apoptotic genes (caspase-3 and BAX) were upregulated with downregulation of the anti-apoptotic gene Bcl-2 as well as the DNA repair OGG1 in cardiac tissues. Furthermore, CaMKII and α-SMA genes were upregulated at cellular levels using cardiac tissue IHC analysis. On the contrary, pretreatment with RHO and CARV alone significantly decreased the cardiac injury markers induced by TIL, inflammatory and anti-inflammatory cytokines, and tissue oxidative-antioxidant parameters. INF-γ, COX-2, OGG1, BAX, and caspase-3 mRNA were downregulated, as observed by real-time PCR, while SOD and Bcl-2 mRNA were upregulated. Furthermore, the CaMKII and α-SMA genes' immune reactivities were significantly decreased in the RHO-pretreated rats.

Keywords: cardiomyopathy; tilmicosin; *Rhodiola rosea*; antioxidants; oxidative stress; apoptosis

1. Introduction

Tilmicosin (TIL) is a popular semi-synthetic macrolide antibiotic in veterinary medicine. It binds to the 50S ribosomal subunit and prevents protein synthesis in bacteria. This compound has 16 atoms. Treatment and prevention of pneumonia caused by *P. multocida*, *Actinobacillus pleuropneumonia*, *Pasteurella haemolytica*, *Streptococcus*, *Staphylococcus*, and *Mycoplasma* are among its approved uses in cattle, sheep, and pigs [1]. In addition, it is used in ruminant animals to avoid and treat mastitis [2]. Low inhibitory concentrations, a high distribution volume, and a long elimination half-life contribute to TIL's effectiveness [3]. TIL has been shown to have a wide range of negative consequences in research, some of which are cardiotoxic [4,5], and the heart is the main target for Tilmicosin action [6].

Several animal studies have shown that high doses of TIL have adverse inotropic and positive chronotropic cardiovascular effects, including ventricular function impairment, abrupt heart failure, and ECG alterations [7]. Many previous studies have not clearly described how TIL induces its cardiotoxicity effect. However, TIL may induce cardiotoxicity through increased production of reactive oxygen species (ROS) or affect the oxidative state of the heart tissues [8] through protein, DNA, and lipid damage, leading to apoptosis [9].

Calmodulin (CaM) is a tiny protein in the cytoplasm with two globular domains that serve as Ca^{2+} binding sites [10]. Muscle contraction in both cardiac and skeletal muscles results from the cytoplasmic release of Ca^{2+} ions through ryanodine receptor calcium release channels (RyRs); RyR2 is prominent in cardiac muscle, while RyR1 is found at high levels in skeletal muscle. Calmodulin in the muscles of the skeleton and the heart is one of several related proteins with RyRs, which are large ion channels made up of four peptide subunits, four tiny 12 kDa FK506 binding proteins (FKBP), and other proteins [11].

Carvedilol is an anti-inflammatory, anti-hypertensive, and anti-ischemic medication for the cardiovascular system with multiple approved uses. In individuals with congestive heart failure, it boosts myocardial function, prolongs life, and reduces mortality [12]. Catecholamines are harmful to the heart. However, this compound can prevent its effects [13]. Carvedilol has recently been shown to reduce the severity of ischemia and reperfusion injury through its antioxidant action, calcium antagonism, anti-arrhythmia, anti-apoptosis, and neutrophil infiltration inhibition in human and animal investigations. In addition, carvedilol can switch myocardial substrates from free fatty acids to glucose oxidation [14]. As a result, there is a greater need to discover new antioxidants derived from natural sources that are efficacious, bio-efficient, and non-toxic for use as safe therapeutic agents [15].

Rhodiola rosea (RHO) is a valuable phytomedicine cultivated in Tibet and China's cold and high altitudes. A member of the Crassulaceae family of plants, "golden root" or "roseroot" is used as a botanical adaptogen all over the world [16]. RHO has been shown to enhance cardiovascular and cerebral blood flow systems. Different physiologically active chemicals found in RHO may have other effects [17]. The glycoside molecule rhodioloside (salidroside) and the class of rosavins primarily found in plant rhizomes, including rosavin, rosarin, and rosin [18], have shown medicinal properties. Multiple investigations have identified salidroside as RHO's most dynamic ingredient [19]. RHO has a complex chemical structure, and its pharmacological effects and therapeutic efficacy for treating a wide range of cardiovascular illnesses are also highly variable among its numerous chemical components [20].

Multiple biological functions, including immune regulation, antioxidants, and cancer cell proliferation inhibition, have been attributed to *Rhodiola rosea* L. preparations by pharmacological researchers in both clinical practice and experimental studies [16]. It has been found that RHO extract has anti-inflammatory properties and protects muscular tissue during exercise [21]. It also has various protective products such as anti-cancer, anti-aging, anti-diabetic, neuroprotective, and cardioprotective effects [16]. The impact of stress on rats' eating, exercise, and reproductive cycles was mitigated by a compound called RHO [22].

Therefore, the current study aimed to outline the protective impacts of RHO that might be involved in ameliorating cardiotoxicity induced by TIL injection. Various biochemical, molecular, and cellular signaling pathways were confirmed to regulate apoptosis, oxidative stress, fibrosis, inflammation, and anti-apoptotic markers.

2. Materials and Methods

2.1. Drugs and Chemicals

A commercial preparation for tilmicosin (Micotil® 300 solution; AH0230) was obtained from (Elanco animal health, Macquarie Park, NSW, Australia)—Dilatrend (25 mg carvedilol, Roche SpA, Segrate MI, Italy). *Rhodiola rosea* extract was obtained from Puritan's Pride® Company (Puritan's Pride, Oakdale, NY, USA) as capsules containing 250 mg of Rhodiola root extract. All additional chemicals and reagents used in this investigation were bought from (Sigma-Aldrich Co., St. Louis, MO, USA) and were of the highest commercial quality.

2.2. Gas Chromatography–Mass Spectrometry (GC-MS) Analysis of *Rhodiola rosea* Extract

The chemical composition analysis of *Rhodiola rosea* extract was conducted using a Trace GC-TSQ mass spectrometer (Thermo Scientific, Austin, TX, USA) equipped with a direct capillary column TG–5MS (30 m × 0.25 mm × 0.25 μm film thickness). Initially, the column oven temperature was maintained at 50 °C and then gradually increased at a rate of 5 °C per minute until it reached 250 °C, where it was held for 2 min. Subsequently, it was further increased to the final temperature of 300 °C at 30 °C per minute and held for 2 min. The injector and MS transfer line temperatures were maintained at 270 °C and 260 °C, respectively. Helium gas was the carrier gas at a constant 1 mL/min flow rate. A solvent delay of 4 min was incorporated, and diluted samples of 1 μL were automatically injected using the Autosampler AS1300 (Thermo Scientific, Austin, TX, USA) in the split mode. Electron ionization (EI) mass spectra were obtained using an ionization voltage of 70 eV, covering the m/z range of 50–650 in full scan mode. The ion source temperature was set at 200 °C. The components present in the extract were identified by comparing their mass spectra with those available in the WILEY 09 and NIST 14 mass spectral databases [23].

2.3. Animal Grouping and Treatment

Thirty-six male Wistar rats (180–200 gm) and two months' age were purchased from an animal breeding farm (Helwan-Cairo, Egypt). Rats were housed in stainless steel cages, monitored for two weeks before the experiment to ensure normal growth and behavior, and kept in a controlled environment with a temperature of 25. 5 °C, a humidity of 55.5%, a 12 h light/dark cycle, and free access to standard rat feed (El-Nasr Co., Cairo, Egypt). Animals were handled manually to ensure complete acclimatization. After acclimatization, rats were distributed randomly into groups of six rats per group. Group I (normal) rats were intubated with 1 mL of distilled water in the stomach tube for 12 consecutive days. Group II (Rhodiola-treated) (RHO): rats received RHO 1 mL orally (400 mg/kg BW daily) [24] for 12 consecutive days. Group III (Carvedilol-treated) (CARV) rats received CARV 1 mL orally at a dose of 10 mg/kg BW [25–27] for 12 consecutive days. Group IV (Tilmicosin-treated; TIL): rats received a single subcutaneous injection (SC) at a dose of 75 mg/kg BW [4,28] on the sixth day of the experiment. Group V (CARV + TIL): rats obtained CARV for 12 consecutive days with a single TIL S/C injection 1 h after CARV administration only on the sixth day of study and continued for another six successive days on CARV with the exact dosage mentioned before. Group VI (RHO + TIL): rats obtained RHO for 12 consecutive days with a single TIL S/C injection 1 h after RHO administration on the sixth day of study only and continued for another six successive days on RHO with the exact dosage mentioned before. Each control, RHO, and CARV rat received a single subcutaneous (S/C) injection of physiological saline (NaCl 0.9%). On day 6, the other three groups only received one S/C injection of TIL. The LD50 of dry R. rosea extract for rats is 3.36 g/kg [29,30], while the oral LD50 doses of carvedilol for mice and rats are over 8000 mg/kg [31].

On the final day of the experiment, the animals had free access to water but were otherwise fasted for 12 h. After that, we used cervical dislocation to end the rats' lives and collected blood samples by puncturing their hearts. Serum samples obtained after centrifuging the collected blood at 3000 rpm for 10 min were stored at 20 °C for further analysis. Next, the heart was taken, rinsed with saline, dried, weighed, and dissected. The heart tissue was divided into three equal longitudinal parts. The heart tissue was frozen and stored at 80 °C in a flash. The frozen materials were mixed in a cold phosphate-buffered saline solution (0.1 M, pH 7.4) in a homogenizer fitted with a Teflon pestle and centrifuged at 5000× *g* for 15 min at 4 °C. For the chemical analysis, aliquots of the supernatant were frozen at 80 °C. Other cardiac tissues were promptly preserved in a neutral buffered formalin 10% solution for histopathological and immunohistochemical analysis [32].

2.4. Assessment of Body Weight and Cardiac Injury Biomarkers

Each rat's total body weight was recorded after the completion of the experiment. As per package directions, we estimated serum LDH and CK-MB concentrations. Serum cTI was also evaluated with an ELISA Plate Reader (Bio-Rad, Hercules, CA, USA) and an ELISA kit specific for the measurement of cTnI. MyBiosource, Inc. (San Diego, CA, USA) kindly provided rat cTnI level ELISA kits. Tsung's method [33] was used to measure creatine kinase-MB (CK-MB), and Lum and Gambino's method [34] was used to calculate LDH.

2.5. Inflammatory and Anti-Inflammatory Cytokine Assessments

Serum concentrations of IL-1 and TNF-alpha were measured using specific ELISA kits (ab255730 and ab46070, respectively) and spectrophotometric analysis according to the manufacturer's instructions. Rat IL-10 ELISA Kit (ab100765), a commercially available kit for measuring IL-10, was purchased from (abcam, Cambridge, MA, USA). The ELISA reader's data was approximated and evaluated according to the kit's instructions.

2.6. Oxidative Stress Biomarker (ROS, PC, MDA, TAC, 8-OHdG, and OGG1) Assessments in Cardiac Tissue Homogenate

The concentration of reactive oxygen species (ROS) in rat heart tissue was determined using an ELISA kit (MBS164653; MyBioSource Co., San Diego, CA, USA) following the manufacturer's instructions. Cayman's Chemical Co., located at 1000 Ellsworth Road, Ann Arbor, MI, USA, produced a colorimetric assay pack (no. 10005020) to measure protein carbonyl (PC) levels. According to Ohkawa et al.'s findings [35], malondialdehyde (MDA) levels were determined by a colorimetric test. The total antioxidant capacity (TAC) was measured using a kit from Biodiagnostic Co., Dokki, Giza, Egypt (catalog no. TA 2513; Egypt).

The My BioSource rat ELISA kits were used to estimate the levels of 8-OHdG (MBS269902) in the heart, following the manufacturer's instructions. To quantify the content of OGG1, a commercially available 96-well ELISA kit (CSB-EL016313RA, Cusabio Biotech Co., Ltd.,

College Park, MD, USA) was used. The ELISA assays were performed following the manufacturer's instructions. In brief, brain tissue was homogenized in a lysis buffer solution. The samples and standards were added to 96-well plates and incubated at 37 °C for 90 min. After washing three times, a biotinylated secondary antibody was added and incubated for 1 h at 37 °C. Following two more washes, an enzyme conjugate liquid was added and incubated for 30 min at 37 °C. After five washes, a color reagent was added to each well and incubated at 37 °C for 30 min in the dark. The absorbance was measured at 450 nm after adding the stop solution.

2.7. qRT-PCR Analysis

The quantitative expression of genes in cardiac tissue was measured and expressed using quantitative real-time PCR. About 100 ng of cardiac tissue was briefly used with Qiazol reagents to extract RNA. cDNA was synthesized with a Multi-Scribe RT-enzyme kit. 7500 Real-Time PCR equipment(Applied Biosystems, Waltham, MA, USA) was used to run qRT-PCR reactions using SYBR Green staining PCR Master Mix. The 2CT technique was used to quantify these genes. The examined genes were normalized to the housekeeping gene actin. The relative cycle threshold (CT) values were used to study the variations in gene expression [36]. Table 1 displays the genes that were analyzed.

Table 1. Primers for gene expression by RT-PCR.

Gene	Direction	Primer Sequence	Annealing Temp.	(bp)	Accession Number
OGG1	F:	ATCTGTTCTTCCAACAACAAC	58 °C	212	NM_030870
	R:	GCCAGCATAAGGTCCCCACAG			
IFN-γ	F:	TGTCATCGAATCGCACCTGATC	57 °C	185	NM_138880
	R:	GACTCCTTTTCCGCTTCCTTAG			
Bax	F:	GGCGAATTGGCGATGAACTG	57 °C	167	NM_017059.2
	R:	ATGGTTCTGATCAGCTCGGG			
Bcl-2	F:	GATTGTGGCCTTCTTTGAGT	57 °C	172	NM_016993.1
	R:	ATAGTTCCACAAAGGCATCC			
SOD	F:	AGGATTAACTGAAGGCGAGCAT	59 °C	173	NM_017050.1
	R:	TCTACAGTTAGCAGGCCAGCAG			
Caspase-3	F:	TGTCAGCTACTCCCAGGTTG	57 °C	146	NM_012922
	R:	TCAAGAAGGTGGTGAAGCAG			
COX-2	F:	CATGGGAGTTGGGCAGTC	56 °C	70	AF159101
	R:	TCAATCTCGGGTGGCTGAACG			
GAPDH	F:	AGGTGGAAGAATGGGAGTTG	55 °C	197	NM_017008.4
	R:	GCTTTCTCCAACCTCTCCTACTACA			

2.8. Cardiac Histopathology and Immunohistochemistry

All cardiac tissue (6 rats per group) was removed, fixed in 10% neutral buffered formalin for 72 h, processed into paraffin blocks, and then sectioned longitudinally at 5 m. The sections were stained with hematoxylin [37]. The previously published free-floating technique was used for immunohistochemistry [38]. Anti-calmodulin ab2860 polyclonal antibody, Abcam (Cambridge, UK), was used as the primary antibody for CaMKII, Ca^{2+}/calmodulin-dependent protein kinase II. Brown cytoplasmic deposits were detected by immunohistochemistry employing monoclonal antibodies directed against alpha-smooth muscle actin (α-SMA) (Lab vision neo markers USA, Catalog No. 1-GE002-07), a sign of active myofibroblasts [39].

2.9. Western Blotting

Tissues were lysed using RIPA lysis buffer (Solarbio, Beijing, China). Protein concentrations were determined using an improved BCA protein assay kit (Beyotime, Shanghai, China). After centrifugation, equal amounts of protein (50 g) were loaded onto a 10% SDS-polyacrylamide gel and transferred to polyvinylidene fluoride membranes (Merck Millipore, Boston, MA, USA). The membranes were blocked in 5% skimmed milk in TBS and incubated overnight at 4 °C with primary antibodies against OGG1 (1:400; Abcam) and β-actin (1:1000; Santa Cruz Biotechnology, Inc., Dallas, TX, USA). Subsequently, the membranes were washed and incubated with horseradish peroxidase-conjugated secondary antibody at 37 °C for 2 h. After further washes, the bands were quantified using (Image-J software 6.1; ImageJ 1.45 s freeware, National Institutes of Health, Rockville, MD, USA), with the optical densities normalized to β-actin.

2.10. Statistical Analysis

SPSS (SPSS Inc., Chicago, IL, USA), version 26 for Windows, tabulated and analyzed the data. The data were presented as a mean + SD. The Shapiro–Wilks test was used to check if the data were normally distributed (t $p > 0.05$). One-way analysis of variance (ANOVA) was used to look for deviations in normally distributed variables. After a statistically significant analysis of variance (ANOVA), we performed post hoc Tuckey HSD multiple comparisons with Bonferroni testing to identify statistically significant pairs. In this study, $p < 0.05$ was indicated as the threshold for statistical significance.

3. Results

3.1. Chromatographic Pattern of GC-MS Spectral Analysis of *Rhodiola rosea* Extract

Table 2 and Figure 1 present the detected compounds and their chemical formulas, peak areas, retention times, and molecular weights. A total of 32 peaks were observed, including n-Hexadecanoic acid (26.47%), á-Sitosterol (5.39%), Stigmast-5-EN-3-ol, (3á,24S) (5.39%), 2-Propenoic acid, 3-phenyl-, methyl ester (5.11%), Linalool (3.62%), 1,6-Octadien-3-ol, 3,7-dimethyl (3.62%), 7,10-Octadecadienoic acid, methyl ester (2.50%), 9,12-octadecadienoic acid (z,z) (1.60%), and 9-Octadecenoic acid (z)-, methyl ester (1.55%), which were the most abundant compounds detected. These compounds were identified in a methanolic extract of *Rhodiola rosea*. In addition, please see the HPLC analysis of *Rhodiola rosea* extract (Figure S1 and Table S1).

Table 2. GC-MS analysis of *Rhodiola rosea* extract.

No.	Retention Time (min)	(Chemical Compound)	Area%	MF	Molecular Formula	Molecular Weight
1	6.85	Linalool	3.62	797	C10H18O	154
2	6.85	1,6-Octadien-3-ol, 3,7-dimethyl	3.62	833	C10H18O	154
3	8.88	Benzene, 1-methoxy-4-(1-propenyl)-	1.41	784	C10H12O	148
4	8.88	Estragole	1.41	866	C10H12O	148
5	12.65	Guaiacol, 6-propenyl-	1.17	831	C10H12O2	164
6	12.65	Phenol, 2-methoxy-6-(1-propenyl)-	1.17	827	C10H12O2	164
7	13.15	2-Propenoic acid, 3-phenyl-, methyl ester	5.11	897	C10H10O2	162
8	15.08	Trans-à-Bergamotene	1.77	919	C15H24	204
9	16.76	á-Copaene	0.92	839	C15H24	204
10	24.16	Ethanol, 2-(9-octadecenyloxy)-, (Z)-	0.73	776	C20H40O2	312
11	25.62	pentadecanoic acid, 14-methyl-, methyl ester	1.16	798	C17H34O2	270
12	25.62	Palmitic Acid methyl ester	1.16	787	C17H34O2	270
13	25.62	Hexadecanoic acid, methyl ester	1.16	789	C17H34O2	270
14	26.44	n-Hexadecanoic acid	26.47	921	C16H32O2	256

Table 2. *Cont.*

No.	Retention Time (min)	(Chemical Compound)	Area%	MF	Molecular Formula	Molecular Weight
15	28.62	7,10-Octadecadienoic acid, methyl ester	2.50	876	C19H34O2	294
16	28.80	9-Octadecenoic acid (z)-, methyl ester	1.55	877	C19H36O2	296
17	29.53	9-Octadecenoic acid (z)-	0.93	819	C18H34O2	282
18	29.63	Oleic Acid	1.15	781	C18H34O2	282
19	29.63	cis-Vaccenic acid	1.15	792	C18H34O2	282
20	29.63	9-Octadecenoic acid (z)-	1.15	809	C18H34O2	282
21	29.63	trans-13-Octadecenoic acid	1.15	796	C18H34O2	282
22	31.94	Alanine	1.16	665	C14H14F3NO4	317
23	35.79	4H-1-Benzopyran-4-one, 2-(3,4-dihydroxyphenyl)-6,8-di-á-d glucopyranosyl-5,7-dihydroxy	1.04	708	C27H30O16	610
24	40.03	17-Pentatriacontene	1.01	677	C35H70	490
25	41.00	4H-1-Benzopyran-4-one, 2-(3,4-dimethoxyphenyl)-3,5-dihydroxy-7-methoxy	0.99	733	C18H16O7	344
26	41.00	Dasycarpidan-1-methanol, acetate (ester)	0.99	730	C20H26N2O2	320
27	41.91	Oleic acid, eicosyl ester	1.56	667	C38H74O2	562
28	43.04	Isochiapin B	1.60	717	C19H22O6	346
29	43.04	9,12-octadecadienoic acid (z,z)	1.60	731	C27H54O4Si2	498
30	43.89	.Psi.,.Psi.-Carotene, 1,1′,2,2′-tetrahydro-1,1′-dimet hoxy	1.45	672	C42H64O2	600
31	44.72	á-Sitosterol	5.39	786	C29H50O	414
32	45.00	Isochiapin B	3.11	725	C19H22O6	346

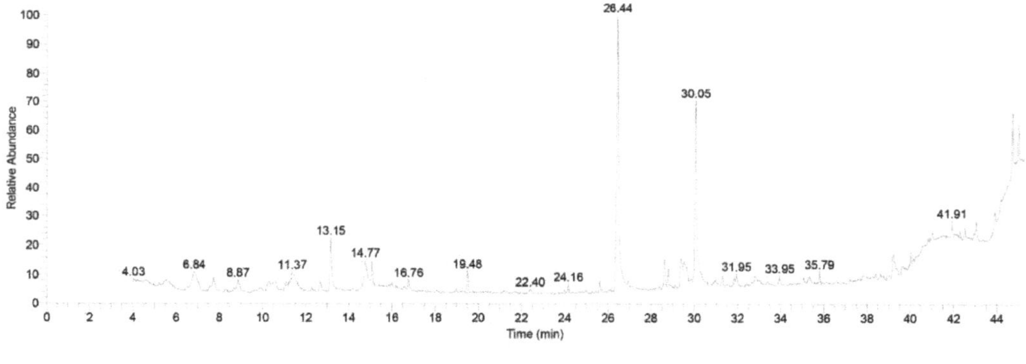

Figure 1. Pattern of GC-MS spectral analysis.

3.2. *Effect of RHO and CARV Alone on Body Weight and Cardiac Injury Markers*

Animals in the control, RHO, and CARV groups appeared healthy with a normal body weight, while the TIL-treated group showed a significantly decreased BW. Pretreatment with RHO and CARV (standard) alone increased BW (Table 2). Regarding cardiac injury markers, serum LDH, CK-MB, and cTI levels were significantly increased in the TIL-treated group. At the same time, pretreatment with RHO and CARV alone considerably decreased compared to the TIL-treated group but was still higher than those recorded in normal rats (Table 3).

Table 3. Impacts of RHO and CARV alone against TIL-induced cardiotoxicity on body weight and cardiac injury markers.

	Normal	RHO	CARV	TIL	CARV + TIL	RHO + TIL
B.W (gm)	195.6 ± 3.1 [a]	199.1 ± 5.1 [a]	198.0 ± 5.0 [a]	154.0 ± 4.8 [c]	170.1 ± 3.5 [b]	168.6 ± 3.1 [b]
LDH (U/L)	707.50 ± 12.6 [b]	721.83 ± 13.2 [b]	717.16 ± 11.7 [b]	1427.33 ± 39.2 [a]	600.16 ± 11.1 [c]	601.83 ± 10.9 [c]
CK-MB (U/L)	701.50 ± 13.03 [b]	697.66 ± 19.38 [b]	700.33 ± 12.19 [b]	1205.33 ± 28.71 [a]	571.16 ± 8.841 [c]	572.33 ± 8.71 [c]
Cardiac cTI (Pg/mL)	0.31 ± 0.04 [c]	0.31 ± 0.05 [c]	0.32 ± 0.06 [c]	1.8367 ± 0.27 [a]	1.060 ± 0.08 [b]	1.06 ± 0.08 [b]

The values represent means ± standard deviations for six separate rats per treatment. [a,b,c] Values with different letters are statistically different at $p < 0.05$.

3.3. Impacts of RHO and CARV Alone on TIL-Induced Changes in Inflammatory Cytokines

The TIL-injected group significantly increased the inflammatory IL-1β and TNF-α cytokines. TIL decreased IL-10 levels (Table 3). The pre-administration of RHO and CARV alone to TIL-injected rats showed a protective effect and significant decreases in inflammatory cytokine levels. In contrast, an apparent increase in IL10 levels was reported compared to other groups (Table 4).

Table 4. Impacts of RHO and CARV alone on TIL-induced cardiotoxicity on inflammatory and anti-inflammatory cytokines.

	Normal	RHO	CARV	TIL	CARV + TIL	RHO + TIL
TNF (pg/mL)	100.16 ± 2.04 [c]	105.33 ± 2.16 [c]	103.00 ± 2.36 [c]	202.66 ± 6.28 [a]	160.33 ± 2.92 [b]	165.02 ± 3.02 [b]
IL-1 B (pg/mL)	119.16 ± 2.85 [c]	121.83 ± 2.92 [c]	120.66 ± 3.07 [c]	230.5 ± 4.27 [a]	177.66 ± 5.92 [b]	176.00 ± 5.79 [b]
IL-10 (pg/mL)	303.33 ± 5.04 [a]	306.83 ± 4.83 [a]	308.33 ± 4.96 [a]	149.33 ± 2.94 [c]	208.83 ± 4.35 [b]	202.33 ± 5.31 [b]

The values represent means ± standard deviations for six separate rats per treatment. [a,b,c] Values with different letters are statistically different at $p < 0.05$.

3.4. Effect of RHO and CARV Alone on Oxidative Stress Biomarkers (ROS, PC, MDA) and TAC in Cardiac Tissue Homogenate

Compared with the control, the oxidative injury markers ROS, PC, and MDA contents in the heart were significantly increased in the TIL-treated rats. At the same time, pretreatment with RHO and CARV alone considerably decreased compared with the TIL-treated group (Figure 2A–C). TAC in TIL-treated rats compared to control was significantly reduced; however, pretreatment of RHO and CARV alone significantly increased the cardiac TAC levels compared to the TIL-treated group (Figure 2D).

In addition, it was shown that the OGG1 protein expression quantity was significantly decreased by the TIL administration in other treated groups. However, the pretreatment of RHO and CARV alone or in combination with the TIL significantly increased the cardiac OGG1 levels compared to the TIL-treated group, and this finding was proved by the OGG1 protein expression using the Western blot analysis. Conversely, the quantitative analysis of the 8-OHdG protein showed that TIL upregulated the expression of 8-OHdG compared with the control, RHO, and CARV groups and downregulated the expression of 8-OHdG. In comparison, the pre-administration of RHO and CARV alone or in combination with TIL protected rats that received TIL, as RHO and CARV significantly increased the OGG1 protein expression and significantly decreased the 8-OHdG protein in relation to the TIL-treated rats. (Please see Supplementary Figure S2).

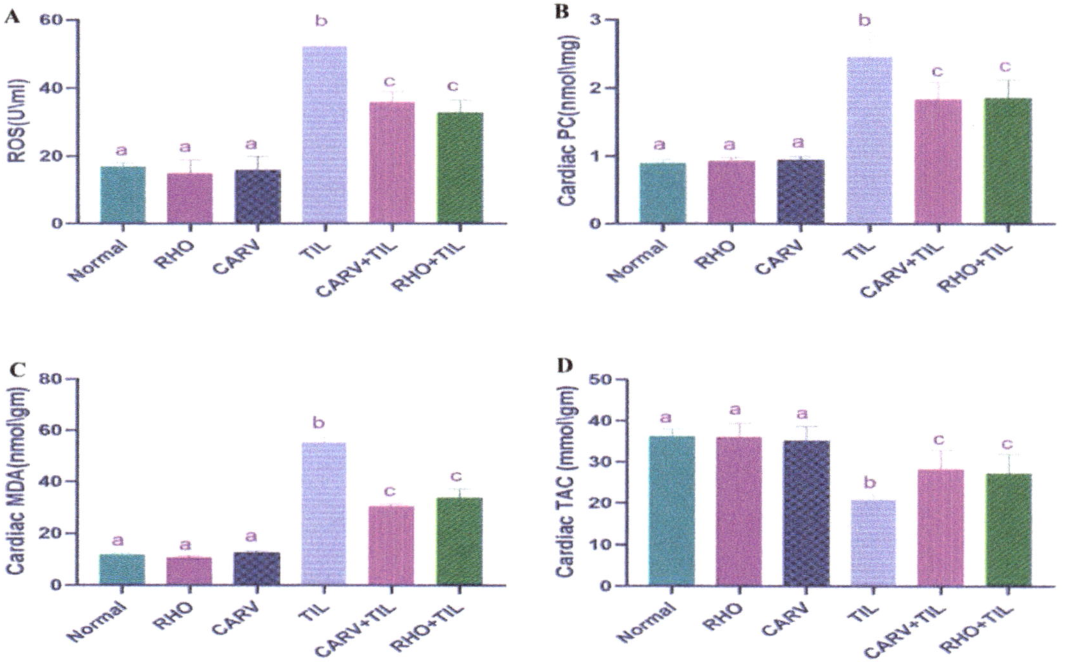

Figure 2. Impacts of Rhodiola against tilmicosin-induced cardiotoxicity on oxidative stress markers. (**A**) ROS, (**B**) Cardiac PC, (**C**). Cardiac MDA, and (**D**) Cardiac TAC. The values represent means ± standard deviations for six separate rats per treatment. [a,b,c] Values with different letters are statistically different at $p < 0.05$.

3.5. Impacts of RHO and CARV Alone against TIL-Induced Alteration in Cardiac mRNA Gene Expression

TIL upregulated the mRNA expression of INF-γ and COX-2 compared with the control, RHO, and CARV groups. In contrast, the pre-administration of RHO and CARV alone in rats protected those that received TIL, as it downregulated the examined genes significantly (Figure 3A–C). TIL upregulated the expression of apoptotic genes caspase-3 and BAX compared with control, RHO, and CARV groups and downregulated the expression of the anti-apoptotic gene BCL-2, antioxidant gene SOD, and OGG1 mRNA expression. In comparison, the pre-administration of RHO and CARV alone protected rats that received TIL, as RHO and CARV significantly decreased the mRNA expression of caspase-3, BAX, INF-γ, and COX-2 while significantly increasing the mRNA expression of BCL-2. SOD and OGG1 resorted to the examined genes (Figures 3A–D and 4A).

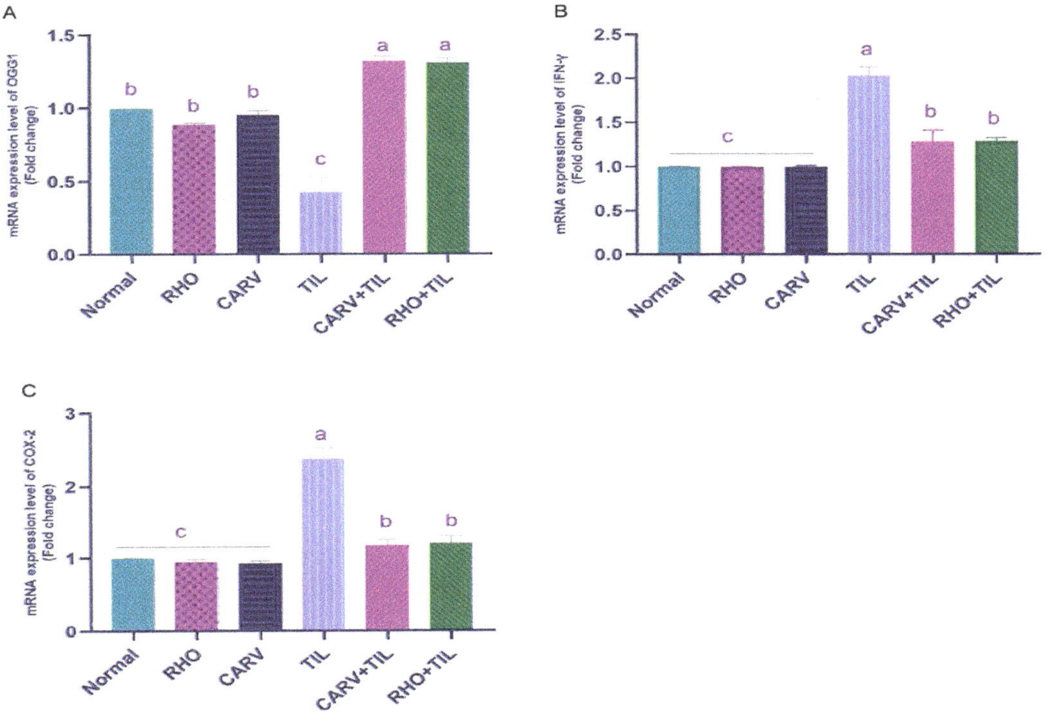

Figure 3. Quantification of mRNA expression of (**A**) OGG1, (**B**) INF-γ, and COX-2 (**C**). Graphic presentations of examined genes were based on qRT-PCR analysis for INF-γ, COX-2, and OGG1 after normalization with GAPDH in different groups. The values represent means ± standard deviations for six separate rats per treatment. [a,b,c] Values with different letters are statistically different at $p < 0.05$.

3.6. Protective Effect of RHO and CARV Alone against TIL-Induced Cardiac Muscle Histopathology

The myocardium of the left ventricle of control rats showed the typical structure of the myocardium, which is formed of branching and anastomosing cardiac muscle fibers with centrally located oval nuclei and acidophilic cytoplasm, and a delicate connective tissue endomysium with dark oval nuclei of fibroblasts that surrounds the cardiac muscle fibers. On the other hand, TIL-treated rats showed loss of the typical architecture of the cardiac muscle fibers with destructed wavy fibers with loss of branching and anastomosing, with some areas showing focal necrosis with intercellular hemorrhage and multiple dilated congested blood vessels surrounded by monocellular infiltration with an increased number of fibroblasts, nuclei of cardiomyocytes appearing with small darkly stained nuclei, areas of hyaline degeneration, and widening of the endomysium. In contrast, CARV + TIL-treated rats and RHO + TIL-treated rats showed the normal structure of the myocardium, formed of branching and anastomosing cardiac muscle fibers with centrally located oval nuclei and dark oval nuclei of fibroblasts. However, there is a widened endomysium with congested blood vessels in the RHO + TIL-treated group (Figure 5A–F).

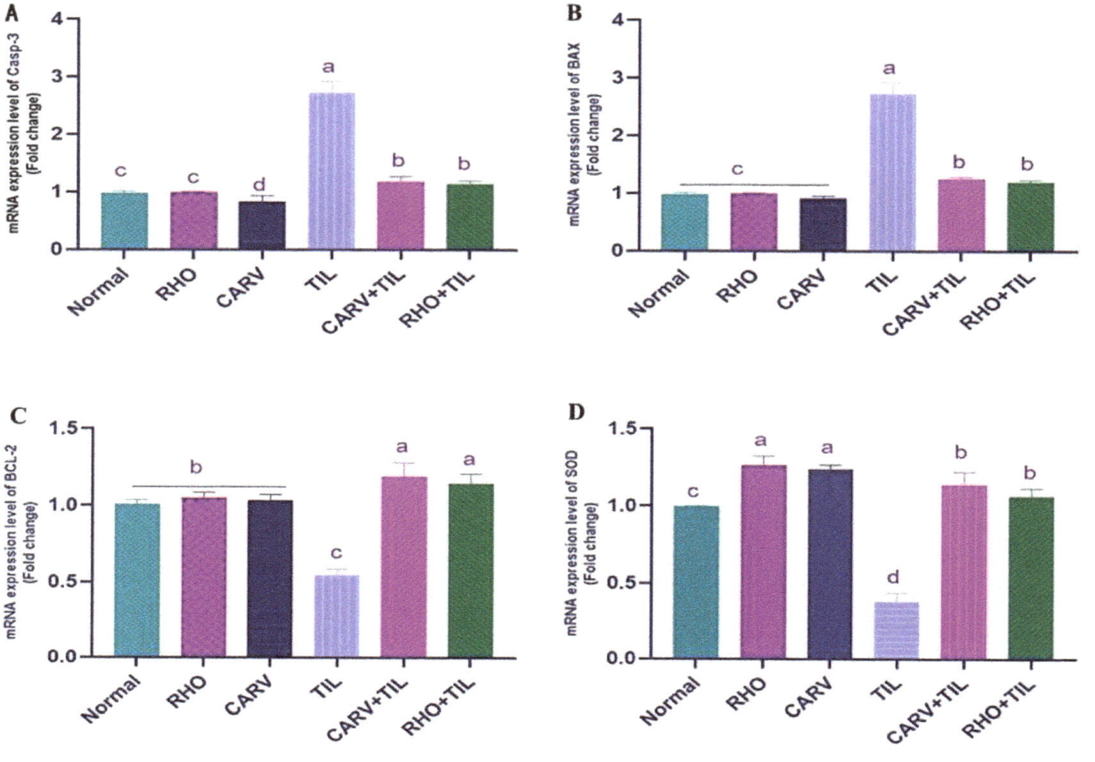

Figure 4. Quantification of mRNA expression of genes. The graphic presentations of examined genes were based on qRT-PCR analysis for (**A**) caspase-3, (**B**) BAX, (**C**) B-cell lymphoma-2 (Bcl-2), and (**D**) superoxide dismutase (SOD) genes after normalization with GAPDH in different groups. The values represent means ± standard deviations for six separate rats per treatment. [a,b,c,d] Values with different letters are statistically different at $p < 0.05$.

3.7. Immunohistochemical Analysis of the Protective Effect of RHO and CARV Alone in TIL-Induced Cardiac Muscle

For the expression of calmodulin in control, CARV, and RHO rats' cardiac cells, normal rats show regular expression of calmodulin (brown color). On the contrary, TIL-intoxicated rats showed a marked increase in calmodulin expression in the form of multiple scattered immune solid-positive foci. In contrast, CARV + TIL-treated rats and RHO + TIL-treated rats showed decreased calmodulin expression (Figure 6A–F) for the expression of alpha-smooth muscle actin (α–SMA) in the cardiac tissue of the control, CARV, and RHO rats' cardiac cells. The control group revealed the normal expression of α–SMA. Conversely, TIL-intoxicated rats showed a marked increase in α–SMA expression in the form of solid immune positivity, while CARV + TIL-treated rats (E) and RHO + TIL-treated rats (F) showed decreased α–SMA expression (Figure 7A–F).

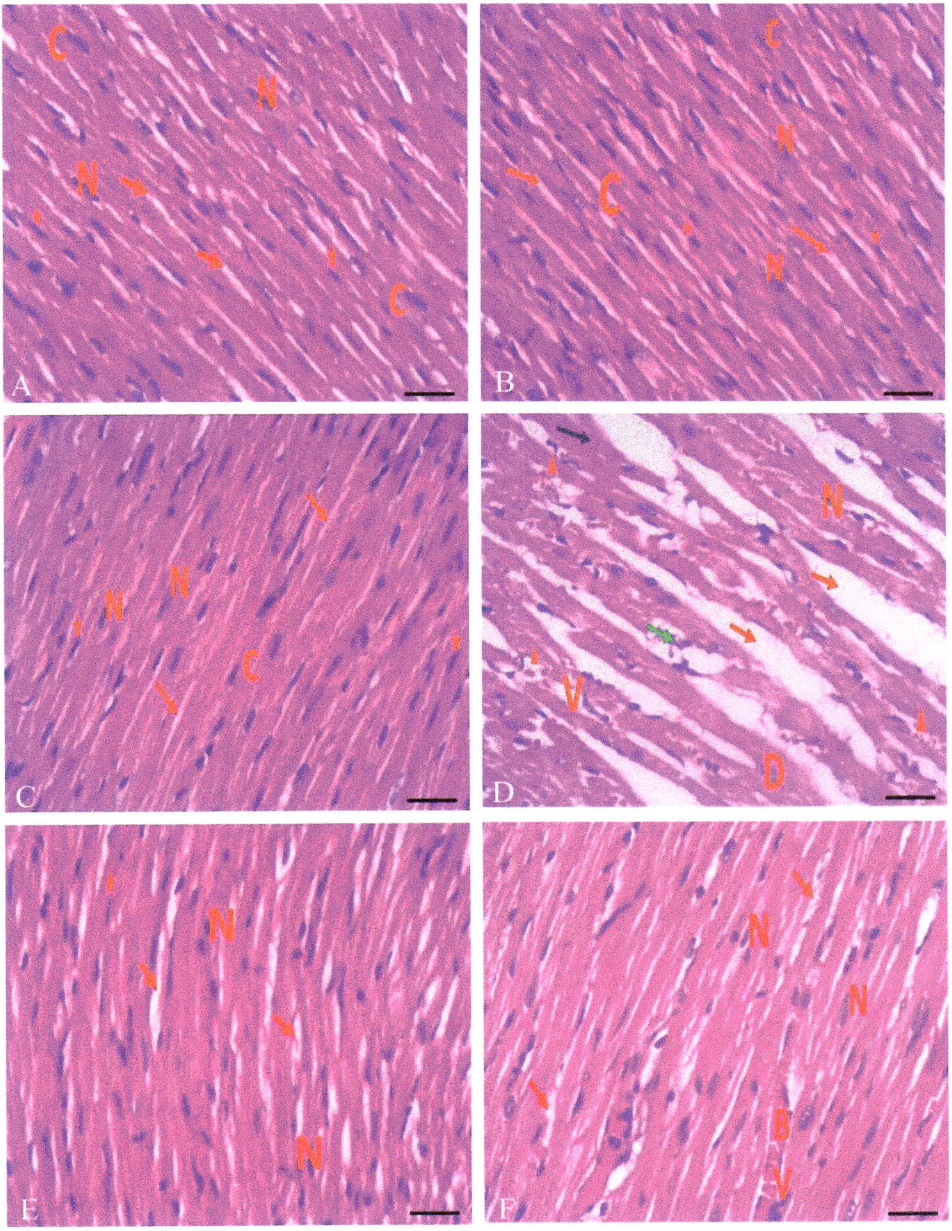

Figure 5. A photomicrograph of a longitudinal section of the myocardium of the left ventricle of control rats (**A–C**) showing the normal structure of the myocardium, formed of branching and anastomosing cardiac muscle fibers with centrally located oval nuclei (N) and acidophilic cytoplasm (**C**) and delicate connective tissue endomysium (red arrows) with dark oval nuclei of fibroblasts (star), surrounds the cardiac muscle fibers. (**D**) TIL-treated rats showed loss of the normal architecture of the cardiac muscle fibers with destructed wavy fibers with loss of branching and anastomosing, with

some areas showing focal necrosis (green arrow) with intercellular hemorrhage (arrowhead) and multiple dilated congested blood vessels (BV) surrounded by monocellular infiltration (CI) with an increased number of fibroblasts (star), nuclei of cardiomyocytes appearing with small darkly stained nuclei (N), areas of hyaline degeneration (blue arrow), and widening of endomysium. (**E**) CARV + TIL-treated rats and (**F**) RHO + TIL-treated rats showed a normal structure of the myocardium, formed of branching and anastomosing cardiac muscle fibers with centrally located oval nuclei (N) and dark oval nuclei of fibroblasts (star). However, there is still a widened endomysium (red arrow) with congested blood vessels (BV) in the RHO + TIL-treated group (H&E X400). Scale bar = 50 μm.

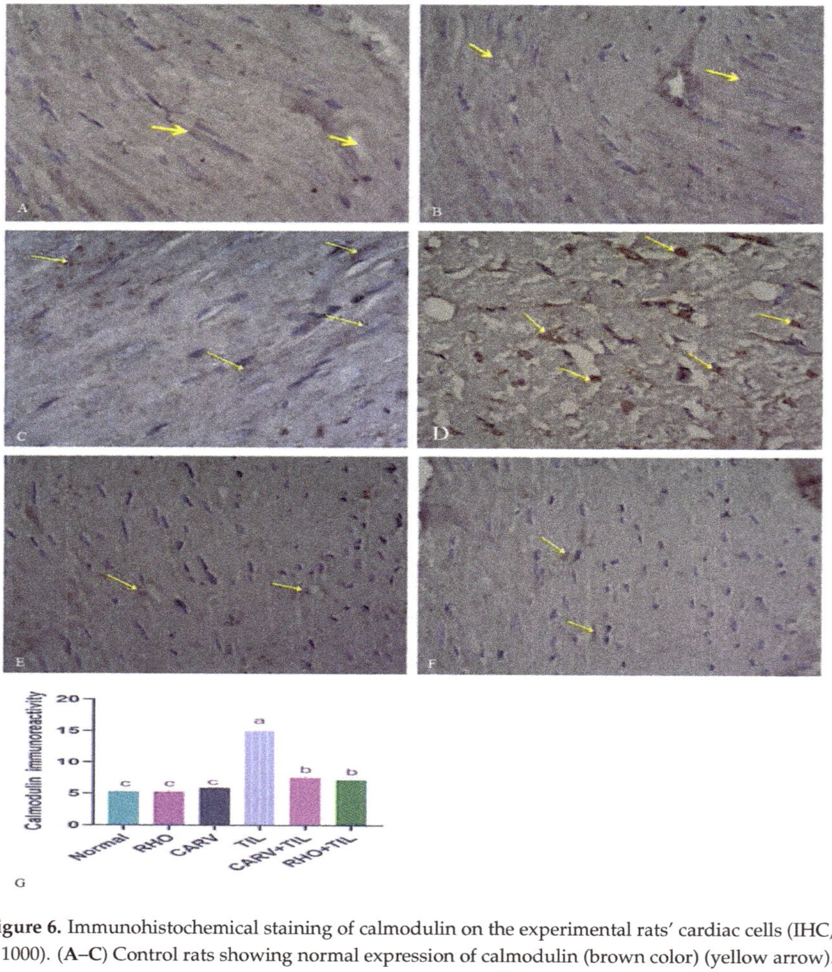

Figure 6. Immunohistochemical staining of calmodulin on the experimental rats' cardiac cells (IHC, ×1000). (**A–C**) Control rats showing normal expression of calmodulin (brown color) (yellow arrow). (**D**) TIL-intoxicated rats showed a marked increase in calmodulin expression in multiple scattered foci of strong immune positivity (yellow arrow). (**E**) CARV + TIL-treated rats and (**F**) RHO + TIL-treated rats showed decreased calmodulin expression (yellow arrow). Scale bar = 50 μm. The level of positive immunoreactivity of calmodulin (**G**) was analyzed using a one-way ANOVA followed by Tukey's multiple comparisons test. HMGB1 immunohistochemical staining was evaluated in 10 individual sections. All data are presented as means + SD, ^{a,b,c} Values with different letters are statistically different at $p < 0.05$.

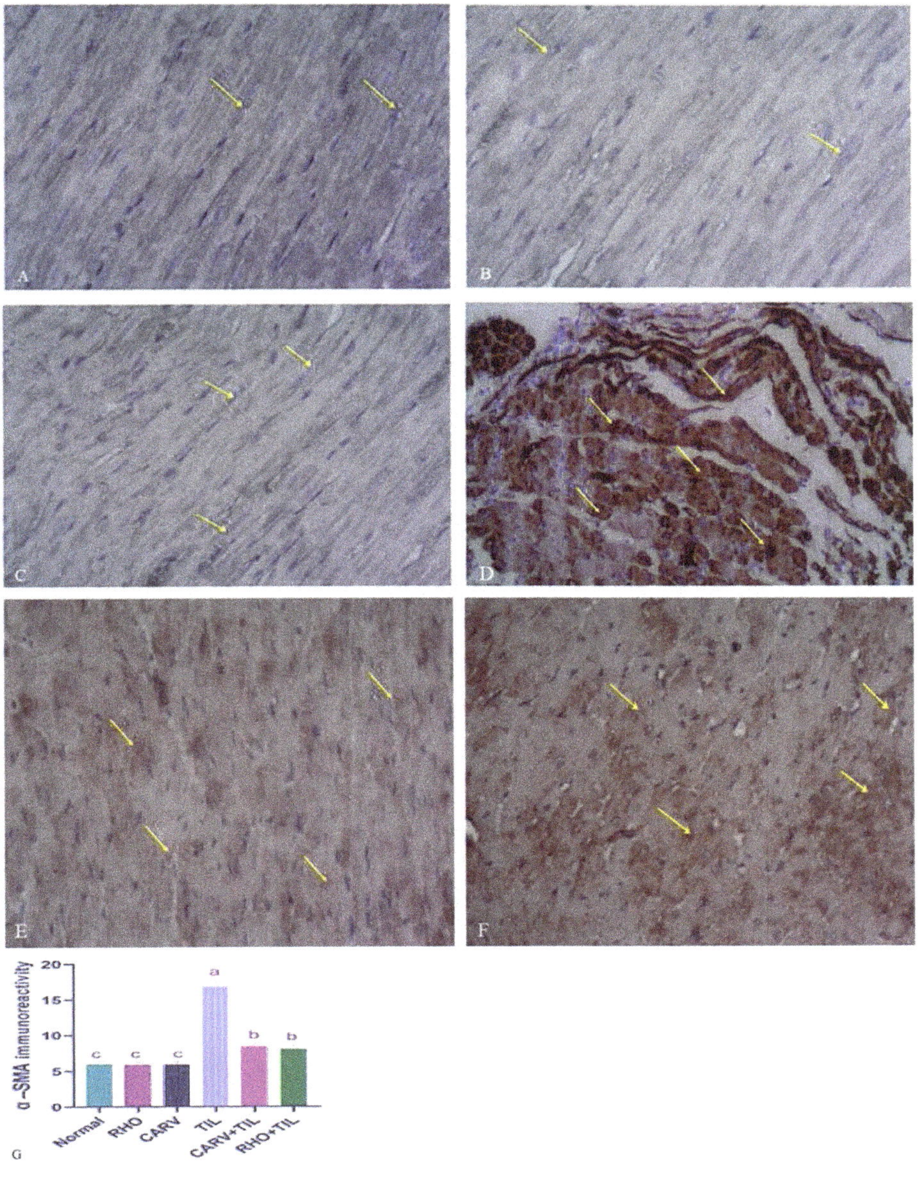

Figure 7. Immunohistochemical staining of alpha-smooth muscle actin (α-SMA) in the experimental rats' cardiac cells (IHC, ×400). (**A–C**) Control rats show normal expression of α-SMA (brown color) (yellow arrow). (**D**) TIL-intoxicated rats showed a marked increase in α-SMA expression in the form of solid immune positivity (yellow arrow). (**E**) CARV + TIL-treated rats and (**F**) RHO + TIL-treated rats showed decreased α-SMA expression (yellow arrow). Scale bar = 50 μm. The level of positive immunoreactivity of α-SMA (**G**) was analyzed using a one-way ANOVA followed by Tukey's multiple comparisons test. HMGB1 immunohistochemical staining was evaluated in 10 individual sections. All data are presented as means + SD, ^{a,b,c} Values with different letters are statistically different at $p < 0.05$.

4. Discussion

This study confirmed RHO's protective impacts against TIL-induced cardiac injury. The serum levels of cardiac injury markers, such as lactate dehydrogenase, increased significantly. Cardiac troponin (cTI) and creatine phosphokinase (CK-MB) showed significant weight loss. In addition, genetic, histopathological, and immunohistochemical alterations were observed in the examined hearts of TIL-receiving rodents [4,37]. However, the mechanisms of cardiotoxicity caused by macrolides, including TIL, have not been fully explained in any previously reported trials. One study found that TIL induced adrenaline production, which could worsen circulatory stress and decrease ventricular function [40]. The parameters mentioned above were ameliorated when CARV or RHO was pretreated in TIL-treated rats, in line with others [41].

GC-MS analysis was performed to investigate the chemical constituents of the *Rhodiola rosea* extract. The results revealed significant amounts of various compounds, including n-Hexadecenoic acids (26.47%), á-Sitosterol (5.39%), 2-Propenoic acid, 3-phenyl-, methyl ester (5.11%), Linalool (3.62%), 1,6-Octadien-3-ol, 3,7-dimethyl (3.62%), 7,10-Octadecadienoic acid, methyl ester (2.50%), 9,12-octadecadienoic acid (z,z) (1.60%), and 9-Octadecenoic acid (z)-, methyl ester (1.55%). These compounds are known for their anti-inflammatory and antioxidant properties [42–44]. N-hexadecenoic acids, also known as palmitoleic acids, are monounsaturated fatty acids with multiple bodily functions. They exhibit anti-inflammatory properties and play a role in maintaining skin health. In terms of metabolism, n-hexadecenoic acids regulate lipid levels in the body. Additionally, they may improve insulin sensitivity. Furthermore, as signaling molecules, these fatty acids can influence cellular processes such as gene expression, cell proliferation, and differentiation [45,46]. á-Sitosterol, a plant sterol found in plants, has various health benefits. It helps manage cholesterol by reducing its absorption and improving lipid profiles. á-Sitosterol also possesses anti-inflammatory properties. Additionally, á-Sitosterol has immunomodulatory effects, boosting immune function [47]. Linalool possesses anti-inflammatory and analgesic properties. Furthermore, it exhibits antimicrobial activity and antioxidant properties [48].

TIL-induced oxidative tissue damage is associated with elevated ROS, a reduction in the activities of GSH and CAT [49], and DNA damage [37]. Antioxidants are the first line of defense against oxidative stress because they neutralize ROS and other free radicals [50]. The current study showed an increase in protein carbonyl (PC), total reactive oxygen species (ROS), and malondialdehyde (MDA) with an apparent decrease in OGG-1 mRNA expression, total antioxidant capacity (TAC), and SOD mRNA in the cardiac homogenate of TIL-treated rats. Our findings and those of Awad et al. [37] reported that lipid peroxidation and redox dysregulation resulting from TIL have been found to induce cardiac injury. Pre-administration of RHO increased TAC in cardiac homogenate while lowering oxidative stress and DNA damage biomarkers if compared with CARV-pretreated rats (standard). This indicates a crucial function for RHO in controlling cardiac oxidative stress. Previous research [41,51] aligns with ours, documenting that salidroside has an antioxidant effect on cardiac damage.

Damage to the structure and physiology of cells due to oxidative stress can affect things like RNA processing, transcription, translation, the structure and function of the cell membrane, and metabolism [5]. The enhanced ROS generation from TIL-induced cellular damage attracts more inflammatory cells and fibroblasts to the wounded tissue and promotes the release of cytokines like tumor necrosis factor-alpha [5]. TNF-α initiates the induction of other cytokines, such as IL-1β and interferon-γ (IFN-γ) [52]. IFN-γ is a pleiotropic cytokine crucial to pro-apoptotic potentials and immune responses [53]. The commencement of a cascade of pro-inflammatory reactions is the result of a complex interplay between immune cell activity and IFN-γ, requiring the coordinated integration of signals from various pathways involving cytokines and pattern recognition receptors (PRRs) such as interleukins, TNFα, and IFNs [54]. Janus kinases JAK1 and JAK2 are phosphorylated in response to IFN-γ. In response, the activated JAKs phosphorylate the dormant STAT1 (signal transducer and activator of transcription) transcription factor located in the cytoplasm [54]. In our study, TIL treatment significantly showed a rise in TNF-α and IL-1β [55], upregulated IFN-γ and COX-2 mRNA expression in cardiac tissues [55], and a significant decrease in anti-inflammatory IL-10 cytokine [55], leading to an increase in the inflammatory process. In contrast, co-administration of CARV + TIL, or RHO + TIL, resulted in a decrease in serum inflammatory cytokines, IFN-γ, and COX-2 mRNA expression in cardiac tissues with an elevation of IL-10, which can be attributed to the anti-inflammatory effect of RHO. Previous experimental studies reported that RHO has an anti-inflammatory impact on myocardial ischemia by inhibiting the PI3K/Akt/mTOR pathway in vivo and the TLR4/NF-κB signaling pathway [51].

The active compounds contained in *Rhodiola rosea* showed anti-oxidation and anti-inflammation (inhibition in TNF-α, IL-1β, and IL-6) via the MAPK signaling pathway [56]. Salidroside restores mitochondrial structure and function by preventing the buildup of reactive oxygen species. Salidroside's antioxidant and protective effects come from its ability to selectively inhibit the activation of genes involved in the Mapk7 pathway that induces oxidative stress. These genes include growth arrest and DNA-damage-inducible 45 a (Gadd45a), mitogen-activated protein kinase 7 (Mapk7), and related RAS viral oncogene homolog 2 (Rras2) [57]. Also, salidroside activates the Nrf2 signaling pathway [58].

Changes in Ca^{2+} handling proteins can be attributed to elevated levels of cardiac ROS in human and animal models of cardiac dysfunction. Diastolic dysfunction is exacerbated by Ca^{2+} overload, and Ca^{2+}/calmodulin-dependent protein kinase II (CaMKII) changes contribute to this problem [59]. However, oxidative stress and intracellular Ca^{2+} overload, leading to cardiac failure, may be sustained by a vicious loop generated by abnormal Ca^{2+} processing in sick myocytes [60,61]. Reduced cardiac systolic potential and an increase in sudden cardiac mortality may result from excess Ca^{2+} buildup via calcium transporters. The involvement of Ca^{2+}/calmodulin-dependent protein kinase II-gamma (CaMKII) in the etiology of cardiotoxicity is primarily attributable to an increase in cytosolic Ca^{2+} levels. Apoptosis is facilitated by CaMKII activation [62]. Here, TIL administration showed significant upregulation of CaMKII expression by IHC, while co-administration of CARV + TIL, or RHO + TIL, decreased its expression. RHO can stabilize the cell membrane's ion pumps and calcium channels, rebuild the atrial potential in heart failure, inhibit atrial fibrillation, eliminate ectopic rhythm, and reduce atrial fibrosis. Calcium channel proteins in the heart, including myocardial sarcoplasmic reticulum Ca^{2+}-ATP Enzyme 2a (SERCA2a), may be responsible for *Rhodiola rosea*'s favorable electrogenic effects [63].

In addition, pro-apoptotic (Bax, caspase-3) and anti-apoptotic (Bcl-2) proteins are significant players in the apoptosis family. Bax activation guarantees cell damage during apoptosis by creating a pore in the mitochondrial membrane. Proteolytic breakdown of cellular components and eventual cell death are triggered by the activation of caspase-3 after the induction of mitochondrial cytochrome-c [32]. Also, the associated proteins, DNA, oxidative stress, and lipid damage have been reported to be strongly linked to the incidence of apoptosis [9]. Our results clarified that TIL caused upregulation of caspase-3, Bax and downregulation of Bcl-2 expression, which reflects the apoptotic role in TIL-induced cardiomyopathy [37]. Also, several studies documented cardio-toxic effects and apoptosis in response to TIL toxicity [64]. RHO's anti-apoptotic action may help preserve myocardial integrity and reduce myocardial damage because pretreatment with RHO decreased caspase-3 and Bax gene levels while increasing Bcl-2 expression compared to CARV pretreatment. Consistent with prior experimental research and our findings [41,58].

Furthermore, α-smooth muscle actin (α-SMA) is the hallmark of mature myofibroblast expression in cardiac tissues [65]. Numerous cells that resemble fibroblasts are seen in the myocardium. These interstitial cells are quiescent without injury, but various noxious stimuli can trigger cardiac fibroblast activation [66]. Myocardial infarction in humans and animals consistently results in cardiac fibroblasts transforming into α-SMA-positive myofibroblasts [65]. Myofibroblast activation and deactivation are critical for proper tissue repair; nevertheless, prolonged or excessive myofibroblast activity can increase stiffness and cause heart failure [67]. In our work, TIL administration exhibited significant upregulation of α-SMA expression; conversely, pre-administration of CARV + TIL or RHO + TIL showed downregulation of its expression in cardiac tissues by IHC, denoting that RHO has an anti-fibrotic effect. A recent experimental study by Yang et al. [68] reported that SAL significantly reduced α-SMA expression in renal fibrosis. This is in line with our findings.

The p38 MAPK pathway strongly influences myocardial fibrosis. Inflammatory cytokines and other stimuli at the receptor level (TGF-R, TLR4) trigger p38 signaling. These actions will trigger the activation of the upstream activator of the p38 MAPK kinases, mitogen-activated protein kinase MKK3/6. Cross-talk between p38 and IKK-NFkB signaling drives the pro-fibrotic (collagen 1 and 3 and α-SMA) and pro-inflammatory (IL-1, IL-6, and TNF-α) gene programs following p38 activation [66]. Molecular studies of how various cells in the body react to environmental cues should focus on the downstream signaling cascades that regulate transcription. OGG-1 is a DNA glycosylase that removes a mutagenic base after it has been damaged by reactive oxygen species (ROS), and its expression levels can be used to gauge how well RHO protects against the DNA damage caused by Tilmicosin. Oxo guanine-DNA glycosylase-1 (OGG1), a major DNA glycosylase that hydrolyzes oxidized-guanine (8-oxo-dG) to guanine, was also found to be at lower levels during this time. DNA damage was significantly elevated in TIL-induced toxicity, correlated with decreased OGG1 levels. These results thus demonstrated that RHO treatment protects the heart from TIL toxicity by maintaining OGG1 levels and protecting mitochondria from DNA damage; this result was in harmony with [69], in which they showed the maintenance of OGG1 levels and protecting mitochondria from DNA damage. OGG1 is a vital DNA glycosylase enzyme that plays a crucial role in removing a specific type of DNA damage produced by 8-OHdG. This implies that OGG1 protects DNA from damage and reduces the risk of carcinogenesis [70,71]. The findings from our research support the notion that Ogg1 and 8-OHdG are interconnected and provide evidence for the antioxidant abilities of RHO and CARV. Pre-treatment with RHO and TIL reduced the levels of inflammatory cytokines, specifically IFN-γ, COX-2, and mRNA expression in heart tissues. Additionally, it resulted in an increase in IL-10 levels, which could be due to the powerful anti-inflammatory effect of RHO. Previous reports cleared the anti-inflammatory effect of RHO in nonalcoholic fatty liver disease [72,73] and acute lung injury [74]. Also, Abou-Zeid et al. [75] showed that Moringa oleifera possesses anti-inflammatory properties that may shield the kidney from damage caused by Til-induced inflammation. Protein, DNA, and lipid damages leading to apoptosis have been linked to TIL exposure, although the mechanisms behind TIL-induced cardiotoxicity are yet unknown [76]. Furthermore, TIL may deplete antioxidant defense mechanisms in treated animals' livers, kidneys, and hearts, leading to oxidative stress [77–79]. In previous studies [23,24], TIL injections have been demonstrated to result in significant tissue damage via oxidative stress and apoptosis; this knowledge was used to determine the dose employed in the present study [4,28]. When comparing TIL with the control, RHO, and CARV groups, we find that TIL downregulates the expression of the anti-apoptotic gene BCL-2 and the antioxidant gene SOD, while upregulating the expression of the apoptotic gene's caspase-3 and BAX. Pretreatment with RHO and CARV alone dramatically reversed the studied genes toward resistance to TIL in rats. These results were in harmony with others [66–68], confirming the antioxidant, anti-inflammatory, and anti-apoptotic effects of RHO administration. RHO aids in the removal and detoxification of external toxins and plays a crucial function in boosting stressed cells' antioxidant defense [80].

5. Conclusions

The current study has shown that RHO can mitigate the side effects of TIL on the heart. These results demonstrate for the first time that RHO is superior to CARV in preventing cardiac damage caused by TIL toxicity. Cardiac gene expression was also modified, and inflammatory cytokine levels and antioxidant activity in the blood were both affected by TIL toxicity. All biochemical and tissue markers that had been changed were retrieved by pre-administration of RHO. The expression of genes involved in DNA damage, inflammation, antioxidants, and apoptosis was controlled by RHO with downregulation of the inflammatory cytokines' genes and increased antioxidant and antiapoptotic gene expression. In contrast to CARV-pretreated animals, cellular immunological reactivity to calmodulin and α-SMA was decreased. These findings provided further evidence for using RHO to shield the heart from TIL's side effects. They opened the door for future research to look for additional signaling pathways involved in such regulation in organs other than the heart.

6. Limitations and Future Directions

The study fails to address using an animal model that mimics human cardiac responses in future experiments. Although no animal model can perfectly mimic human cardiac responses, pigs and specific dog breeds are often used in research due to their anatomical and functional similarities to humans. These models provide valuable insights into cardiac physiology and pathologies. It is important to note that while each model has its benefits and limitations regarding their resemblance to human cardiac responses, combining different models and experimental approaches is necessary to understand the complexities of human cardiac responses comprehensively.

The study should incorporate hemodynamic measurements to address the current lack of information. Additionally, further investigations should involve a wide range of age distributions to allow for more accurate disease staging. It would also be valuable to study changes in blood pressure and conduct proteome analysis as practical tools for identifying new proteins associated with various heart diseases and understanding alterations in heart structure.

Supplementary Materials: The following supporting information can be downloaded at: https://www.mdpi.com/article/10.3390/toxics11100857/s1. Figure S1: HPLC peaks; Figure S2: Quantification of protein content of (A) Ogg1, (B) 8-OHdG, and (C) Western blotting and densitometric quantification of the protein expression of OGG1 following different treatment; Table S1: Rhodiola Rosea Extract (compound names).

Author Contributions: S.A.E., M.M.S., H.I.G. and M.S. supervised, designed, performed experiments, and revised the manuscript. L.A.M. performed experiments, analyzed data, and wrote and finalized the manuscript. M.S., L.A.M., H.E.N., A.O., B.S.A. and M.S. performed experiments. I.J., S.S., A.O., H.A.E. and S.A.E. performed experiments, wrote, interpreted data, and finalized the manuscript. All authors have read and agreed to the published version of the manuscript.

Funding: The authors would like to thank King Khalid University's Deanship of Scientific Research for the Large Group Research Project (Grant number RGP2/435/44). Moreover, we appreciate the resources provided by Princess Nourah bint Abdulrahman University Researchers Supporting Project number (PNURSP2023R73), Princess Nourah bint Abdulrahman University, Riyadh, Saudi Arabia.

Institutional Review Board Statement: All experiments were performed according to the Kafrelsheikh University guidelines for the care and use of lab animals, and all of the steps meant to keep the animals from suffering were taken in code KFS-IACUC/131/2023.

Informed Consent Statement: Not applicable.

Data Availability Statement: Upon request.

Acknowledgments: The authors thank the Pharmacology, Pathology, and Clinical Pathology Departments, Faculty of Medicine at Benha University, Egypt, for their invaluable technical and administrative support throughout this work. The authors would like to thank King Khalid University's Deanship of Scientific Research for the Large Group Research Project (Grant number RGP2/435/44). We appreciate the resources provided by Princess Nourah bint Abdulrahman University Researchers Supporting Project number (PNURSP2023R73), Princess Nourah bint Abdulrahman University, Riyadh, Saudi Arabia.

Conflicts of Interest: The authors declare no conflict of interest.

References

1. Naccari, F.; Pellegrino, M.; Calò, M.; Licata, P.; Giofré, F.; Carli, S. Effectiveness and kinetic behaviour of tilmicosin in the treatment of respiratory infections in sheep. *Vet. Rec.* **2001**, *148*, 773–776. [CrossRef] [PubMed]
2. Dingwell, R.; Duffield, T.; Leslie, K.; Keefe, G.P.; DesCoteaux, L.; Kelton, D.; Lissemore, K.; Schukken, Y.; Dick, P.; Bagg, R. The efficacy of intramammary tilmicosin at drying-off, and other risk factors for the prevention of new intramammary infections during the dry period. *J. Dairy Sci.* **2002**, *85*, 3250–3259. [CrossRef] [PubMed]
3. Zhu, L.; Cao, X.; Xu, Q.; Su, J.; Li, X.; Zhou, W. Evaluation of the antibacterial activity of tilmicosin-SLN against Streptococcus agalactiae: In vitro and in vivo studies. *Int. J. Nanomed.* **2018**, *13*, 4747. [CrossRef] [PubMed]
4. Khalil, S.R.; Abdel-Motal, S.M.; Abd-Elsalam, M.; Abd El-Hameed, N.E.; Awad, A. Restoring strategy of ethanolic extract of Moringa oleifera leaves against Tilmicosin-induced cardiac injury in rats: Targeting cell apoptosis-mediated pathways. *Gene* **2020**, *730*, 144272. [CrossRef]
5. Gheith, I.; El-Mahmoudy, A.; Elmajdoub, A.; Awidat, S. Pharmacovigilance of tilmicosin in mice. *Acta Sci. Vet.* **2015**, *43*, 1318.
6. Oztekin, E.; Sivrikaya, A.; Col, R.; Elmas, M.; Bas, A. Effects of different doses of tilmicosin on malondialdehyde and glutathione concentrations in mice. *Acta Vet. Brno* **2004**, *73*, 69–72.
7. Youssef, M.A.; Ibrahim, H.M.; Farag, E.-S.M.; El-Khodery, S.A. Effects of tilmicosin phosphate administration on echocardiographic parameters in healthy donkeys (*Equus asinus*): An experimental study. *J. Equine Vet. Sci.* **2016**, *38*, 24–29. [CrossRef]
8. Kart, A.; Karapehlivan, M.; Yapar, K.; Citil, M.; Akpinar, A. Protection through L-Carnitine on tissue oxidant status and sialic acid content in tilmicosin-induced alterations in BALB/c Mice. *Acta Vet. Brno* **2007**, *76*, 203–207. [CrossRef]
9. Aboubakr, M.; Elsayd, F.; Soliman, A.; Fadl, S.E.; El-Shafey, A.; Abdelhiee, E.Y. L-Carnitine and vitamin E ameliorate cardiotoxicity induced by tilmicosin in rats. *Environ. Sci. Pollut. Res.* **2020**, *27*, 23026–23034. [CrossRef]
10. Meissner, G.; Pasek, D.A.; Yamaguchi, N.; Ramachandran, S.; Dokholyan, N.V.; Tripathy, A. Thermodynamics of calmodulin binding to cardiac and skeletal muscle ryanodine receptor ion channels. *Proteins Struct. Funct. Bioinform.* **2009**, *74*, 207–211. [CrossRef]
11. Meissner, G. Regulation of mammalian ryanodine receptors. *Front. Biosci. Landmark* **2002**, *7*, 2072–2080. [CrossRef] [PubMed]
12. Shaddy, R.E.; Boucek, M.M.; Hsu, D.T.; Boucek, R.J.; Canter, C.E.; Mahony, L.; Ross, R.D.; Pahl, E.; Blume, E.D.; Dodd, D.A. Carvedilol for children and adolescents with heart failure: A randomized controlled trial. *JAMA* **2007**, *298*, 1171–1179. [CrossRef] [PubMed]
13. Soma, K.; Yao, A.; Saito, A.; Inaba, T.; Ishikawa, Y.; Hirata, Y.; Komuro, I. Regular treatment strategy with a large amount of carvedilol for heart failure improves biventricular systolic failure in a patient with repaired tetralogy of Fallot. *Int. Heart J.* **2018**, *59*, 1169–1173. [CrossRef] [PubMed]
14. Onay-Besikci, A.; Suzmecelik, E.; Ozcelikay, A.T. Carvedilol suppresses fatty acid oxidation and stimulates glycolysis in C2C12 cells. *Can. J. Physiol. Pharmacol.* **2012**, *90*, 1087–1093. [CrossRef] [PubMed]
15. Khalil, S.R.; Salem, H.F.; Metwally, M.M.; Emad, R.M.; Elbohi, K.M.; Ali, S.A. Protective effect of Spirulina platensis against physiological, ultrastructural and cell proliferation damage induced by furan in kidney and liver of rat. *Ecotoxicol. Environ. Saf.* **2020**, *192*, 110256. [CrossRef]
16. Pu, W.-L.; Zhang, M.-Y.; Bai, R.-Y.; Sun, L.-K.; Li, W.-H.; Yu, Y.-L.; Zhang, Y.; Song, L.; Wang, Z.-X.; Peng, Y.-F. Anti-inflammatory effects of Rhodiola rosea L.: A review. *Biomed. Pharmacother.* **2020**, *121*, 109552. [CrossRef]
17. Kucinskaite, A.; Briedis, V.; Savickas, A. Experimental analysis of therapeutic properties of Rhodiola rosea L. and its possible application in medicine. *Medicina* **2004**, *40*, 614–619.
18. Tolonen, A.; Pakonen, M.; Hohtola, A.; Jalonen, J. Phenylpropanoid glycosides from Rhodiola rosea. *Chem. Pharm. Bull.* **2003**, *51*, 467–470. [CrossRef]
19. Mao, Y.; Li, Y.; Yao, N. Simultaneous determination of salidroside and tyrosol in extracts of *Rhodiola* L. by microwave assisted extraction and high-performance liquid chromatography. *J. Pharm. Biomed. Anal.* **2007**, *45*, 510–515. [CrossRef]
20. Chou Lin, S.S.; Chin, L.W.; Chao, P.C.; Lai, Y.Y.; Lin, L.Y.; Chou, M.Y.; Chou, M.C.; Wei, J.C.C.; Yang, C.C. In vivo Th1 and Th2 cytokine modulation effects of Rhodiola rosea standardised solution and its major constituent, salidroside. *Phytother. Res.* **2011**, *25*, 1604–1611. [CrossRef]
21. Abidov, M.; Grachev, S.; Seifulla, R.; Ziegenfuss, T. Extract of Rhodiola rosea radix reduces the level of C-reactive protein and creatinine kinase in the blood. *Bull. Exp. Biol. Med.* **2004**, *138*, 63–64. [CrossRef] [PubMed]
22. Mattioli, L.; Funari, C.; Perfumi, M. Effects of Rhodiola rosea L. extract on behavioural and physiological alterations induced by chronic mild stress in female rats. *J. Psychopharmacol.* **2009**, *23*, 130–142. [CrossRef]
23. Abd El-Kareem, M.S.; Rabbih, M.A.E.F.; Selim, E.T.M.; Elsherbiny, E.A.E.-m.; El-Khateeb, A.Y. Application of GC/EIMS in combination with semi-empirical calculations for identification and investigation of some volatile components in basil essential oil. *Int. J. Anal. Mass Spectrom. Chromatogr.* **2016**, *4*, 14–25. [CrossRef]
24. Mahamat, I.A.; Mehanna, E.S.E.; Oda, S.S.; Khafaga, A.F.; Gad El-Karim, D.R. Lead-Induced Hepato-Renal Toxicosis and the Ameliorative Role of Rhodiolarosea Extract in Albino Rats. *Alex. J. Vet. Sci.* **2021**, *70*, 76–85. [CrossRef]
25. Kim, Y.-H.; Park, S.-M.; Kim, M.; Kim, S.H.; Lim, S.-Y.; Ahn, J.-C.; Song, W.-H.; Shim, W.-J. Cardioprotective effects of rosuvastatin and carvedilol on delayed cardiotoxicity of doxorubicin in rats. *Toxicol. Mech. Methods* **2012**, *22*, 488–498. [CrossRef] [PubMed]
26. El-Demerdash, E.; Abdel-Sattar, S.A.; El-Bakly, W.M.; Mohamed, E.A. Antifibrotic effects of carvedilol and impact of liver fibrosis on carvedilol pharmacokinetics in a rat model. *Eur. J. Drug Metab. Pharmacokinet.* **2017**, *42*, 767–779. [CrossRef] [PubMed]

27. Hamdy, N.M.; El-Demerdash, E. New therapeutic aspect for carvedilol: Antifibrotic effects of carvedilol in chronic carbon tetrachloride-induced liver damage. *Toxicol. Appl. Pharmacol.* **2012**, *261*, 292–299. [CrossRef] [PubMed]
28. Kart, A.; Yapar, K.; Karapehlivan, M.; Citil, M. The Possible Protective effect of L-carnitine on Tilmicosin-induced Cardiotoxicity in Mice. *J. Vet. Med. Ser. A* **2007**, *54*, 144–146. [CrossRef]
29. Tao, H.; Wu, X.; Cao, J.; Peng, Y.; Wang, A.; Pei, J.; Xiao, J.; Wang, S.; Wang, Y. Rhodiola species: A comprehensive review of traditional use, phytochemistry, pharmacology, toxicity, and clinical study. *Med. Res. Rev.* **2019**, *39*, 1779–1850. [CrossRef]
30. Chiang, H.-M.; Chen, H.-C.; Wu, C.-S.; Wu, P.-Y.; Wen, K.-C. Rhodiola plants: Chemistry and biological activity. *J. Food Drug Anal.* **2015**, *23*, 359–369. [CrossRef]
31. Murthy, T.; Sowjanya, G. Development of discriminatory method for dissolution of carvedilol marketed formulations. *Int. J. ChemTech Res.* **2010**, *2*, 1047–1050.
32. Elblehi, S.S.; El-Sayed, Y.S.; Soliman, M.M.; Shukry, M. Date palm pollen extract avert doxorubicin-induced cardiomyopathy fibrosis and associated oxidative/nitrosative stress, inflammatory cascade, and apoptosis-targeting Bax/Bcl-2 and Caspase-3 signaling pathways. *Animals* **2021**, *11*, 886. [CrossRef]
33. Tsung, S.H. Creatine kinase activity and isoenzyme pattern in various normal tissues and neoplasms. *Clin. Chem.* **1983**, *29*, 2040–2043. [CrossRef] [PubMed]
34. Lum, G.; Gambino, S.R. A comparison of serum versus heparinized plasma for routine chemistry tests. *Am. J. Clin. Pathol.* **1974**, *61*, 108–113. [CrossRef]
35. Ohkawa, H.; Ohishi, N.; Yagi, K. Assay for lipid peroxides in animal tissues by thiobarbituric acid reaction. *Anal. Biochem.* **1979**, *95*, 351–358. [CrossRef]
36. Livak, K.J.; Schmittgen, T.D. Analysis of relative gene expression data using real-time quantitative PCR and the $2-\Delta\Delta CT$ method. *Methods* **2001**, *25*, 402–408. [CrossRef]
37. Awad, A.; Khalil, S.R.; Hendam, B.M.; Abd El-Aziz, R.M.; Metwally, M.M.; Imam, T.S. Protective potency of Astragalus polysaccharides against tilmicosin-induced cardiac injury via targeting oxidative stress and cell apoptosis-encoding pathways in rat. *Environ. Sci. Pollut. Res.* **2020**, *27*, 20861–20875. [CrossRef]
38. Lee, J.C.; Cho, Y.J.; Kim, J.; Kim, N.; Kang, B.G.; Cha, C.I.; Joo, K.M. Region-specific changes in the immunoreactivity of vasoactive intestinal peptide and pituitary adenylate cyclase-activating polypeptide receptors (VPAC2, and PAC1 receptor) in the aged rat brains. *Brain Res.* **2010**, *1351*, 32–40. [CrossRef] [PubMed]
39. Bancroft, J.D.; Layton, C. The hematoxylins and eosin. *Bancroft's Theory Pract. Histol. Tech.* **2012**, *7*, 173–186.
40. Tamargo, J.; De Miguel, B.; Tejerina, M. A comparison of josamycin with macrolides and related antibiotics on isolated rat atria. *Eur. J. Pharmacol.* **1982**, *80*, 285–293. [CrossRef]
41. Jin, P.; Li, L.-H.; Shi, Y.; Hu, N.-B. Salidroside inhibits apoptosis and autophagy of cardiomyocyte by regulation of circular RNA hsa_circ_0000064 in cardiac ischemia-reperfusion injury. *Gene* **2021**, *767*, 145075. [CrossRef] [PubMed]
42. Aparna, V.; Dileep, K.V.; Mandal, P.K.; Karthe, P.; Sadasivan, C.; Haridas, M. Anti-inflammatory property of n-hexadecanoic acid: Structural evidence and kinetic assessment. *Chem. Biol. Drug Des.* **2012**, *80*, 434–439. [CrossRef]
43. Paniagua-Pérez, R.; Flores-Mondragón, G.; Reyes-Legorreta, C.; Herrera-López, B.; Cervantes-Hernández, I.; Madrigal-Santillán, O.; Morales-González, J.A.; Alvarez-González, I.; Madrigal-Bujaidar, E. Evaluation of the anti-inflammatory capacity of beta-sitosterol in rodent assays. *Afr. J. Tradit. Complement. Altern. Med.* **2017**, *14*, 123–130. [CrossRef]
44. Khatri, P.; Rana, J.; Jamdagni, P.; Sindhu, A. Phytochemical screening, GC-MS and FT-IR analysis of methanolic extract leaves of Elettaria cardamomum. *Int. J. Res.* **2017**, *5*, 213–224. [CrossRef]
45. de Souza, C.O.; Vannice, G.K.; Rosa Neto, J.C.; Calder, P.C. Is palmitoleic acid a plausible nonpharmacological strategy to prevent or control chronic metabolic and inflammatory disorders? *Mol. Nutr. Food Res.* **2018**, *62*, 1700504. [CrossRef]
46. Hernández-Saavedra, D.; Stanford, K.I. The regulation of lipokines by environmental factors. *Nutrients* **2019**, *11*, 2422. [CrossRef]
47. Ibrahim, K.A.; Abdelgaid, H.A.; El-Desouky, M.A.; Fahmi, A.A.; Abdel-Daim, M.M. Linseed ameliorates renal apoptosis in rat fetuses induced by single or combined exposure to diesel nanoparticles or fenitrothion by inhibiting transcriptional activation of p21/p53 and caspase-3/9 through pro-oxidant stimulus. *Environ. Toxicol.* **2021**, *36*, 958–974. [CrossRef] [PubMed]
48. Ola, O.S.; Sofolahan, T.A. A monoterpene antioxidant, linalool, mitigates benzene-induced oxidative toxicities on hematology and liver of male rats. *Egypt. J. Basic Appl. Sci.* **2021**, *8*, 39–53. [CrossRef]
49. Ibrahim, A.E.; Abdel-Daim, M.M. Modulating effects of Spirulina platensis against tilmicosin-induced cardiotoxicity in mice. *Cell J.* **2015**, *17*, 137.
50. Nepali, S.; Ki, H.-H.; Lee, J.-H.; Cha, J.-Y.; Lee, Y.-M.; Kim, D.-K. Triticum aestivum sprout-derived polysaccharide exerts hepatoprotective effects against ethanol-induced liver damage by enhancing the antioxidant system in mice. *Int. J. Mol. Med.* **2017**, *40*, 1243–1252. [CrossRef]
51. Chen, L.; Liu, P.; Feng, X.; Ma, C. Salidroside suppressing LPS-induced myocardial injury by inhibiting ROS-mediated PI 3K/Akt/mTOR pathway in vitro and in vivo. *J. Cell. Mol. Med.* **2017**, *21*, 3178–3189. [CrossRef] [PubMed]
52. Oku, H.; Nakazato, H.; Horikawa, T.; Tsuruta, Y.; Suzuki, R. Pirfenidone suppresses tumor necrosis factor-α, enhances interleukin-10 and protects mice from endotoxic shock. *Eur. J. Pharmacol.* **2002**, *446*, 167–176. [CrossRef]
53. Dunn, G.P.; Ikeda, H.; Bruce, A.T.; Koebel, C.; Uppaluri, R.; Bui, J.; Chan, R.; Diamond, M.; Michael White, J.; Sheehan, K.C. Interferon-γ and cancer immunoediting. *Immunol. Res.* **2005**, *32*, 231–245. [CrossRef] [PubMed]

54. Kak, G.; Raza, M.; Tiwari, B.K. Interferon-gamma (IFN-γ): Exploring its implications in infectious diseases. *Biomol. Concepts* **2018**, *9*, 64–79. [CrossRef] [PubMed]
55. Cao, X.-Y.; Dong, M.; Shen, J.-Z.; Wu, B.-B.; Wu, C.-M.; Du, X.-D.; Wang, Z.; Qi, Y.-T.; Li, B.-Y. Tilmicosin and tylosin have anti-inflammatory properties via modulation of COX-2 and iNOS gene expression and production of cytokines in LPS-induced macrophages and monocytes. *Int. J. Antimicrob. Agents* **2006**, *27*, 431–438. [CrossRef]
56. Pan, J.; Zhang, L.; Xu, S.; Cheng, X.; Yu, H.; Bao, J.; Lu, R. Induction of apoptosis in human papillary-thyroid-carcinoma BCPAP cells by diallyl trisulfide through activation of the MAPK signaling pathway. *J. Agric. Food Chem.* **2018**, *66*, 5871–5878. [CrossRef]
57. Lin, S.-Y.; Xu, D.; Du, X.-X.; Ran, C.-L.; Xu, L.; Ren, S.-J.; Tang, Z.-T.; Yin, L.-Z.; He, C.-L.; Yuan, Z.-X. Protective effects of salidroside against carbon tetrachloride (CCl4)-induced liver injury by initiating mitochondria to resist oxidative stress in mice. *Int. J. Mol. Sci.* **2019**, *20*, 3187. [CrossRef]
58. Zhu, Y.; Zhang, Y.-J.; Liu, W.-W.; Shi, A.-W.; Gu, N. Salidroside suppresses HUVECs cell injury induced by oxidative stress through activating the Nrf2 signaling pathway. *Molecules* **2016**, *21*, 1033. [CrossRef]
59. Erickson, J.R.; He, B.J.; Grumbach, I.M.; Anderson, M.E. CaMKII in the cardiovascular system: Sensing redox states. *Physiol. Rev.* **2011**, *91*, 889–915. [CrossRef]
60. Kohlhaas, M.; Liu, T.; Knopp, A.; Zeller, T.; Ong, M.F.; Böhm, M.; O'Rourke, B.; Maack, C. Elevated cytosolic Na+ increases mitochondrial formation of reactive oxygen species in failing cardiac myocytes. *Circulation* **2010**, *121*, 1606–1613. [CrossRef]
61. Cappetta, D.; Esposito, G.; Coppini, R.; Piegari, E.; Russo, R.; Ciuffreda, L.P.; Rivellino, A.; Santini, L.; Rafaniello, C.; Scavone, C. Effects of ranolazine in a model of doxorubicin-induced left ventricle diastolic dysfunction. *Br. J. Pharmacol.* **2017**, *174*, 3696–3712. [CrossRef] [PubMed]
62. Tscheschner, H.; Meinhardt, E.; Schlegel, P.; Jungmann, A.; Lehmann, L.H.; Müller, O.J.; Most, P.; Katus, H.A.; Raake, P.W. CaMKII activation participates in doxorubicin cardiotoxicity and is attenuated by moderate GRP78 overexpression. *PLoS ONE* **2019**, *14*, e0215992. [CrossRef] [PubMed]
63. Hsiao, Y.-W.; Tsai, Y.-N.; Huang, Y.-T.; Liu, S.-H.; Lin, Y.-J.; Lo, L.-W.; Hu, Y.-F.; Chung, F.-P.; Lin, S.-F.; Chang, S.-L. Rhodiola crenulata reduces ventricular arrhythmia through mitigating the activation of IL-17 and inhibiting the MAPK signaling pathway. *Cardiovasc. Drugs Ther.* **2021**, *35*, 889–900. [CrossRef] [PubMed]
64. Oda, S.S.; Derbalah, A.E. Impact of diclofenac sodium on tilmicosin-induced acute cardiotoxicity in rats (tilmicosin and diclofenac cardiotoxicity). *Cardiovasc. Toxicol.* **2018**, *18*, 63–75. [CrossRef]
65. Shinde, A.V.; Humeres, C.; Frangogiannis, N.G. The role of α-smooth muscle actin in fibroblast-mediated matrix contraction and remodeling. *Biochim. Biophys. Acta (BBA)-Mol. Basis Dis.* **2017**, *1863*, 298–309. [CrossRef]
66. Umbarkar, P.; Ejantkar, S.; Tousif, S.; Lal, H. Mechanisms of fibroblast activation and myocardial fibrosis: Lessons learned from FB-specific conditional mouse models. *Cells* **2021**, *10*, 2412. [CrossRef]
67. Kong, P.; Christia, P.; Frangogiannis, N.G. The pathogenesis of cardiac fibrosis. *Cell. Mol. Life Sci.* **2014**, *71*, 549–574. [CrossRef]
68. Yang, S.; Pei, T.; Wang, L.; Zeng, Y.; Li, W.; Yan, S.; Xiao, W.; Cheng, W. Salidroside Alleviates Renal Fibrosis in SAMP8 Mice by Inhibiting Ferroptosis. *Molecules* **2022**, *27*, 8039. [CrossRef]
69. Pillai, V.B.; Bindu, S.; Sharp, W.; Fang, Y.H.; Kim, G.; Gupta, M.; Samant, S.; Gupta, M.P. Sirt3 protects mitochondrial DNA damage and blocks the development of doxorubicin-induced cardiomyopathy in mice. *Am. J. Physiol. Heart Circ. Physiol.* **2016**, *310*, H962–H972. [CrossRef]
70. Cooke, M.S.; Olinski, R.; Evans, M.D. Does measurement of oxidative damage to DNA have clinical significance? *Clin. Chim. Acta* **2006**, *365*, 30–49. [CrossRef]
71. Evans, M.D.; Dizdaroglu, M.; Cooke, M.S. Oxidative DNA damage and disease: Induction, repair and significance. *Mutat. Res./Rev. Mutat. Res.* **2004**, *567*, 1–61.
72. Sun, P.; Song, S.Z.; Jiang, S.; Li, X.; Yao, Y.L.; Wu, Y.L.; Lian, L.H.; Nan, J.X. Salidroside Regulates Inflammatory Response in Raw 264.7 Macrophages via TLR4/TAK1 and Ameliorates Inflammation in Alcohol Binge Drinking-Induced Liver Injury. *Molecules* **2016**, *21*, 1490. [CrossRef]
73. Zheng, T.; Yang, X.; Li, W.; Wang, Q.; Chen, L.; Wu, D.; Bian, F.; Xing, S.; Jin, S. Salidroside Attenuates High-Fat Diet-Induced Nonalcoholic Fatty Liver Disease via AMPK-Dependent TXNIP/NLRP3 Pathway. *Oxidative Med. Cell. Longev.* **2018**, *2018*, 8597897. [CrossRef] [PubMed]
74. Song, D.; Zhao, M.; Feng, L.; Wang, P.; Li, Y.; Li, W. Salidroside attenuates acute lung injury via inhibition of inflammatory cytokine production. *Biomed. Pharmacother. Biomed. Pharmacother.* **2021**, *142*, 111949. [CrossRef] [PubMed]
75. Abou-Zeid, S.M.; Ahmed, A.I.; Awad, A.; Mohammed, W.A.; Metwally, M.M.; Almeer, R.; Abdel-Daim, M.M.; Khalil, S.R. Moringa oleifera ethanolic extract attenuates tilmicosin-induced renal damage in male rats via suppression of oxidative stress, inflammatory injury, and intermediate filament proteins mRNA expression. *Biomed. Pharmacother.* **2021**, *133*, 110997. [CrossRef] [PubMed]
76. Tsai-Turton, M.; Luong, B.T.; Tan, Y.; Luderer, U. Cyclophosphamide-induced apoptosis in COV434 human granulosa cells involves oxidative stress and glutathione depletion. *Toxicol. Sci. Off. J. Soc. Toxicol.* **2007**, *98*, 216–230. [CrossRef]
77. Farag, M.R.; Elhady, W.M.; Ahmed, S.Y.A.; Taha, H.S.A.; Alagawany, M. Astragalus polysaccharides alleviate tilmicosin-induced toxicity in rats by inhibiting oxidative damage and modulating the expressions of HSP70, NF-kB and Nrf2/HO-1 pathway. *Res. Vet. Sci.* **2019**, *124*, 137–148. [CrossRef]

78. Cetin, N.; Boyraz, U.; Cetin, E. Ghrelin alleviates tilmicosin-induced myocardial oxidative stress in rats. *J. Anim. Vet. Adv.* **2011**, *10*, 2038–2042. [CrossRef]
79. Seddik, M.; El-Borai, N.; El-Bialy, B.; Elsabbagh, H. The Potential Teratogenic Effect of Tilmicosin in Rats: Visceral Malformations and Histomorphological Alterations in Fetal Internal Organs. *J. Curr. Vet. Res.* **2021**, *3*, 50–62. [CrossRef]
80. Lu, H.; Lei, X.; Zhang, Q. Moderate activation of IKK2-NF-kB in unstressed adult mouse liver induces cytoprotective genes and lipogenesis without apparent signs of inflammation or fibrosis. *BMC Gastroenterol.* **2015**, *15*, 94. [CrossRef]

toxics

MDPI

Article

The Role of Vitamin E in Protecting against Oxidative Stress, Inflammation, and the Neurotoxic Effects of Acute Paracetamol in Pregnant Female Rats

Alaa M. Hammad [1,*], Baraa Shawaqfeh [1], Suhair Hikmat [1,*], Tariq Al-Qirim [1], Lama Hamadneh [1,2], Sameer Al-Kouz [1], Mariam M. Awad [1] and Frank S. Hall [3]

1 Department of Pharmacy, College of Pharmacy, Al-Zaytoonah University of Jordan, Amman 11733, Jordan
2 Department of Basic Medical Sciences, Faculty of Medicine, Al-Balqa Applied University, Al-Salt 19117, Jordan
3 Department of Pharmacology and Experimental Therapeutics, College of Pharmacy and Pharmaceutical Sciences, University of Toledo, Toledo, OH 43606, USA; frank.hall@utoledo.edu
* Correspondence: alaa.hammad@zuj.edu.jo (A.M.H.); suhair.jasim@zuj.edu.jo (S.H.)

Abstract: Paracetamol (acetaminophen, APAP) is the most common non-prescription analgesic drug used during pregnancy. The aim of this study was to investigate the effect of vitamin E on acute APAP toxicity in pregnant rats. Toxicity in the liver, kidney, and brain (hippocampus, cerebellum, and olfactory bulb) was examined. Twenty pregnant female Wistar rats at gestational day 18 were used. Pregnant rats were divided into four groups: Control, APAP, E + APAP, and APAP + E. The Control group was treated with 0.5 mL p.o. corn oil. The APAP group received 3000 mg/kg p.o. APAP. The E + APAP group received 300 mg/kg p.o. vitamin E one hour before 3000 mg/kg APAP. The APAP + E group received 3000 mg/kg paracetamol one hour before 300 mg/kg p.o. vitamin E. Twenty-four hours after the last treatment administration, rats were euthanized and blood, brain, liver, and kidney samples were collected. Alanine aminotransferase (ALT), aspartate aminotransferase (AST), blood urea nitrogen (BUN), creatinine levels, uric acid (UA), and superoxide dismutase (SOD) levels, as well as the relative mRNA expression of *Cyp1a4*, *Cyp2d6*, and *Nat2*, were determined. Acute APAP treatment upregulated ALT, AST, BUN, and creatinine levels. APAP treatment downregulated UA and SOD levels. APAP treatment upregulated the relative mRNA expression of *Cyp1a4* and *Cyp2d6*, but downregulated *Nat2* expression. Vitamin E treatment, either before or after APAP administration, attenuated the toxic effects of APAP. In conclusion, the results showed that an acute toxic APAP dose in late pregnancy can cause oxidative stress and dysregulation in Cyp isoform expression, and that vitamin E treatment attenuates these effects.

Keywords: APAP; *Cyp1a2*; *Cyp2d6*; *Nat2*; ALT; AST

Citation: Hammad, A.M.; Shawaqfeh, B.; Hikmat, S.; Al-Qirim, T.; Hamadneh, L.; Al-Kouz, S.; Awad, M.M.; Hall, F.S. The Role of Vitamin E in Protecting against Oxidative Stress, Inflammation, and the Neurotoxic Effects of Acute Paracetamol in Pregnant Female Rats. *Toxics* **2023**, *11*, 368. https://doi.org/10.3390/toxics11040368

Academic Editor: Matthew J. Winter

Received: 10 March 2023
Revised: 29 March 2023
Accepted: 6 April 2023
Published: 12 April 2023

1. Introduction

Acetaminophen, also known as paracetamol (N-acetyl-para-aminophenol; APAP), is an over-the-counter painkiller and antipyretic that is safe when used at the recommended levels (no more than 4 gm/day) for adults, including pregnant women, as well as children (no more than 75 mg/kg/day). Self-medication with APAP is commonly used for fever and pains associated with common upper respiratory infections, e.g., rhinoviruses, influenza viruses, and many others, including COVID-19, for which it has been widely used during the pandemic in Jordan [1,2]. APAP overdoses, both accidental and purposeful, are common and frequently result in hepatotoxicity [3–5]. Nephrotoxicity and neurotoxicity are less common in APAP overdoses, but can occur independently or concurrently with hepatotoxicity [6,7]. Importantly, a previous study showed that APAP is the most common over the counter (OTC) drug associated with toxicity during pregnancy [8,9], so it is especially important to examine APAP toxicity under these conditions. Thus, in this study, we used pregnant female rats to investigate the effect of an acute toxic dose on toxicity in different organs during pregnancy.

APAP directly alters hormone-dependent processes in vivo, in vitro, and ex vivo, thereby affecting neural and reproductive development in both sexes [10]. In rodents, fetal APAP exposure has been experimentally demonstrated to produce male urogenital tract reproductive problems, including abnormalities in testicular function, sperm, and sexual behavior [11–13]. According to several studies, female ovarian development is also disrupted by APAP, resulting in fewer oocytes, early ovarian insufficiency, and decreased fertility [14,15]. It has been shown that exposure to fetal APAP alters neurotransmission in the brain, resulting in changes to cognition, behavior, and movement [16]. The results of these investigations have demonstrated that the impact of APAP is influenced by the timing of exposure in relation to particular developmental processes, as well as length of the exposure. For further information, see [10].

APAP is metabolized in the liver through either phase I or phase II metabolism [17]. Phase II metabolism is the major pathway, starting with UDP-glucuronosyl transferases (UGT) which conjugate glucuronic acid to APAP to form a glucuronic acid conjugate (a nontoxic metabolite) [18]. The sulfuric acid conjugate is the second nontoxic metabolite that is formed by the action of sulfotransferase (SULT). The phase I pathway involves CYP450 enzymes, including *Cyp1a2* and *Cyp2d6*, that de-acetylate APAP to form N-acetyl-p-benzoquinone imine (NAPQI) (a toxic metabolite) that is normally detoxified by glutathione (GSH) to form mercaptate conjugates (nontoxic metabolites) at non-toxic APAP doses [19,20]. Furthermore, since APAP shares structural similarities with endogenous acetylated products, it decreases N-acetyltransferase 2 (NAT2) function, which has been shown both in vitro and in vivo [21]. Acute APAP toxicity overproduction of NAPQI depletes hepatic GSH reserves, resulting in the accumulation of reactive oxygen species (ROS), oxidative stress, mitochondrial dysfunction, DNA fragmentation, and ultimately, hepatocyte apoptosis or necrosis [22–24].

Many studies have also reported increased levels of creatinine and blood urea nitrogen (BUN) after APAP toxicity [25–27]. These outcomes are potential indicators of nephrotoxicity. Paracetamol-induced nephrotoxicity also results in metabolism of APAP to form NAPQI and then p-aminophenol (PAP) in the kidney by de-acetylation. PAP converts to a reactive quinone by the action of prostaglandin endoperoxide synthase [26,28], thereby resulting in nephrotoxicity.

Although not a common outcome reported for APAP overdose, neurotoxicity might also result. Several studies have detected the presence of CYP450 enzymes within the brain that would most likely metabolize APAP to the toxic metabolite NAPQI. These enzymes are mainly found in the olfactory bulbs, olfactory cortex, hippocampus, cerebellum, and brain stem [29–31]. APAP metabolism in the brain to the toxic metabolite NAPQI leads to an increase in *Cyp2e1* levels and neuronal cell death [32]. Nevertheless, the currently available studies on the effects of acute APAP toxicity on the brain are sparse and none of them have measured CYP450 expression in many brain regions. Other studies have shown that during APAP toxicity, ROS generation and GSH depletion occurs [33].

The primary purpose of this work was to assess the potential effects of acute APAP toxicity in pregnant rats. Several measures of toxicity were examined, including the production of ROS in the liver, kidney, and brain, as well as the expression of CYP450 isozymes in the liver, kidney, and brain. This study also investigated the therapeutic effects of vitamin E against APAP-induced toxicity in the liver, kidney, and brain in pregnant rats. Vitamin E (α-tocopherol) is a lipid-soluble small molecule antioxidant that belongs to the class of tocopherol antioxidants. Vitamin E is the most potent form that can neutralize reactive nitrogen species (RNS), including nitric oxide (NO), nitrogen dioxide (NO_2), and peroxy-nitrite. Vitamin E has many sources, including common food oils produced from corn, peanuts, and soybeans [34]. Consequently, we investigated the pre- and post-treatment protective effects of vitamin E on an acute toxic APAP dose in pregnant rats.

2. Materials and Methods

2.1. Animals

This study was conducted on 20 female Wistar rats weighing 200 ± 10 g at the start of the study. Wistar rats were obtained from the Al-Zaytoonah University of Jordan animal program. The animals were kept in a 12/12 hr light/dark cycle at room temperature (25 ± 2 °C) and humidity (45 ± 5%), with free access to food and water. The animal protocol (protocol #18/06/2018–2019) for this work was approved by the Animal Care and Use Committee of Al-Zaytoonah University of Jordan and all work was conducted in accordance with the Helsinki guidelines for animal research [35] and all applicable Jordanian governmental rules and guidelines.

2.2. Drugs

APAP powder (UC448) (≥97%) and vitamin E (T3251) (89.0–102.0%) were purchased from Sigma Aldrich, Burghausen, Germany.

2.3. Experimental Procedure

Twenty female Wistar rats, average weight 200 ± 10 g, were housed in cages alone after mating. Mating consisted of placing two virgin females and one male together prior to the end of the daily light cycle. The following morning, each female was examined for the presence of a vaginal plug. Females were separated after vaginal plug detection and housed alone. Pregnant females were placed in separate cages and kept on a 12:12 light/dark cycle. The day after mating was taken as day zero of pregnancy (gestational day zero, GD0). Abdominal enlargement of female rats was observed on day 16 after mating, which was taken as a positive indicator of pregnancy. GD19 was chosen to take samples for this experiment because pregnancy is visually apparent at this stage and GD19 is representative of late-stage pregnancy in rats, equivalent to the second trimester in humans. On GD17, all animals were fasted overnight. On GD18, rats were divided into 4 groups (5 rats per group): Group I (Control) received 0.5 mL p.o. corn oil (vehicle), Group II (APAP) received a single dose of 3000 mg/kg p.o. APAP dissolved in corn oil, Group III (E + APAP) received a single dose of 3000 mg/kg p.o. APAP dissolved in corn oil one hour after 300 mg/kg p.o. vitamin E dissolved in corn oil, and Group IV (APAP + E) received a single dose of 3000 mg/kg p.o. APAP dissolved in corn oil one hour before 300 mg/kg p.o. vitamin E dissolved in corn oil. All groups were euthanized twenty-four hours after the treatments (GD19). Blood was collected in heparinized tubes and stored at −20 °C, and the brain, kidney, and liver were collected and snap frozen for further analysis. High doses of APAP were used to ensure toxicity in liver, kidney, and brain, as shown by previous studies [36–38]. Additionally, high doses of vitamin E were used to measure pre- and post-treatment protective effects as reported previously [39–41].

2.4. Blood Biochemistry: Hepatotoxicity

Blood samples (around 1 mL) were taken from the aorta of each rat, collected in heparinized tubes, and were centrifuged at 4000 rpm for 10 min. Plasma was obtained from the samples using a calibrated micropipette and stored at −20 °C until subsequent analysis for the levels of ALT, AST, serum creatinine (SCr), and BUN.

2.4.1. Alanine Aminotransferase (ALT)

ALT analyses were conducted using the manual procedure of the Biolabo ALT analysis kit (REF#80027, Biolabo S.A.S., Les Hautes Rives, France). A reduction in absorbance in the samples is due to the conversion of NADH to NAD^+ and is proportional to ALT activity in the specimen. The absorbance was measured at 340 nm after 1, 2, and 3 min. The absorbance at each time point was subtracted from the previous measurement (e.g., the value for time-point 2 minus the value for time-point 1, and the value for time-point 3 minus the value for time-point 2). The absolute values of these two measurements were averaged and converted to a measure of absorbance rates (ΔAbs/min). ALT levels were calculated using the following equation by comparing the absorbance in the sample to the absorbance in a calibration standard:

$$\text{ALT levels (IU/L)} = \frac{\text{Sample Absorbance}(\Delta\text{abs/min})}{\text{Standard Absorbance}(\Delta\text{abs/min})} \times \text{standard concentration}$$

2.4.2. Aspartate Aminotransferase (AST)

AST analyses were conducted using the manual procedure of the Biolabo AST analysis kit (REF#80025, Biolabo S.A.S., Les Hautes Rives, France). Decreases in absorbance in the sample reflect the conversion of NADH to NAD^+ and are proportional to AST activity in the sample. The absorbance was measured at 340 nm after 1, 2, and 3 min. The absorbance for each time point was subtracted from the previous measurement (e.g., the value for time-point 2 minus the value for time-point 1, and the value for time-point 3 minus the value for time-point 2). The absolute values of these two measurements were averaged and converted to a measure of absorbance rates (ΔAbs/min). The AST levels were calculated using the following equation by comparing the absorbance in the sample to the absorbance in a calibration standard:

$$\text{AST levels (IU/L)} = \frac{\text{Sample Absorbance}(\Delta\text{abs/min})}{\text{Standard Absorbance}(\Delta\text{abs/min})} \times \text{standard concentration}$$

2.5. Blood Biochemistry: Nephrotoxicity

2.5.1. Serum Creatinine (SCr)

Serum creatinine analyses were conducted using the manual procedure of the Bioloabo creatinine kinetic method kit (REF#80107, Biolabo S.A.S., Les Hautes Rives, France). Creatinine levels were measured by an analysis of the absorbance reduction. Creatinine reacts with picrate to form a colored complex in alkaline solution. Creatinine levels were measured at 490 nm at 30 (A_1) and 150 (A_2) seconds. Serum creatinine levels were calculated using the following equation:

$$\text{Serum Creatine levels (mg/dL)} = \frac{\text{Sample Absorbance}(A_2 - A_1)}{\text{Standard Absorbance}(A_2 - A_1)} \times \text{standard concentration}$$

2.5.2. Blood Urea Nitrogen (BUN)

BUN levels were analyzed using the manual procedure of the Biolabo urea kit (REF#80221, Biolabo S.A.S., Les Hautes Rives, France). BUN analyses depend on the hydrolysis of urea to ammonium ions and carbon dioxide by the action of urease. The ammonium ions form chloride and salicylate complexes that are blue-green in color, which correlate with the urea concentration in the specimen. The absorbance is determined at 600 nm. BUN levels were calculated using the following equation:

$$\text{Serum BUN levels (mg/dL)} = \frac{\text{Sample Absorbance}}{\text{Standard Absorbance}} \times \text{standard concentration}$$

2.6. Brain Tissue Harvesting

Twenty-four hours after treatments, rats were euthanized by diethyl ether (#673811, Sigma Aldrich, Burghausen, Germany) overdose and decapitated using a guillotine. Brains were extracted and frozen in liquid nitrogen prior to storage at −80 °C. Using a cryostat set to −20 °C, tissue samples from the olfactory bulbs, hippocampus, and cerebellum were anatomically identified and dissected according to the rat brain atlas of Paxinos and Watson [42]. Samples were divided for analysis by quantitative polymerase chain reaction (qPCR) for mRNA, uric acid levels, and superoxide dismutase inhibition. All samples were frozen until later analysis.

2.7. Tissue Preparation

Kidney, liver, olfactory bulb, hippocampus, and cerebellum tissue samples were isolated and washed with ice-cold isotonic saline (0.9%) and then stored at −80 °C until later processing. Tissue samples were homogenized in 50 mM phosphate buffer (PH = 7.4) using an electronic homogenizer polytron PT1200 Kinematica AG®, Eschbach, Germany) to prepare 10% (*w*/*v*) homogenates. After centrifugation, supernatants were isolated and aliquots were separated into Eppendorf tubes and stored at −80 °C until assays were conducted for PCR, superoxide dismutase (SOD), and uric acid (UA) measurements.

2.8. Antioxidant Status

2.8.1. Uric Acid (UA)

UA levels in liver, kidney, and brain samples were detected using the Biosystem uric acid kit (REF#80351, Biosystem®, Barcelona, Spain). The kit forms a colored complex (red color) that can be measured at 520 nm. Uric acid levels were calculated using the following equation:

$$\text{Uric acid levels (mg/L)} = \frac{\text{Sample Absorbance}}{\text{Standard Absorbance}} \times \text{standard concentration}$$

2.8.2. Superoxide Dismutase (SOD)

SOD activities in liver, kidney, and brain samples were detected using the Abcam superoxide dismutase activity kit (ab65354; Abcam, Cambridge, UK). Analyses of SOD activity depend on a rection with the tetrazolium salt WST-1 that produces a water-soluble formazan dye upon reduction with a superoxide anion. The activity is determined as the difference between the activity of the sample and control (blank) samples. Blanks 1, 2, and 3, as well as the samples, were prepared according to Table 1.

Table 1. Components of the working solutions in SOD activity procedure.

Component	Sample (μL)	Blank 1 (μL)	Blank 2 (μL)	Blank 3 (μL)
Sample solution	20	0	20	0
ddH_2O	0	20	0	20
WST working solution	200	200	200	200
Enzyme working solution	20	20	0	0
Dilution Buffer	0	0	20	20

The absorbance was measured at 450 nm. SOD activity, expressed as % inhibition, was calculated using the following equation:

$$\%\ \text{SOD Inhibition} = \frac{[\{\text{Absorbance}(\text{Blank 1}) - \text{Absorbance}(\text{Blank 3})\} - \{\text{Absorbance}(\text{sample}) - \text{Absorbance}(\text{Blank 2})\}]}{[\{\text{Absorbance}(\text{Blank 1}) - \text{Absorbance}(\text{Blank 3})\}]} \times 100$$

2.9. Real-Time, Quantitative PCR

Total RNA was isolated from the tissue samples using the Thermo Scientific, GeneJet RNA Purification Kit (cat# K0732) according to the manufacturer's protocol. Ten micrograms of total RNA from each sample was used for cDNA synthesis. Reverse transcription (RT) was conducted using a Thermo Scientific verso cDNA synthesis kit (cat#AB1453B) according to the manufacturer's protocol. RNA and cDNA concentrations were measured using a Nanodrop Quawell DNA/Protein Analyzer (Thermo Fisher Scientific, Sunnyvale, CA, USA). cDNA samples were diluted to a total concentration of 100 ng/μL, and 1 μL of each diluted cDNA sample was used for a quantitative real-time polymerase chain reaction (qRT-PCR) according to the manufacturer's protocol. qRT-PCR was performed using a reaction mixture of TAKARA Cyber Green as a fluorescent dye (TP Green Premix Ex Taq II), a 1/20 volume of cDNA as a template, and the appropriate primers for the genes of interest. A threshold cycle number (CT) for each sample was obtained from an iCycler thermal cycler (Bio-Rad Laboratories, München, Germany) and used to compare the relative amount of target mRNA in experimental groups with those of controls, using the $2^{-\Delta\Delta CT}$ method [43]. Each sample was run in triplicate. To calculate ΔCT, the mean CT value for the control gene (GAPDH) was subtracted from the mean CT value of the gene of interest. The ΔCT values for group C were then averaged and subtracted from ΔCT for each measurement from the experimental groups to obtain ΔΔCT. The relative fold changes from the control were then expressed by calculating $2^{-\Delta\Delta CT}$ for each sample.

2.10. Statistical Analysis

Graphpad Prism 9.0 software was used for statistical analyses. The results were expressed as means ± standard deviation (SD). Values between groups were compared using a one-way ANOVA followed by Tukey post hoc tests to determine statistical significance between individual means.

3. Results

3.1. Blood Biochemistry: Hepatotoxicity

The hepatic function was assessed by measuring the levels of ALT and AST.

3.1.1. Alanine Aminotransferase (ALT)

APAP increased the ALT levels compared to the control group. Either pre-treatment or post-treatment administration of vitamin E attenuated this effect of APAP (Figure 1A). This effect was confirmed by a one-way ANOVA showing a significant main effect of treatment ($F(3, 16) = 57.37$, $p < 0.0001$). Tukey multiple comparisons showed a significant elevation in ALT after treatment with 3000 mg/kg p.o. APAP in comparison to the control group, while pre-treatment and post-treatment vitamin E administration prevented this increase. These results indicate that single dose of APAP caused liver toxicity and vitamin E reversed this effect.

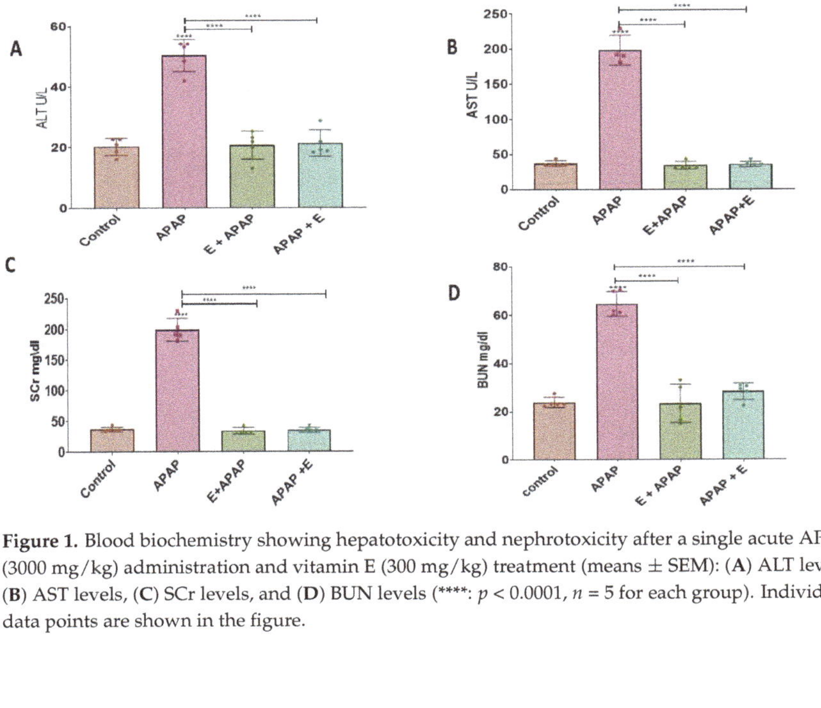

Figure 1. Blood biochemistry showing hepatotoxicity and nephrotoxicity after a single acute APAP (3000 mg/kg) administration and vitamin E (300 mg/kg) treatment (means ± SEM): (**A**) ALT levels, (**B**) AST levels, (**C**) SCr levels, and (**D**) BUN levels (****: $p < 0.0001$, $n = 5$ for each group). Individual data points are shown in the figure.

3.1.2. Aspartate Aminotransferase (AST)

APAP (3000 mg/kg p.o.) increased AST levels compared to the control group, while pre-treatment or post-treatment administration with vitamin E attenuated this effect (Figure 1B). This effect was confirmed by a one-way ANOVA showing a significant main effect of treatment (F (3, 15) = 261.8, $p < 0.0001$). Tukey multiple comparisons showed a significant elevation in AST after treatment with 3000 mg/kg p.o. compared to the control group, while pre-treatment or post-treatment vitamin E administration prevented this increase. These results indicate that APAP administration caused lever toxicity and vitamin E reversed this effect.

3.2. *Blood Biochemistry: Nephrotoxicity*

The kidney function was assessed by measuring the levels of creatinine and BUN.

3.2.1. Serum Creatinine (SCr)

APAP (3000 mg/kg) increased SCr levels compared to the control group, while pre-treatment or post-treatment administration of vitamin E attenuated this effect (Figure 1C). This effect was confirmed by a one-way ANOVA showing a significant main effect of treatment (F (3, 16) = 331.4, $p < 0.0001$). Tukey multiple comparisons showed a significant elevation in SCr after treatment with 3000 mg/kg p.o. in comparison to the control group, while pre-treatment or post-treatment vitamin E administration prevented this increase. These results suggest that APAP caused nephrotoxicity and vitamin E reversed this effect.

3.2.2. Blood Urea Nitrogen (BUN)

APAP (3000 mg/kg) increased BUN levels compared to the control group, while pre-treatment or post-treatment administration of vitamin E attenuated this effect (Figure 1D). This effect was confirmed by a one-way ANOVA showing a significant main effect of treatment (F (3, 16) = 75.45, $p < 0.0001$). Tukey multiple comparisons showed a significant elevation in BUN after treatment with 3000 mg/kg p.o. APAP in comparison to the control group, while pre-treatment or post-treatment administration of vitamin E prevented this increase. These results suggest that APAP caused nephrotoxicity and vitamin E reversed this effect.

3.3. *Antioxidant Status*

The results for the determination of uric acid (UA) and superoxide dismutase (SOD) activities in liver, kidney, and brain samples are shown below.

3.3.1. Uric Acid (UA)

APAP (3000 mg/kg p.o.) decreased UA concentrations in the liver and kidney compared to the control group, while pre-treatment or post-treatment vitamin E administration attenuated this effect (Figures 2A and 2B, respectively). These effects were confirmed by a one-way ANOVA, which showed the main effect of treatment in the liver (F (3, 16) = 11.95, $p = 0.0002$) and the kidney (F (3, 16) = 20.65, $p < 0.0001$). Tukey multiple comparison tests showed a significant decrease in UA after treatment with 3000 mg/kg p.o. APAP in comparison to the control group, while pre-treatment and post-treatment vitamin E administration prevented this increase. In addition, APAP decreased UA concentrations in the hippocampus and cerebellum compared to the control group (Figures 2C and 2D, respectively). This effect was confirmed by a one-way ANOVA, which showed significant main effects of treatment in the hippocampus (F (3, 16) = 3.555, $p = 0.0383$) and the cerebellum (F (3, 16) = 3.347, $p = 0.0456$). Tukey multiple comparisons showed a significant decrease in UA after treatment with 3000 mg/kg p.o. APAP in comparison to the control group. These decreases were very modest and comparisons to the vitamin E-treated groups did not demonstrate a significant difference from APAP treatment alone. However, in the olfactory bulbs APAP decreased the UA concentration compared to the control group. This effect was more obvious than the changes seen in the other brain regions and pre-treatment or post-treatment administration of vitamin E significantly attenuated this effect (Figure 2E).

This effect was confirmed by a one-way ANOVA, which showed the significant main effect of treatment (F (3, 16) = 50.49, $p < 0.0001$). Tukey multiple comparisons showed a significant decrease in UA after treatment with 3000 mg/kg p.o. APAP in comparison to the control group, while vitamin E administration prevented this decrease.

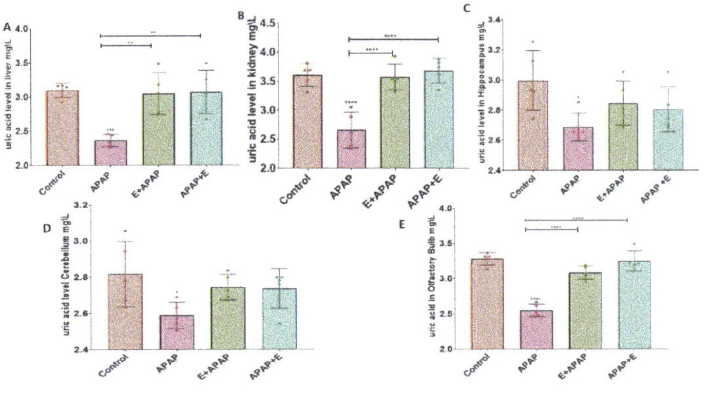

Figure 2. UA levels in different organs and brain regions after a single acute APAP (3000 mg/kg) administration and vitamin E (300 mg/kg) treatment (means ± SEM): (**A**) liver, (**B**) kidney, (**C**) hippocampus, (**D**) cerebellum, and (**E**) olfactory bulbs (*: $p < 0.05$, **: $p < 0.01$, ***: $p < 0.001$, ****: $p < 0.0001$, $n = 5$ for each group). Individual data points are shown in the figure.

3.3.2. Superoxide Dismutase (SOD)

APAP (3000 mg/kg p.o.) decreased the SOD concentration in the liver, kidneys, hippocampus, cerebellum, and olfactory bulb compared to the control group, while pre-treatment or post-treatment vitamin E administration attenuated this effect (Figures 3A, 3B, 3C, 3D and 3E, respectively). These effects were confirmed by a one-way ANOVA, which showed the significant main effects of treatment in the liver (F (3, 16) = 65.34, $p < 0.0001$), the kidneys (F (3, 16) = 52.65, $p < 0.0001$), the hippocampus (F (3, 16) = 41.10, $p < 0.0001$), the cerebellum (F (3, 16) = 52.11, $p < 0.0001$), and the olfactory bulb (F (3, 16) = 31.28, $p < 0.0001$). Tukey multiple comparisons showed a significant decrease in SOD concentration after treatment with 3000 mg/kg p.o. APAP in comparison to the control group, while vitamin E administration eliminated or reduced this decrease in all organs/brain regions that were compared.

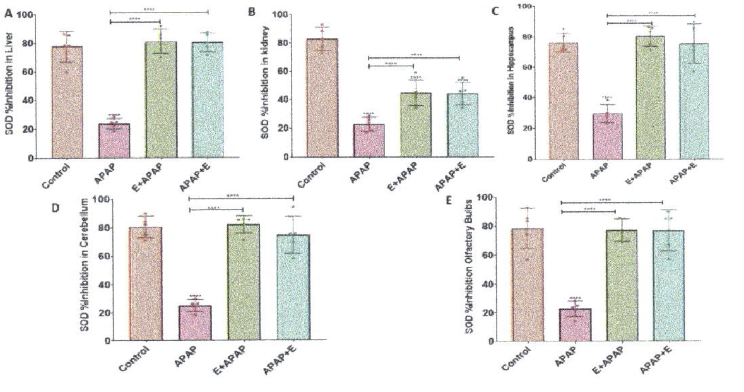

Figure 3. SOD inhibition (%) in different organs and brain regions after a single acute APAP (3000 mg/kg) administration and vitamin E (300 mg/kg) treatment (mean ± SEM): (**A**) liver, (**B**) kidney, (**C**) hippocampus, (**D**) cerebellum, and (**E**) olfactory bulbs (****: $p < 0.0001$, $n = 5$ for each group). Individual data points are shown in the figure.

3.4. Real-Time Quantitative PCR Results:

The results for determination of the relative expressions of *Cyp1a4*, *Cyp2d6*, and *Nat2* mRNA in liver, kidney, and brain samples are shown below.

3.4.1. *Cyp1a4*

APAP (3000 mg/kg p.o.) increased the relative *Cyp1a4* expression in the liver, kidneys, and hippocampus compared to the control group, while pre-treatment or post-treatment vitamin E attenuated this effect (Figure 4A,B and Figure 3C, respectively). These effects were confirmed by a one-way ANOVA, showing the main effects of treatment in the liver (F (3, 17) = 10.54, $p = 0.0004$), the kidney (F (3, 16) = 6.063, $p = 0.0059$), and the hippocampus (F (3, 17) = 21.86, $p < 0.0001$). Tukey multiple comparisons found significant increases in the relative *Cyp1a4* expression after treatment with 3000 mg/kg p.o. APAP in comparison to the control group, while pre-treatment and post-treatment vitamin E administration prevented these increases. However, the relative *Cyp1a4* expression in the cerebellum was not significantly different (Figure 4D), although it did show a similar pattern.

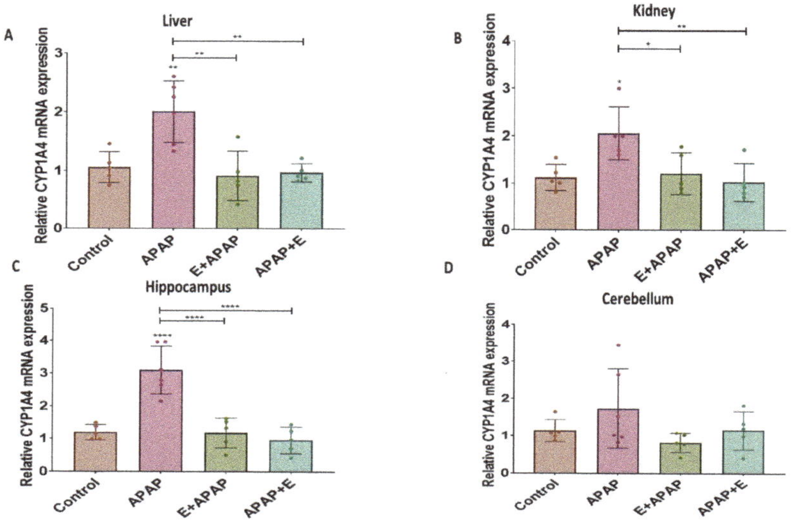

Figure 4. Relative *Cyp1a4* mRNA expression in different organs and brain regions after single acute APAP (3000 mg/kg) administration and vitamin E (300 mg/kg) treatment (mean ± SEM): (**A**) liver, (**B**) kidney, (**C**) hippocampus, and (**D**) cerebellum (*: $p < 0.05$, **: $p < 0.01$, ****: $p < 0.0001$, $n = 5$ for each group). Individual data points are shown in the figure.

3.4.2. *Cyp2d6*

APAP (3000 mg/kg p.o.) increased the relative *Cyp2d6* expression in the liver, kidneys, and hippocampus compared to the control group. Vitamin E pre-treatment attenuated this effect in the liver and the hippocampus, while vitamin E post-treatment attenuated this effect in the hippocampus (Figures 5A, 5B and 5C, respectively). These effects were confirmed by a one-way ANOVA, which found the main effects of treatment in the liver (F (3, 17) = 6.41, $p = 0.0042$), the kidney (F (3, 16) = 4.84, $p= 0.0139$), and the hippocampus (F (3, 16) = 8.11, $p < 0.0016$). Tukey multiple comparison tests found significant increases in the relative *Cyp2d6* expression after treatment with 3000 mg/kg p.o. APAP in comparison to the control group. Pre-treatment vitamin E administration prevented this increase in the liver, while pre-treatment and post-treatment vitamin E administration prevented this increase in the hippocampus. However, the relative *Cyp2d6* expression in the cerebellum was not significantly affected by the treatments (Figure 5D).

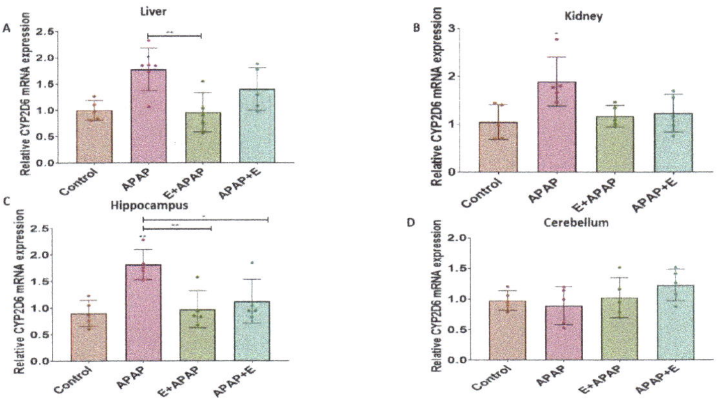

Figure 5. Relative *Cyp2d6* mRNA expression in different organs and brain regions after single acute APAP (3000 mg/kg) administration and vitamin E (300 mg/kg) treatment (mean ± SEM): (**A**) liver, (**B**) kidney, (**C**) hippocampus, and (**D**) cerebellum (*: $p < 0.05$, **: $p < 0.01$, $n = 5$ for each group). Individual data points are shown in the figure.

3.4.3. Nat2

APAP (3000 mg/kg p.o.) decreased the relative *Nat2* expression in the liver, hippocampus, and cerebellum compared to the control group, while pre-treatment or post-treatment vitamin E attenuated this effect in the liver and the cerebellum (Figures 6A, 6C and 6D, respectively). These effects were confirmed by a one-way ANOVA, showing the main effects of treatment in the liver ($F(3, 17) = 11.29$, $p = 0.0003$), the hippocampus ($F(3, 16) = 4.38$, $p = 0.0198$), and the cerebellum ($F(3, 16) = 8.36$, $p = 0.0014$). Tukey multiple comparison tests found significant decreases in the relative *Nat2* expression after treatment with 3000 mg/kg p.o APAP in comparison to control group, while pre-treatment and post-treatment vitamin E administration in the liver and the cerebellum normalized the *Nat2* levels. A similar pattern was observed in the hippocampus, but the post hoc comparisons did not find significant differences. The relative *Nat2* expression in the kidney was not significantly different between treatment groups (Figure 6B).

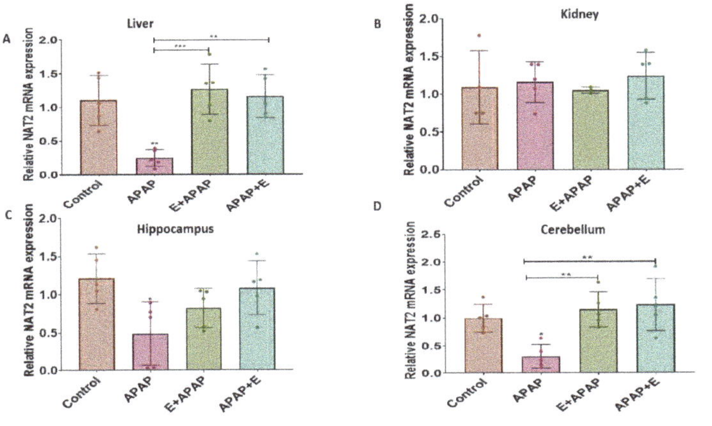

Figure 6. Relative *Nat2* mRNA expression in different organs and brain regions after a single acute APAP (3000 mg/kg) administration and vitamin E (300 mg/kg) treatment (mean ± SEM): (**A**) liver, (**B**) kidney, (**C**) hippocampus, and (**D**) cerebellum (*: $p < 0.05$, **: $p < 0.01$, ***: $p < 0.001$, $n = 5$ for each group). Individual data points are shown in the figure.

4. Discussion

This study examined acute APAP toxicity in the liver, kidney, and brain. These studies focused on pregnant rats, since APAP is the most commonly used analgesic during pregnancy. Moreover, this study investigated the potential benefits of prophylactic and therapeutic vitamin E treatment for preventing APAP-induced hepatic, renal, and neural toxicity. Acute APAP poisoning is frequently linked to hepatic and, to a lesser extent, renal inflammation, which occurs primarily as a result of cellular oxidative stress and pro-inflammatory immunological responses [44]. Our results showed that the levels of ALT and AST were significantly increased in the APAP-treated group in comparison to the control group, indicating that APAP induced hepatic injury. These results are consistent with a previous study that found a significant elevation in ALT and AST levels after administration of 750 mg/kg APAP to male rats [45]. Another study also showed that 3000 mg/kg p.o. APAP increased the levels of ALT and AST in male rats [46]. This elevation was explained by impaired transport function in hepatocytes, resulting in leakage of the plasma membrane, thereby increasing enzyme levels in the serum [45]. In order to investigate the in vivo protective effects of vitamin E, ALT and AST levels were determined in the present study and a significant improvement in both biochemical parameters were observed after either pre-treatment or post-treatment with vitamin E. Similarly, a previous study showed a restoration in ALT and AST levels after treatment with 50 mg/kg vitamin E for 6 weeks before treatment with 3000 mg/kg p.o. APAP [47].

In this study, the levels of creatinine and BUN were significantly increased in rats treated with APAP compared to the control group, indicative of kidney injury. These results agree with observations in previous studies that showed that levels of serum creatinine and BUN were elevated 24 h after administration of 1000 mg/kg i.p. APAP to male rats [48,49]. Moreover, similar to the observations here, in pregnant female rats, elevated creatinine and BUN levels were reduced by vitamin E administration either pre-treatment or post-treatment. Similarly, a previous study in mice also showed that pretreatment with 30 mg/kg p.o. vitamin E before 900 mg/kg i.p. APAP restored the creatinine and BUN levels to normal [50].

APAP produced other evidence of hepatic, renal, and neural (hippocampus, cerebellum, and olfactory bulb) toxicity as shown by a significant decline in UA and SOD levels, suggesting the presence of oxidative stress. Similar results were reported in a previous study, which examined the level of oxidative biomarkers such as catalase and SOD in rat kidneys following administration of 750 mg/kg/day APAP for 7 days [51]. Furthermore, another study found reduced hepatic antioxidant markers after administration of 2 g/kg p.o. APAP to mice in terms of catalase and SOD functions [52]. Other studies have demonstrated significant reductions in catalase and SOD in the entire brain of male mice after administration of 600 mg/kg i.p. APAP [53,54].

To investigate the protective effects of vitamin E treatment, given either before or after APAP in pregnant female rats, the present study examined uric acid and SOD levels in the liver, kidney, hippocampus, cerebellum, and olfactory bulb. Vitamin E attenuated the harmful effects of acute APAP administration on hepatic, renal, and neural (hippocampus and cerebellum) toxicity, as shown by the measures of oxidative stress. Moreover, significant improvements in uric acid and SOD levels were shown in the liver, kidney, hippocampus, cerebellum, and olfactory bulb after vitamin E treatment. This is consistent with one study showing that pretreatment with 30 mg/kg p.o. vitamin E before 900 mg/kg i.p APAP in BALB/c mice of both sexes restored the antioxidant status to normal levels [50].

APAP causes acute toxicity primarily through the metabolism of APAP to NAPQI by CYP enzymes. APAP is metabolized by the CYP isoforms *Cyp2e1*, *Cyp1a4*, and *Cyp2d6* [55–57]. An additional pathway for APAP-induced acute toxicity is through deacetylation of APAP by N-deacetylase enzymes to PAP, which is a toxic metabolite [58]. Following the production of PAP, it can be reconverted into APAP by NAT2 [59]. In the present study, the relative mRNA expressions of *Cyp1a4*, *Cyp2d6*, and *Nat2* were measured. A single dose of 3000 mg/kg p.o. APAP to pregnant female rats upregulated the relative mRNA expressions of *Cyp1a4* and *Cyp2d6*, while the expression of *Nat2* was reduced in all organs except in the kidney. *Nat2* RNA expression is low in the kidney compared to the liver and even lower in female kidneys compared to male kidneys, which might make detecting reductions in *Nat2* expression in the kidney more difficult [60]. More studies are warranted to evaluate this finding. These changes in enzyme expression are consistent with previous studies that reported that the relative mRNA expressions of some CYP isoforms are upregulated after a single oral dose of APAP [44,61]. The present study showed that pre-treatment or post-treatment vitamin E administration attenuated CYP mRNA dysregulation caused by APAP. This result is in line with several reports on the protective effects of vitamin E through effects that normalize CYP isoform function and/or expression [62–64].

The present study examined pregnant female subjects because APAP is commonly used by pregnant women, and APAP overdose occurs at a fairly high rate, perhaps because users perceive it to be relatively safe drug. The present study expanded upon our knowledge of the mechanisms underlying APAP overdoses, both in terms of the role of alterations in the expression of metabolizing enzymes and the effects on measures of neurotoxicity. The present study did not examine non-pregnant female rats or male rats, so it is difficult to say whether pregnancy or sex influences the toxicity of APAP or the protective effects of vitamin E. Comparisons across previous studies do not provide answers because the conditions (dose and treatment regimens and other factors) differ substantially between studies. Thus, answers to these questions will require further, specific study. It should also be noted that the effects of these drugs on in utero toxicity were not explored here but given the findings here on toxicity in pregnant dams, in utero toxicity might be of some concern, and, at the very least, worthy of study as well.

There are a few additional limitations to this study. In future studies, a positive control group treated with N-acetyl cysteine (NAC), the standard treatment for acetaminophen toxicity [50], should be added to the experimental design in order to compare it to the efficacy of vitamin E. This would be an important comparison for evaluation of vitamin E as an alternative for NAC, which has a high rate of serious anaphylaxis associated with its use [65]. A study by Şener, Şehirli and Ayanoğlu-Dülger [50] showed that NAC was more efficacious than vitamin E, but it remains to be seen whether that was due to the treatment parameters used in that single-dose study. Other parameters of the experiment certainly need additional examination, including the time between exposure and treatment. The present study also showed that pre-treatment with vitamin E reduces APAP toxicity, which should also be investigated, particularly in the context of APAP use in pregnancy. The present studies investigated such pre-treatments under a limited set of circumstances. Future studies should address potential prophylactic uses of vitamin E under more clinically relevant circumstances. Another limitation involves additional investigations of the underlying mechanism, including the effect of APAP on *Cyp2e1*, which is also involved in acetaminophen detoxification. Vitamin E effects on post APAP toxicity survival should also be evaluated. More studies are warranted in order to address these limitations. Additionally, future research should be focused on the effect of acute APAP doses and vitamin E treatment on fetal survival and health.

In summary, this study clearly demonstrated the toxic effects of an acute oral dose of APAP in pregnant female rats. This addresses an important limitation of previous studies that have not addressed overdose in pregnant females. Furthermore, the toxic effects of APAP were shown in the liver and kidney, and for the first time in the brain (cerebellum, hippocampus, and olfactory bulbs). APAP-induced toxic effects were shown for oxidative stress biomarkers, including SOD and UA. Additionally, and for the first time, we showed the effects of APAP on the expression of minor metabolic enzymes, often not examined in toxicity studies, that could contribute to the production of toxic metabolites after APAP administration. Moreover, the many outcomes detailed in the current study reveal the protective effect of vitamin E treatment, either when given prophylactically or after exposure to an acute toxic dose of APAP. These actions of vitamin E protected against hepatotoxicity, nephrotoxicity, and neurotoxicity. The potential for prophylactic use of vitamin E may be especially worthy of future investigation as it would limit the negatives effects associated with APAP use and might potentially prevent the toxic effects before they occur.

Author Contributions: Conceptualization, T.A.-Q.; and S.H.; methodology, B.S.; software, A.M.H.; validation, A.M.H., F.S.H. and S.H.; formal analysis, B.S.; investigation, A.M.H.; data curation, B.S., L.H., S.A.-K. and M.M.A.; writing—original draft preparation, A.M.H.; writing—review and editing, F.S.H.; visualization, S.H.; supervision, A.M.H.; project administration, S.H.; funding acquisition, T.A.-Q. All authors have read and agreed to the published version of the manuscript.

Funding: The work was supported by funds provided by Al-Zaytoonah University of Jordan (18/06/2018-2019).

Institutional Review Board Statement: The animal protocol (protocol #18/06/2018-2019) for this work was approved by the Animal Care and Use Committee of Al-Zaytoonah University of Jordan and all work was conducted in accordance with the Helsinki guidelines for animal research [28] and all applicable Jordanian governmental rules and guidelines.

Informed Consent Statement: Not applicable.

Data Availability Statement: Not applicable.

Conflicts of Interest: The authors declare no conflict of interest.

References

1. Elayeh, E.; Akour, A.; Haddadin, R.N. Prevalence and predictors of self-medication drugs to prevent or treat COVID-19: Experience from a Middle Eastern country. *Int. J. Clin. Pract.* **2021**, *75*, e14860. [CrossRef] [PubMed]
2. Malak, M.-Z.; AbuKamel, A. Self-medication Practices among University Students in Jordan. *Malays. J. Med. Health Sci.* **2019**, *15*, 112–119.
3. Breu, A.C.; Patwardhan, V.R.; Nayor, J.; Ringwala, J.N.; Devore, Z.G.; Ganatra, R.B.; Hathorn, K.E.; Horton, L.; Iriana, S.; Tapper, E.B. A multicenter study into causes of severe acute liver injury. *Clin. Gastroenterol. Hepatol.* **2019**, *17*, 1201–1203. [CrossRef] [PubMed]
4. Mikhail, A.; Tanoli, O.; Légaré, G.; Dubé, P.-A.; Habel, Y.; Lesage, A.; Low, N.C.; Lamarre, S.; Singh, S.; Rahme, E. Over-the-counter drugs and other substances used in attempted suicide presented to emergency departments in Montreal, Canada. *Crisis* **2018**, *40*, 166–175. [CrossRef]
5. Moore, N.; Duret, S.; Grolleau, A.; Lassalle, R.; Barbet, V.; Duong, M.; Thurin, N.; Droz-Perroteau, C.; Gulmez, S.E. Previous drug exposure in patients hospitalised for acute liver injury: A case-population study in the French National Healthcare Data System. *Drug Saf.* **2019**, *42*, 559–572. [CrossRef]
6. Yan, M.; Huo, Y.; Yin, S.; Hu, H. Mechanisms of acetaminophen-induced liver injury and its implications for therapeutic interventions. *Redox Biol.* **2018**, *17*, 274–283. [CrossRef] [PubMed]
7. Ndetan, H.; Evans, M.W., Jr.; Singal, A.K.; Brunner, L.J.; Calhoun, K.; Singh, K.P. Light to moderate drinking and therapeutic doses of acetaminophen: An assessment of risks for renal dysfunction. *Prev. Med. Rep.* **2018**, *12*, 253–258. [CrossRef]
8. Wilkes, J.M.; Clark, L.E.; Herrera, J.L. Acetaminophen Overdose in Pregnancy. *South. Med. J.* **2005**, *98*, 1118–1122. [CrossRef]
9. Mokhtari, V.; Afsharian, P.; Shahhoseini, M.; Kalantar, S.M.; Moini, A. A Review on Various Uses of N-Acetyl Cysteine. *Cell J.* **2017**, *19*, 11. [CrossRef]
10. Bauer, A.Z.; Swan, S.H.; Kriebel, D.; Liew, Z.; Taylor, H.S.; Bornehag, C.-G.; Andrade, A.M.; Olsen, J.; Jensen, R.H.; Mitchell, R.T. Paracetamol use during pregnancy—A call for precautionary action. *Nat. Rev. Endocrinol.* **2021**, *17*, 757–766. [CrossRef]
11. Skakkebaek, N.E.; De Meyts, E.R.; Main, K.M. Testicular dysgenesis syndrome: An increasingly common developmental disorder with environmental aspects. *Apmis* **2001**, *109*, S22–S30. [CrossRef]
12. Albert, O.; Desdoits-Lethimonier, C.; Lesné, L.; Legrand, A.; Guillé, F.; Bensalah, K.; Dejucq-Rainsford, N.; Jégou, B. Paracetamol, aspirin and indomethacin display endocrine disrupting properties in the adult human testis in vitro. *Hum. Reprod.* **2013**, *28*, 1890–1898. [CrossRef] [PubMed]
13. Kristensen, D.M.; Mazaud-Guittot, S.; Gaudriault, P.; Lesné, L.; Serrano, T.; Main, K.M.; Jégou, B. Analgesic use—Prevalence, biomonitoring and endocrine and reproductive effects. *Nat. Rev. Endocrinol.* **2016**, *12*, 381–393. [CrossRef] [PubMed]
14. Arendrup, F.S.; Mazaud-Guittot, S.; Jégou, B.; Kristensen, D.M. EDC IMPACT: Is exposure during pregnancy to acetaminophen/paracetamol disrupting female reproductive development? *Endocr. Connect.* **2018**, *7*, 149. [CrossRef]
15. Rossitto, M.; Ollivier, M.; Déjardin, S.; Pruvost, A.; Brun, C.; Marchive, C.; Nguyen, A.L.; Ghettas, A.; Keime, C.; De Massy, B. In utero exposure to acetaminophen and ibuprofen leads to intergenerational accelerated reproductive aging in female mice. *Commun. Biol.* **2019**, *2*, 310. [CrossRef]

16. Thiele, K.; Solano, M.E.; Huber, S.; Flavell, R.A.; Kessler, T.; Barikbin, R.; Jung, R.; Karimi, K.; Tiegs, G.; Arck, P.C. Prenatal acetaminophen affects maternal immune and endocrine adaptation to pregnancy, induces placental damage, and impairs fetal development in mice. *Am. J. Pathol.* **2015**, *185*, 2805–2818. [CrossRef]
17. Cederbaum, A.I. Molecular mechanisms of the microsomal mixed function oxidases and biological and pathological implications. *Redox Biol.* **2015**, *4*, 60–73. [CrossRef]
18. Bertolini, A.; Ferrari, A.; Ottani, A.; Guerzoni, S.; Tacchi, R.; Leone, S. Paracetamol: New Vistas of an Old Drug. *CNS Drug Rev.* **2006**, *12*, 250–275. [CrossRef]
19. McGill, M.R.; Sharpe, M.R.; Williams, C.D.; Taha, M.; Curry, S.C.; Jaeschke, H.J. The mechanism underlying acetaminophen-induced hepatotoxicity in humans and mice involves mitochondrial damage and nuclear DNA fragmentation. *J. Clin. Investig.* **2012**, *122*, 1574–1583. [CrossRef]
20. Pingili, R.B.; Pawar, A.K.; Challa, S.R. Effect of chrysin on the formation of N-acetyl-p-benzoquinoneimine, a toxic metabolite of paracetamol in rats and isolated rat hepatocytes. *Chem.-Biol. Interact.* **2019**, *302*, 123–134. [CrossRef]
21. Rothen, J.-P.; Haefeli, W.; Meyer, U.A.; Todesco, L.; Wenk, M. Acetaminophen is an inhibitor of hepatic N-acetyltransferase 2 in vitro and in vivo. *Pharmacogenetics* **1998**, *8*, 553–560. [CrossRef] [PubMed]
22. Guo, C.; Xie, G.; Su, M.; Wu, X.; Lu, X.; Wu, K.; Wei, C. Characterization of acetaminophen-induced cytotoxicity in target tissues. *Am. J. Transl. Res.* **2016**, *8*, 4440. [PubMed]
23. Wang, X.; Wu, Q.; Liu, A.; Anadón, A.; Rodríguez, J.-L.; Martínez-Larrañaga, M.-R.; Yuan, Z.; Martínez, M.-A. Paracetamol: Overdose-induced oxidative stress toxicity, metabolism, and protective effects of various compounds in vivo and in vitro. *Drug Metab. Rev.* **2017**, *49*, 395–437. [CrossRef] [PubMed]
24. Iorga, A.; Dara, L.; Kaplowitz, N. Drug-induced liver injury: Cascade of events leading to cell death, apoptosis or necrosis. *Int. J. Mol. Sci.* **2017**, *18*, 1018. [CrossRef]
25. Ozatik, F.Y.; Teksen, Y.; Kadioglu, E.; Ozatik, O.; Bayat, Z. Effects of hydrogen sulfide on acetaminophen-induced acute renal toxicity in rats. *Int. Urol. Nephrol.* **2019**, *51*, 745–754. [CrossRef]
26. Kennon-McGill, S.; McGill, M.R. Extrahepatic Toxicity of Acetaminophen: Critical Evaluation of the Evidence and Proposed Mechanisms. *J. Clin. Transl. Res.* **2018**, *3*, 5. [CrossRef]
27. Moshaei-Nezhad, P.; Hosseini, S.M.; Yahyapour, M.; Iman, M.; Khamesipour, A. Protective effect of ivy leaf extract on paracetamol-induced oxidative stress and nephrotoxicity in mice. *J. Herbmed Pharmacol.* **2019**, *8*, 64–68. [CrossRef]
28. Abu, I.F.; Mat, A.C.; Zulkifli, M.; Juliana, N.; Mohamad, M.H.N. Improvement of kidney histological morphology in nephrotoxic paracetamol-induced rats by Cassia alata treatment. *Int. J. Res. Pharm. Sci.* **2018**, *9*, 6–11.
29. Upadhya, S.C.; Tirumalai, P.S.; Boyd, M.R.; Mori, T.; Ravindranath, V. Cytochrome P4502E (*Cyp2e*) in Brain: Constitutive Expression, Induction by Ethanol and Localization by Fluorescence in Situ Hybridization. *Arch. Biochem. Biophys.* **2000**, *373*, 23–34. [CrossRef]
30. Howard, L.A.; Miksys, S.; Hoffmann, E.; Mash, D.; Tyndale, R.F. Brain *Cyp2e1* is induced by nicotine and ethanol in rat and is higher in smokers and alcoholics. *Br. J. Pharmacol.* **2003**, *138*, 1376–1386. [CrossRef]
31. Joshi, M.; Tyndale, R.F. Induction and recovery time course of rat brain *Cyp2e1* after nicotine treatment. *Drug Metab. Dispos.* **2006**, *34*, 647–652. [CrossRef] [PubMed]
32. Posadas, I.; Santos, P.; Blanco, A.; Muñoz-Fernández, M.; Ceña, V. Acetaminophen induces apoptosis in rat cortical neurons. *PloS ONE* **2010**, *5*, e15360. [CrossRef] [PubMed]
33. Ghanem, C.I.; Pérez, M.J.; Manautou, J.E.; Mottino, A.D. Acetaminophen from liver to brain: New insights into drug pharmacological action and toxicity. *Pharmacol. Res.* **2016**, *109*, 119–131. [CrossRef] [PubMed]
34. El Hadi, H.; Vettor, R.; Rossato, M. Vitamin E as a Treatment for Nonalcoholic Fatty Liver Disease: Reality or Myth? *Antioxidants* **2018**, *7*, 12. [CrossRef]
35. Emanuel, E.J.; Grady, C.C.; Crouch, R.A.; Lie, R.K.; Miller, F.G.; Wendler, D.D. *The Oxford Textbook of Clinical Research Ethics*; Oxford University Press: Oxford, UK, 2008.
36. Henneh, I.T.; Ahlidja, W.; Alake, J.; Kwabil, A.; Ahmed, M.A.; Kyei-Asante, B.; Adinortey, M.B.; Ekor, M.; Armah, F.A. Ziziphus abyssinica root bark extract ameliorates paracetamol-induced liver toxicity in rats possibly via the attenuation of oxidative stress. *Toxicol. Rep.* **2022**, *9*, 1929–1937. [CrossRef]
37. Doğan, N.; Akçam, M.; Koca, T.; Doğuç, D.K.; Özgöçmen, M. The protective effect of Capparis ovata in acute hepatotoxicity induced by paracetamol. *Turk. J. Med. Sci.* **2016**, *46*, 561–566. [CrossRef]
38. Iyanda, A.A.; Adeniyi, F.A.A. Biochemical and histologic presentations of female Wistar rats administered with different doses of paracetamol/methionine. *Niger. J. Physiol. Sci.* **2011**, *26*, 151–160.
39. Tras, B.; Faki, H.E.; Kutahya, Z.O.; Bahcivan, E.; Dik, B.; Uney, K. The effects of dexamethasone and minocycline alone and combined with N-acetylcysteine and vitamin E on serum matrix metalloproteinase-9 and coenzyme Q10 levels in aflatoxin B1 administered rats. *Pol. J. Vet. Sci.* **2022**, *25*, 419–427.
40. Raeeszadeh, M.; Saleh Hosseini, S.M.; Amiri, A.A. Impact of Co-Administration of N-Acetylcysteine and Vitamin E on Cyclophosphamide-Induced Ovarian Toxicity in Female Rats. *J. Toxicol.* **2022**, *2022*, 9073405. [CrossRef]
41. Sudheesh, N.; Ajith, T.; Janardhanan, K. Hepatoprotective effects of DL-α-lipoic acid and α-Tocopherol through amelioration of the mitochondrial oxidative stress in acetaminophen challenged rats. *Toxicol. Mech. Methods* **2013**, *23*, 368–376. [CrossRef]
42. Paxinos, G.; Watson, C. *The Rat Brain in Stereotaxic Coordinates*; Hard Cover Edition; Elsevier: Amsterdam, The Netherlands, 2006.

43. Livak, K.J.; Schmittgen, T.D. Analysis of relative gene expression data using real-time quantitative PCR and the $2^{-\Delta\Delta CT}$ method. *Methods* **2001**, *25*, 402–408. [CrossRef] [PubMed]
44. El-Boshy, M.; BaSalamah, M.A.; Ahmad, J.; Idris, S.; Mahbub, A.; Abdelghany, A.H.; Almaimani, R.A.; Almasmoum, H.; Ghaith, M.M.; Elzubier, M. Vitamin D protects against oxidative stress, inflammation and hepatorenal damage induced by acute paracetamol toxicity in rat. *Free. Radic. Biol. Med.* **2019**, *141*, 310–321. [CrossRef]
45. Jafri, M.; Subhani, M.J.; Javed, K.; Singh, S. Hepatoprotective activity of leaves of Cassia occidentalis against paracetamol and ethyl alcohol intoxication in rats. *J. Ethnopharmacol.* **1999**, *66*, 355–361. [CrossRef]
46. Hamid, Z.A.; Budin, S.B.; Jie, N.W.; Hamid, A.; Husain, K.; Mohamed, J. Nephroprotective effects of Zingiber zerumbet Smith ethyl acetate extract against paracetamol-induced nephrotoxicity and oxidative stress in rats. *J. Zhejiang Univ. Sci. B* **2012**, *13*, 176–185. [CrossRef] [PubMed]
47. Walker, B.; Kelleher, J.; Dixon, M.; Losowsky, M. Vitamin E protection of the liver from paracetamol in the rat. *Clin. Sci. Mol. Med.* **1974**, *47*, 449–459. [CrossRef] [PubMed]
48. Trumper, L.; Girardi, G.; Elías, M.M. Acetaminophen nephrotoxicity in male Wistar rats. *Arch. Toxicol.* **1992**, *66*, 107–111. [CrossRef]
49. Cekmen, M.; Ilbey, Y.; Ozbek, E.; Simsek, A.; Somay, A.; Ersoz, C. Curcumin prevents oxidative renal damage induced by acetaminophen in rats. *Food Chem. Toxicol.* **2009**, *47*, 1480–1484. [CrossRef]
50. Şener, G.; Şehirli, A.Ö.; Ayanoğlu-Dülger, G. Protective effects of melatonin, vitamin E and N-acetylcysteine against acetaminophen toxicity in mice: A comparative study. *J. Pineal Res.* **2003**, *35*, 61–68. [CrossRef]
51. Sathishkumar, T.; Baskar, R. Renoprotective effect of *Tabernaemontana heyneana* Wall. leaves against paracetamol-induced renotoxicity in rats and detection of polyphenols by high-performance liquid chromatography–diode array detector–mass spectrometry analysis. *J. Acute Med.* **2014**, *4*, 57–67. [CrossRef]
52. Hamza, R.Z.; Al-Harbi, M.S. Amelioration of paracetamol hepatotoxicity and oxidative stress on mice liver with silymarin and Nigella sativa extract supplements. *Asian Pac. J. Trop. Biomed.* **2015**, *5*, 521–531. [CrossRef]
53. da Silva, M.H.; da Rosa, E.J.F.; de Carvalho, N.R.; Dobrachinski, F.; da Rocha, J.B.T.; Mauriz, J.L.; González-Gallego, J.; Soares, F.A.A. Acute brain damage induced by acetaminophen in mice: Effect of diphenyl diselenide on oxidative stress and mitochondrial dysfunction. *Neurotox. Res.* **2012**, *21*, 334–344. [CrossRef]
54. Mohammed, E.; Safwat, G. Assessment of the ameliorative role of selenium nanoparticles on the oxidative stress of acetaminophen in some tissues of male albino rats. *Beni-Suef Univ. J. Basic Appl. Sci.* **2013**, *2*, 80–85. [CrossRef]
55. Manyike, P.T.; Kharasch, E.D.; Kalhorn, T.F.; Slattery, J.T. Contribution of *Cyp2e1* and *Cyp3a* to acetaminophen reactive metabolite formation. *Clin. Pharmacol. Ther.* **2000**, *67*, 275–282. [CrossRef]
56. Rumack, B.H. Acetaminophen misconceptions. *Hepatology* **2004**, *40*, 10–15. [CrossRef] [PubMed]
57. Zhao, L.; Pickering, G. Paracetamol metabolism and related genetic differences. *Drug Metab. Rev.* **2011**, *43*, 41–52. [CrossRef] [PubMed]
58. Kim, S.-H. *Acetaminophen Associated Neurotoxicity and Its Relevance to Neurodevelopmental Disorders*; University of South Florida: Tampa, FL, USA, 2017.
59. Dressler, W.E.; Appelqvist, T. Plasma/blood pharmacokinetics and metabolism after dermal exposure to para-aminophenol or para-phenylenediamine. *Food Chem. Toxicol.* **2006**, *44*, 371–379. [CrossRef]
60. Estrada, L.; Kanelakis, K.C.; Levy, G.N.; Weber, W.W. Tissue-and Gender-Specific Expression ofN-Acetyltransferase 2 (Nat2*) during Development of the Outbred Mouse Strain CD-1. *Drug Metab. Dispos.* **2000**, *28*, 139–146.
61. Jiang, Y.; Fan, X.; Wang, Y.; Chen, P.; Zeng, H.; Tan, H.; Gonzalez, F.J.; Huang, M.; Bi, H. Schisandrol B protects against acetaminophen-induced hepatotoxicity by inhibition of CYP-mediated bioactivation and regulation of liver regeneration. *Toxicol. Sci.* **2015**, *143*, 107–115. [CrossRef]
62. Pattarachotanant, N.; Prasansuklab, A.; Tencomnao, T. *Momordica charantia* L. Extract Protects Hippocampal Neuronal Cells against PAHs-Induced Neurotoxicity: Possible Active Constituents Include Stigmasterol and Vitamin E. *Nutrients* **2021**, *13*, 2368. [CrossRef]
63. Nirala, S.K.; Bhadauria, M.; Mathur, R.; Mathur, A. Amelioration of beryllium induced alterations in hepatorenal biochemistry and ultramorphology by co-administration of tiferron and adjuvants. *J. Biomed. Sci.* **2007**, *14*, 331–345. [CrossRef]
64. Landes, N.; Pfluger, P.; Kluth, D.; Birringer, M.; Rühl, R.; Böl, G.F.; Glatt, H.; Brigelius-Flohé, R. Vitamin E activates gene expression via the pregnane X receptor. *Biochem. Pharmacol.* **2003**, *65*, 269–273. [CrossRef] [PubMed]
65. Hodgman, M.J.; Garrard, A.R. A review of acetaminophen poisoning. *Crit. Care Clin.* **2012**, *28*, 499–516. [CrossRef] [PubMed]

toxics

MDPI

Article

Time- and Concentration-Dependent Adverse Effects of Paclitaxel on Non-Neuronal Cells in Rat Primary Dorsal Root Ganglia

Amira Elfarnawany [1,2] and Faramarz Dehghani [1,*]

[1] Department of Anatomy and Cell Biology, Medical Faculty, Martin Luther University Halle-Wittenberg, Grosse Steinstrasse 52, 06108 Halle (Saale), Germany; amira.elfarnawany@science.tanta.edu.eg
[2] Zoology Department, Faculty of Science, Tanta University, Tanta 31527, Egypt
* Correspondence: faramarz.dehghani@medizin.uni-halle.de

Abstract: Paclitaxel is a chemotherapeutic agent used to treat a wide range of malignant tumors. Although it has anti-tumoral properties, paclitaxel also shows significant adverse effects on the peripheral nervous system, causing peripheral neuropathy. Paclitaxel has previously been shown to exert direct neurotoxic effects on primary DRG neurons. However, little is known about paclitaxel's effects on non-neuronal DRG cells. They provide mechanical and metabolic support and influence neuronal signaling. In the present study, paclitaxel effects on primary DRG non-neuronal cells were analyzed and their concentration or/and time dependence investigated. DRGs of Wister rats (6–8 weeks old) were isolated, and non-neuronal cell populations were separated by the density gradient centrifugation method. Different concentrations of Paclitaxel (0.01 μM–10 μM) were tested on cell viability by MTT assay, cell death by lactate dehydrogenase (LDH) assay, and propidium iodide (PI) assay, as well as cell proliferation by Bromodeoxyuridine (BrdU) assay at 24 h, 48 h, and 72 h post-treatment. Furthermore, phenotypic effects have been investigated by using immunofluorescence techniques. Paclitaxel exhibited several toxicological effects on non-neuronal cells, including a reduction in cell viability, an increase in cell death, and an inhibition of cell proliferation. These effects were concentration- and time-dependent. Cellular and nuclear changes such as shrinkage, swelling of cell bodies, nuclear condensation, chromatin fragmentation, retraction, and a loss in processes were observed. Paclitaxel showed adverse effects on primary DRG non-neuronal cells, which might have adverse functional consequences on sensory neurons of the DRG, asking for consideration in the management of peripheral neuropathy.

Keywords: peripheral neuropathy; DRG non-neuronal cells; paclitaxel; MTT assay; LDH assay; BrdU assay

Citation: Elfarnawany, A.; Dehghani, F. Time- and Concentration-Dependent Adverse Effects of Paclitaxel on Non-Neuronal Cells in Rat Primary Dorsal Root Ganglia. *Toxics* **2023**, *11*, 581. https://doi.org/10.3390/toxics11070581

Academic Editor: Youssef Sari

Received: 23 May 2023
Revised: 27 June 2023
Accepted: 1 July 2023
Published: 4 July 2023

1. Introduction

Many chemotherapeutic agents may trigger chemotherapy-induced peripheral neuropathy (CIPN), which manifests as tingling, numbness, and burning pain in both hands and feet [1]. The high incidence of CIPN [2] frequently results in dose reduction or the discontinuation of chemotherapy regimens [2–4]. Additionally, CIPN symptoms can continue for a very long time after chemotherapy, significantly lowering patients' quality of life [5].

Sensory neurons are more vulnerable to the toxic effects of anticancer drugs, and patients with CIPN typically experience more sensory symptoms than those in the motor or autonomic systems [6,7]. Chemotherapeutic drugs cause toxicity in myelin sheaths (myelopathy), sensory cell bodies (neuronopathy), and axonal compartments (axonopathy) in the DRG by affecting ion channels, microtubules, mitochondria, and associated capillaries [7,8]. DRG explants have thus been demonstrated to be a good, simple, and well-accepted model for studying peripheral neuropathy caused by antineoplastic agents [9–11]. Peripheral sensory (somatic) neurons can easily be reached by chemotherapy drugs as they are located outside the central nervous system without a brain–blood barrier and show strong vascularization due to fenestrated capillaries [12]. Additionally, chemotherapeutic drugs accumulate more in the sensory ganglia than in the peripheral nerves [13,14]. DRG explants' ability to outgrow neurites in vitro, as well as their response to toxic substances with neurite shortening, make them a reliable model in drug neurotoxicity assessment [9,15–17].

Primary DRG cultures consist of a diverse population of cells, including differentiated sensory post-mitotic neuronal cells (neurons) and proliferative non-neuronal cells (Satellite glial cells (SGCs), Schwann cells (SCs), and other glial cells) [18–21]. In parallel to the valuable impact of neurons, DRG non-neuronal cells are increasingly recognized as important in the development and maintenance of neuropathic pain [22,23]. SGCs, for instance, provide mechanical and metabolic support for neurons by forming an envelope surrounding their cell bodies [14,24]. Therefore, they closely monitor neuronal functions and interact with neurons using both diffusible (e.g., the paracrine release of glial modulators) and non-diffusive mechanisms (e.g., gap junctions) [25–28]. After nerve injury, SGCs become activated and contribute to the development of neuropathic pain [22,29]. SCs aid in myelinating axons, eliminate cellular debris [30], and play an important role in the outgrowth and guidance of re-growing peripheral axons [31]. SCs not only physically support the long axons, but they also have several growth factors that nourish and myelinate the large associated axons [32–34].

Paclitaxel is one of a wide range of commonly used chemotherapeutic agents. Although it has anti-tumoral properties, it also has significant adverse effects on the peripheral nervous system, causing peripheral neuropathy [2,17,35,36]. Paclitaxel shows neurotoxic effects on DRG neurons, including a significant reduction in neurite length and an increase in neuronal cell bodies at different investigated time points, as reported earlier [17,37]. The effects of paclitaxel on neuronal survival and neurite length in the DRG are shown to be dose- and time-dependent [17,37,38]. However, little is known about the effects of paclitaxel on primary DRG non-neuronal cells. The question is still open as to whether similar paclitaxel toxicity in primary DRG non-neuronal cells exists.

Previous research measured the process areas of non-neuronal cells of the DRG inside the mixed culture of neuronal and non-neuronal cells after 24 h of exposure to paclitaxel and found a decrease in the process areas of the non-neuronal cells [39]. In addition, paclitaxel has been shown to reduce cell viability and change the phenotype of SCs isolated from the sciatic nerve at 24 h and 48 h [31]. A recent study also investigated the impact of paclitaxel on the viability and proliferation of SGCs and found no effect on viability but a decrease in cell proliferation [14]. However, more research is needed to fully understand paclitaxel toxicity in the entire culture of non-neuronal cells (SCs, SGCs, and other glial cells). These outcomes may shed more light on the potential functional consequences of paclitaxel on primary DRG sensory neurons and the therapeutic interventions for peripheral neuropathy.

Therefore, the aim of this study was to investigate the effects of paclitaxel on primary DRG non-neuronal cells and determine the time course of those changes. DRG non-neuronal cells were isolated and treated with different concentrations of paclitaxel at different time points. Effects on viability, morphology, and proliferation were analyzed. We applied approaches such as the MTT assay to study cell viability [40], the lactate dehydrogenase (LDH) assay [41], and the propidium iodide (PI) assay to study cell death [42], as well as Bromodeoxyuridine (BrdU), to study cell proliferation [43]. These approaches are frequently employed in related studies [14,44–46]. We hypothesized that paclitaxel exposure would have severe toxic effects on DRG non-neuronal cells, which might be dose- or/and time-dependent.

2. Materials and Methods

2.1. Ethics Statement

All research involving animal material was carried out in accordance with the ethics policy and the policy on animal use in neuroscience research as outlined in Directive 2010/63/EU of the European Parliament and of the Council of the European Union on the protection of animals used for scientific purposes and was approved by local authorities for laboratory animal care and use (State of Saxony-Anhalt, Germany, permission number: I11M27).

2.2. Materials

Paclitaxel was used and administered into culture media in accordance with the treatment protocol (Taxol equivalent, Invitrogen, cat No. P3456-5 mg, Schwerte, Germany). Dimethyl sulfoxide (DMSO, Sigma–Aldrich, cat. No. D4540-500 mL, Lyon, France) was used to dissolve paclitaxel to obtain stock solutions of 1 mM and stored at −20 °C.

2.3. Isolation and Preparation of Primary DRG Co-Culture

DRG tissues were isolated from Wister rats aged 6–8 weeks. Rats were deeply anesthetized before scarification by isoflurane (Florene, 100% (*v*/*v*), 250 mL, Abcam, cat No. B506, France). Under aseptic conditions, the vertebral column was isolated, and all surrounding muscle, fat, and soft tissue were carefully removed. The spinal cord was exposed, and after that, DRGs were located, removed, and collected from intervertebral foramina on both sides in a sterile dish containing 3 mL of Hanks balanced salt solutions without Mg^{2+}/Ca^{2+} (HBSS, Invitrogen, REF. 24020-091, Schwerte, Germany) (Figure S1). The culture of non-neuronal cells was conducted in accordance with a previously published protocol [47], with some modifications. In brief, isolated DRGs were enzymatically digested in the first enzymatic solution containing 60 U/mL papain (Sigma–Aldrich, cat No. P4762-100 mg, St. Louis, MO, USA), 3 μL of 80 mg/mL saturated sodium hydrogen carbonate solution ($NaHCO_3$, Merck, cat No. k22399729, Darmstadt, Germany), and 0.6 mg/mL L-Cysteine (L-Cys, Sigma–Aldrich, Cat No. C7352-25 g, St. Louis, MO, USA) dissolved in 1.5 mL of HBSS without Mg^{2+}/Ca^{2+}. DRGs were then incubated for 15 min in a 37 °C water bath before being incubated in the second solution containing 4 mg/mL collagenase type II solution (CLS2, Sigma–Aldrich, Cat No. C6885-1 gm, St. Louis, MO, USA) and 4.6 mg/mL dispase type II (Dispase II, Sigma–Aldrich, Cat No. D4693-1 gm, St. Louis, MO, USA) solution in 3 mL HBSS without Mg^{2+}/Ca^{2+}. The DRGs were gently mixed with collagenase solution and incubated for an additional 15 min at 37 °C.

The resulting cell suspension underwent a one-minute centrifugation at 200 g. After carefully aspirating the collagenase solution, the DRGs were triturated 10–15 times with 1 mL of F12 medium (1X, Invitrogen, REF.11765-054, Schwerte, Germany) supplemented with 10 % of heat-inactivated Fetal Bovine Serum (FBS, Invitrogen, REF. 10270-106, Schwerte, Germany) and 1 % of 0.1 mg/mL streptomycin/penicillin (Sigma–Aldrich, cat No. P4333/100 mL, Darmstadt, Germany) by using 1000 μL pipette tips till the cell suspension became cloudy.

2.4. Seeding and Growth of Primary DRG Co-Culture

Circular coverslips were pre-coated for at least 1 h or overnight at 4 °C with 2 mg/mL Poly-D-lysine (PDL, Sigma–Aldrich, cat No. P6407, St. Louis, MO, USA) and 0.2 mg/mL laminin (Sigma–Aldrich, cat No. L2020-1 mg, St. Louis, MO, USA), then washed once with dist. H_2O and added directly before seeding cells in the culture medium. DRGs (50,000 cells) co-cultured in 50 μL culture medium were then pre-seeded on the coated coverslips for 2 h in an incubator at 37 °C and with 5% CO_2. One mL of warm culture medium adjusted to pH 7.4 was gently added to cells per well and maintained at 37 °C with 5% CO_2. Growth and morphology of co-cultivation of neurons and non-neurons were observed after 24 h, 72 h, 7 days, and 10 days (Figure 1a).

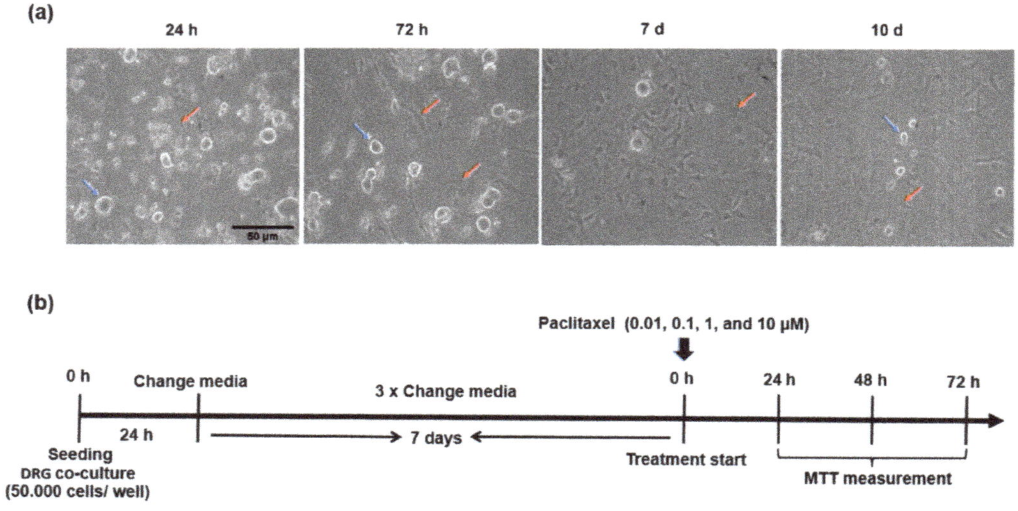

Figure 1. Morphological features and treatment protocol of primary DRG co-culture. (**a**) Representative images show the morphology and growth of DRG co-culture at different time points, blue arrows indicate neuronal populations, while red arrows indicate different subpopulations of DRG non-neuronal cells, Scale bar = 50 µm. (**b**) Treatment protocol for studying the effects of paclitaxel on DRG co-culture viability by using MTT assays at 24 h, 48 h, and 72 h post-treatment.

2.5. Effects on Cell Viability of Primary DRG Co-Culture (MTT Assay)

DRG co-cultured cells (5×10^4 cells/well) were treated 8 days after seeding with different concentrations of paclitaxel (0.01–10 µM) at 24 h, 48 h, and 72 h post-treatment in 96 well plates to study the effects on cell viability (Figure 1b). Four concentrations were then chosen that were as close to clinically applied doses as possible. Furthermore, the selected paclitaxel concentrations are in line with earlier reports from the literature [37–39,48–51]. Cell viability (%) was measured at 24 h, 48 h, and 72 h post-treatment using MTT assay. Four hours prior to the end of the experiments at various time points, 3-(4,5-dimethylthiazol-2-yl)-2,5-diphenyltetrazolium bromide solution (MTT, Invitrogen, cat. No M6494, 5 mg/mL, Eugene, OR, USA) was added. After an additional 4 h of incubation, the MTT solution was removed from the cells, and formazan crystals were dissolved in 100 µL of DMSO. Absorbance values were determined at two wavelengths (540 nm and 720 nm) by a microplate reader (SynergyTMMx, BioTek Instruments, Winooski, VT, USA) after another 20 min. Co-cultures maintained in standard media without paclitaxel were used as the control group. To rule out any effects of the solvent on cell viability, controls had DMSO at the same highest concentration (0.1%) as those used in other groups. For each treatment, three technical replicas were used in three biologically independent experiments.

2.6. Separation of Primary DRG Non-Neuronal Cells

To separate non-neuronal cells, density gradient centrifugation was applied by using bovine serum albumin (BSA, Sigma Aldrich, cat No.A7906-10 G, St. Louis, MO, USA) (15% (*w/v*) BSA solution) for purification [52]. The DRGs were triturated 10–15 times in 1 mL of high-glucose Dulbecco's Modified Eagle Medium (DMEM; Invitrogen; Ref. 41965-039; Schwerte, Germany) supplemented with 10 % FBS. Non-neuronal cells were separated from the DRG mixed culture by centrifuging single-cell suspensions through a 15% (*w/v*) BSA solution in DMEM. One milliliter of cell suspension was added to three milliliters of 15% BSA solution in a 15 mL conical tube and centrifuged at 300 g for 8 min at room temperature (RT) (Figure S1b). Thereafter, the layer of non-neuronal cells was carefully transferred to a 15 mL conical tube by using 1000 μL pipette tips. Then, 1 mL of warmed F12 medium supplemented with 10% FBS and 1% of 0.1 mg/mL streptomycin/penicillin was added, and the DRG non-neurons were suspended. A 40 μm cell strainer (SARSTEDT, cat. no. D-51588, Schwerte, Germany) was then used to filter the cell suspension to remove cell debris and undigested tissue fragments.

2.7. Seeding and Growth of Primary DRG Non-Neuronal Cells

Sterilized 12 mm circular coverslips were used, and they were washed and dried once with dist. H_2O. 50,000 cells resuspended in 50 μL culture medium were then pre-seeded on the sterilized coverslips for 2 h in an incubator at 37 °C and with 5% CO_2. One mL of warm culture medium adjusted to pH 7.4 was gently added to the cells, which were then preserved at 37 °C with 5% CO_2. Growth and morphology of DRG non-neuronal cells were observed after 24 h, 48 h, 72 h, 96 h, and 7, 10 days.

2.8. Effects of Paclitaxel on DRG Non-Neuronal Cells

2.8.1. Cell Viability (MTT Assay)

In 96 well plates, non-neuronal cells (15×10^3 cells/well) were seeded for 7 days, followed by treatment with four different concentrations of paclitaxel (0.01 μM, 0.1 μM, 1 μM, and 10 μM) at three different time points: 24 h, 48 h, and 72 h post treatment (Figure 2a). The effects of paclitaxel on the cell viability of non-neuronal cells were measured by MTT assay, as described above in Section 2.5.

2.8.2. Determination of Cytotoxicity (LDH Assay)

In 24 well plates, DRG non-neuronal cells (7×10^3 cells/well) were seeded in DMEM/F12 free phenol red medium (1X, Gibco, REF.21041-025, Schwerte, Germany) supplemented with 10 % inactivated FBS and 1% of 0.1 mg/mL streptomycin/penicillin for 7 days, followed by treatment with four different concentrations of paclitaxel (0.01 μM, 0.1 μM, 1 μM, and 10 μM) prepared in culture media supplemented with 1% FBS at different time points: 24 h, 48 h, and 72 h post treatment (Figure 2b). Additional wells were filled without cells for culture media control (blank). For determination of maximum LDH release (positive LDH control, 100 % cell death), 1:10 of the LDH lysis kit (LDH, Sigma Aldrich, cat. No. TOX7, St. Louis, MO, USA) was added to some wells and incubated for 45 min. According to the manufacturer's instructions, culture media samples from cells or controls at certain time points were transferred to 1.5 mL tubes and then centrifuged at $250\times g$ for 4 min to pellet cells. Afterward, 40 μL of the supernatant of different samples was added in 5 replicates to a clean flat-bottom 96-well plate and proceeded with enzymatic analysis. The LDH assay mixture was prepared at the time of use by adding 20 μL per well. The plates were covered with aluminum foil for light protection and incubated at room temperature for 30 min. The reaction was then stopped by adding 6 μL of 1 N Hydrochloric acid (HCl, Sigma Aldrich, cat. No. H9892, St. Louis, MO, USA) to each well. Absorbance values of samples were measured at a wavelength of 490 nm and the background absorbance of multi-well plates at 690 nm. Background absorbance values were subtracted from the primary wavelength measurements (490 nm). Finally, all controls, samples, and maximal measurements were normalized with blank measurements. Then the percent of cytotoxicity was calculated according to the below equation [53].

$$\%\ Cell\ death = \frac{(\text{sample absorbance value} - \text{mean control value})}{(\text{mean complete kill result} - \text{mean control value})} \times 100$$

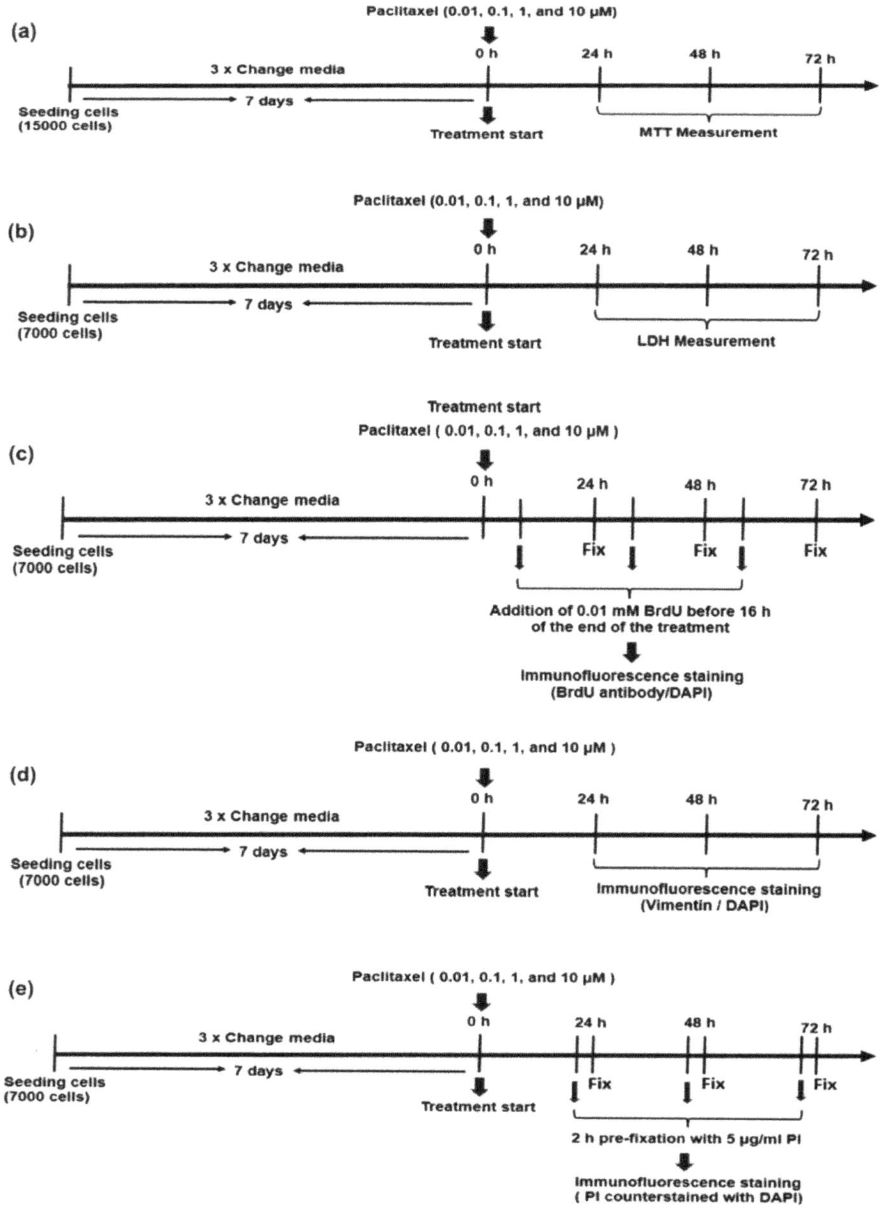

Figure 2. Various treatment protocols investigate the effects of different paclitaxel concentrations on primary DRG non-neuronal cells after 24 h, 48 h, and 72 h of the application. (**a**) The MTT assay was used for cell viability determination; (**b**) the LDH assay for cytotoxicity measurements; (**c**) the BrdU assay was used to detect cell proliferation; (**d**) treatment protocol for studying the effects of paclitaxel on cellular morphology through immunofluorescence staining; (**e**) detection of cell death by using the PI assay.

2.8.3. Detection of Cell Proliferation by BrdU Assay

To investigate the effects of paclitaxel on cell proliferation, DRG non-neuronal cells (7×10^3 cells/well) were seeded on 12 mm sterile coverslips in a 24 well plate, cultured for 7 days, and treated with various concentrations of paclitaxel at different time windows. Four µL of 0.01 mM 5-bromo-2′-deoxyuridine (BrdU, Sigma Aldrich, cat No. B5002-1G, St. Louis, MO, USA) was added to each well 16 h before fixation (Figure 2c). Cells were either immediately subjected to immunofluorescence or stored in 0.02 M PBS at 4 °C pending further use after fixation with 4% paraformaldehyde (PFA, AppliChem, cat No. 141451.1211, Darmstadt, Germany) for 15 min at room temperature. For labeling, non-specific bindings were blocked with normal goat serum (NGS, Sigma–Aldrich, cat. No. G9023-10 mL, Taufkirchen, Germany, 1:20) in 0.02 M PBS/0.3% (*v/v*) plus triton X-100 for 30 min. Thereafter, cells were washed three times with 0.02 M PBS for ten minutes each and incubated with a monoclonal mouse anti-BrdU antibody (Dako, cat. No. M0744-1 mL, Glostrup, Denmark, 1:200) overnight. Coverslips were then incubated with the goat anti-mouse Alexa Fluor® 488 conjugated secondary antibody (Life Technologies, cat. no. 2066710, Darmstadt, Germany, 1:200) for 1 h washed three times with PBS/triton for ten minutes. By using DAPI (4′,6-Diamin-2-phenylindol, Sigma–Aldrich, Munich, Germany, cat No. D9542), nuclei were visualized, and coverslips were mounted with DAKO fluorescence mounting medium (DAKO, Agilent Technologies, Inc., Santa Clara, CA 95051, USA). A confocal laser scanning microscope (Leica DMi8, Wetzlar, Germany) was used to take photomicrographs from five to eight randomly chosen areas. BrdU-positive cells were manually counted with Image J's multipoint tool (version 1.46r, National Institutes of Health, Laboratory for Optical and Computational Instrumentation, University of Wisconsin, Madison, WI, USA), and the percentage of proliferating cells was determined by dividing the number of $BrdU^+$ cells by the total number of DAPI-stained nuclei. To obtain the data, three independent experiments were conducted.

2.8.4. Determination of Paclitaxel Effects on Cellular Morphology

To study the effects of paclitaxel on the morphology of DRG non-neuronal cells, cells (7×10^3 cells/well) were seeded on 12 mm sterile coverslips in a 24 well plate, cultured for 7 days to allow nearly all cells to proliferate, and then treated with various concentrations of paclitaxel at different time windows (Figure 2d). After fixation, the immunofluorescence staining procedure was followed as described in Section 2.8.3. Chicken anti-vimentin polyclonal primary antibody (Abcam, cat No. ab24525, Cambridge, UK, 1:1000) combined with goat anti-chicken IgY Alexa Fluor® 488 conjugated (Invitrogen, REF. A11039-0.5 mL, Eugene, OR, USA, 1:200) as secondary antibody was used for labeling the cytoskeleton of non-neuronal cells. Then the procedure is completed as previously described in Section 2.8.3. Images were taken with a Leica confocal laser scanning microscope (Leica DMi8, Wetzlar, Germany), and five to eight areas were randomly captured per coverslip in three independent experiments.

2.8.5. Analysis of Apoptosis by Assessment of Nuclear Morphology

DRG non-neuronal cells were stained with the DNA dye DAPI to visualize nuclear morphology. The percentage of apoptosis (early and late apoptosis) was quantitated by scoring the percentage of apoptotic cells in the adherent cell population. Stained nuclei with a uniform and regular morphology were scored manually as healthy and viable cells. Cells with condensed, fragmented, or blubber nuclei were scored as apoptotic cells. The total number of nuclei in non-neuronal cells was counted automatically using Fiji software (https://imagej.net/Fiji/Downloads). After converting DAPI images into 8-bit gray scale images, the threshold of nuclei was adjusted manually, and the separation of attached nuclei was performed by applying a binary watershed. Finally, the analyzing particles option was applied, and the total number of nuclei was determined per image (Figure S2). Photomicrographs were captured using a Leica (DMi8, Wetzlar, Germany) confocal laser scanning microscope, and five to eight areas were recorded per each coverslip randomly in three independent experiments.

2.8.6. Detection of Cell Death by Propidium Iodide Staining

For detection of degenerating non-neuronal nuclei of dead cells by late apoptosis or necrosis, cells (7×10^3 cells/well) were seeded on 12 mm sterile coverslips in a 24 well plate, cultured for 7 days, and then treated with various concentrations of paclitaxel at 24 h, 48 h, and 72 h after treatment. Then, 5 µg/mL propidium iodide (PI, Merk, cat No. 537059-50 mg, Darmstadt, Germany) was added 2 h before fixation. Afterwards, cells were washed three times with PBS and then fixed with 4% PFA for 15 min (Figure 2e). Coverslips were washed three times with PBS/triton and incubated with DAPI. All stained slides were washed with aqua distilled water before being covered with a DAKO fluorescence mounting medium. Images were captured using a Leica (DMi8, Wetzlar, Germany) confocal laser scanning microscope, and five to eight areas were recorded per each coverslip randomly in three independent experiments. For the detection of PI-labeled dead cells, monochromatic light at 543 nm and an emission bandpass filter of 585–615 nm was used. PI-positive cells were counted manually using the multipoint tool of Image J software version v1.46r.

2.9. Statistical Analysis

GraphPad Prism 8.0.1 for Windows (GraphPad Software, La Jolla, CA, USA, www.graphpad.com, accessed on 22 May 2023) was used for data analysis and visualization. All the data were checked for normality using the Kolmogorov–Smirnov test. Statistics were performed using a one-way ANOVA (analysis of variance) followed by a Bonferroni post-test, with significance set at $p < 0.05$. All tests had an alpha level of 0.05.

3. Results

3.1. Characterization of Primary DRG Co-Culture

The growth of DRG co-culture was checked at different timelines (1, 3, 7, and 10 days) by a light microscope. DRG co-culture is a heterogeneous population of neuronal and non-neuronal cells. DRG neurons were characterized by refractile and bright cell bodies, and three different subpopulations were observed according to the size of their somata (small, $\leq$599 µm^2; medium, 600–1199 µm^2 and large, 1200–1300 µm^2), which represented 67%, 31%, and 2% of neurons in culture, respectively [37]. Additionally, three different subpopulations of DRG non-neuronal cells were observed in the culture (SCs, SGCs, and fibroblasts) (Figure 1a).

3.2. Effects of Paclitaxel on Viability of Primary DRG Co-Culture by MTT Assay

DRG co-cultures (neurons and non-neuronal cells) were treated with different concentrations of paclitaxel for 24 h, 48 h, and 72 h post-treatment. At 24 h post-treatment, the four different concentrations of paclitaxel showed no significant effects on the viability of DRG co-culture in comparison with the control group ($p > 0.05$) (Figure 3a). However, all paclitaxel concentrations demonstrated a significant suppression in the viability of cells in DRG co-culture compared to the control group at 48 h and 72 h post-treatment ($p < 0.0001$) (Figure 3b,c).

3.3. Characterization of Primary DRG Non-Neuronal Cells

DRG non-neuronal cells were examined under a light microscope at different time points (1, 2, 3, 4, 7, and 10 days) to analyze their growth and morphology. DRG non-neuronal cells are divided into three different subpopulations. The first population are SCs, which represent the majority of DRG non-neuronal cells [20,21]. They are distinguished by a single, small, spindle-shaped nucleus. These cells have a thin layer of cytoplasm surrounding the nucleus and bipolar cell bodies with long, thin projections or processes extending from each side. These long processes can either form a dense bundle of fibers or travel in a single thread of fibers away from the cell body (Figure 4). The population of

SGCs shows small, round, and flat cell bodies with wide cytoplasmic projections (Figure 4). These cells play a crucial role in the formation of an enveloping layer around DRG neurons for protection and metabolism. Lastly, fibroblasts are found under SCs with a large flat cell body and are pyramidal in shape with multipolar wide projections that are not associated with any other fibers. These cells are secretory active and form the connective tissue that supports cells in the culture (Figure 4).

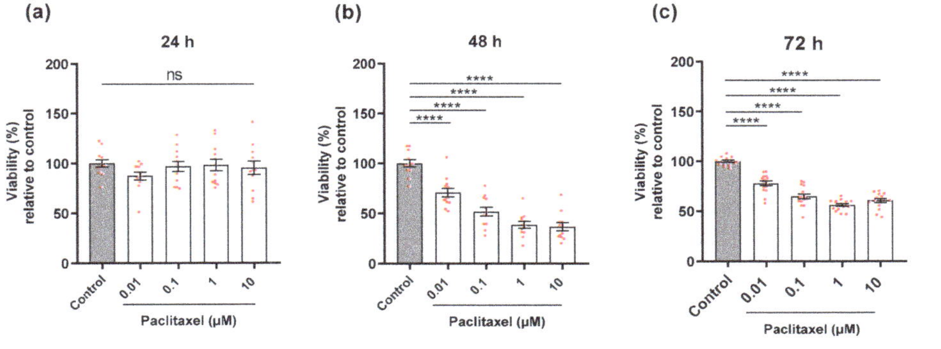

Figure 3. Effects of different paclitaxel concentrations on viability (%) of DRG co-culture at 24 h, 48 h, and 72 h post-treatment by MTT assay. (**a**) No significant effect on viability was found in co-cultures compared to controls at 24 h post-treatment ($p > 0.05$). (**b**) 48 h, and (**c**) 72 h post-treatment, paclitaxel displayed a significant reduction in the viability of cells compared to the control (**** $p < 0.0001$). The asterisks depict statistically significant results regarding the respective measurement indicated with the bar. Values are served as the mean ± SEM of three independent experiments performed in triplicate. ns, non-significant.

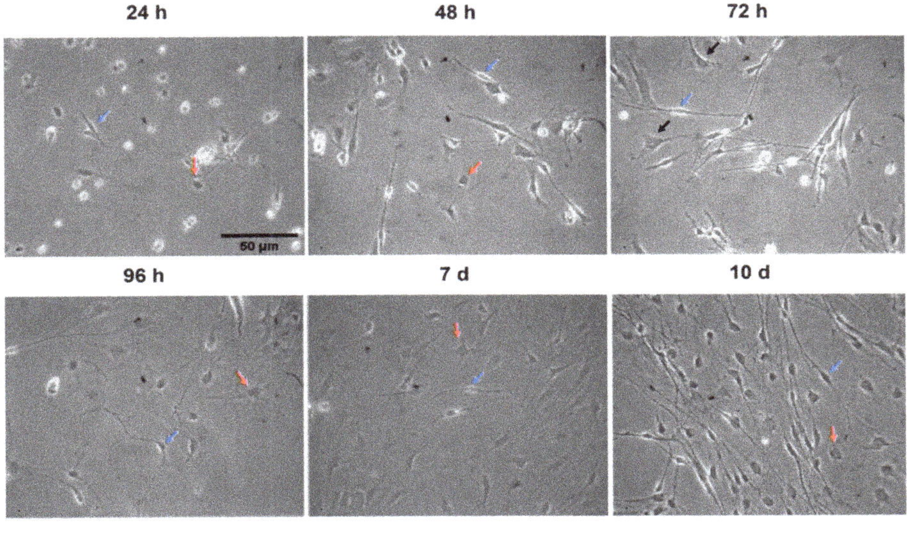

Figure 4. Representative phase contrast images show the morphology and growth of primary DRG non-neuronal cells at various time points. Blue arrows indicate Schwann cells, red arrows satellite glial cells, and black arrows represent fibroblasts, Scale bar = 50 μm.

3.4. Effects of Paclitaxel on Primary DRG Non-Neuronal Cells

3.4.1. Cell Viability (MTT Assay)

DRG non-neuronal cells were exposed to various concentrations of paclitaxel for 24 h, 48 h, and 72 h post-treatment. Only 10 µM of paclitaxel showed a significant reduction in the viability of cells compared to the control group at 24 h post-treatment ($p < 0.05$) (Figure 5a). While, at 48 h and 72 h post-treatment, different paclitaxel concentrations showed a significant reduction in the viability of non-neuronal cells compared to the untreated control group ($p < 0.05$) (Figure 5b,c). At 72 h post-treatment, the effects of paclitaxel on the viability of non-neuronal cells were clearly concentration-dependent (Figure S3a). Notably, the effects of 10 µM paclitaxel on the viability of non-neuronal cells were only time- but not concentration-dependent (Figure S3b).

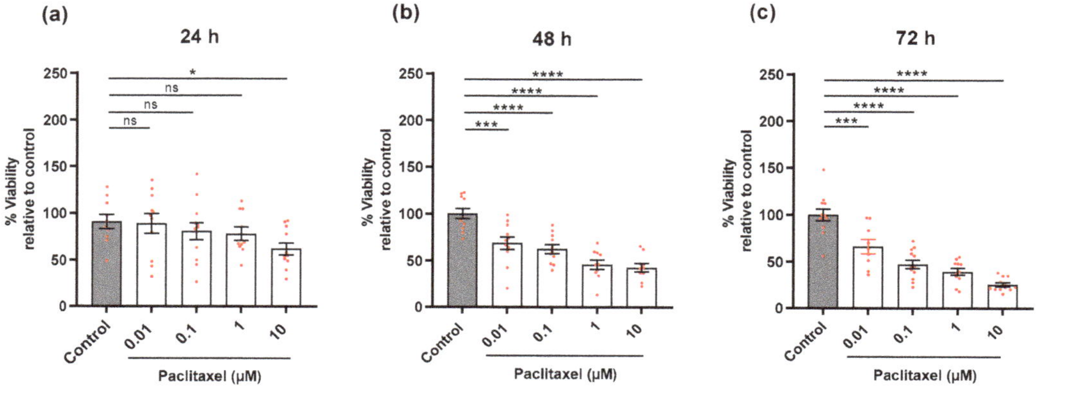

Figure 5. Effects of different concentrations of paclitaxel on the viability (%) of DRG non-neuronal cultures at 24 h, 48 h, and 72 h post-treatment by using MTT assay. (**a**) 10 µM of paclitaxel was the only concentration that showed a significant effect on the viability of DRG non-neuronal cells compared to control at 24 h post-treatment (* $p < 0.05$). (**b**,**c**), Different concentrations of paclitaxel elucidated a significant reduction in the viability of cells compared to the control at 48 h and 72 h post-treatment (*** $p < 0.001$, **** $p < 0.0001$). The asterisk denotes significant results regarding the respective measurement indicated with the bar. Values are served as mean ± SEM of three independent experiments performed in triplicate, ns: non-significant.

3.4.2. Determination of Cytotoxicity (LDH Assay)

The treatment of DRG non-neuronal cells with different paclitaxel concentrations (0.01 µM, 0.1 µM, 1 µM, and 10 µM) resulted in a significant increase in the number of damaged or dead cells that was proportional to the amount of LDH released in the cell culture media compared to non-treated cells ($p < 0.0001$) at 24 h after treatment (Figure 6a). After 48 h of treatment, the cytotoxicity of the four concentrations of paclitaxel increased remarkably compared to the control ($p < 0.0001$) (Figure 6b). The increase in the number of dead cells in response to the exposure of non-neuronal cells to paclitaxel continued in comparison to the control group ($p < 0.0001$) at 72 h post-treatment (Figure 6c). It was obvious that the effects of different paclitaxel concentrations on cytotoxicity were dose-dependent at only 72 h post-treatment (Figure S4a). Furthermore, a considerable difference was observed between different investigated time points for all applied concentrations, indicating time-dependent effects (Figure S4b).

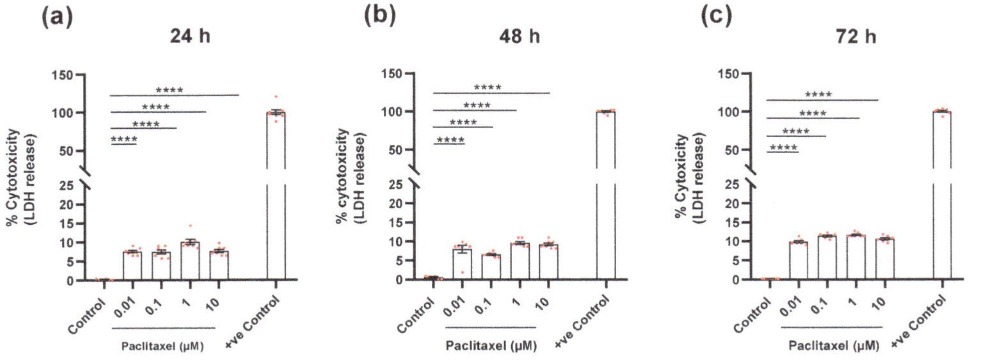

Figure 6. Effects of different concentrations of paclitaxel on cytotoxicity of DRG non-neuronal cultures using lactate dehydrogenase (LDH) assay. Levels of released LDH were quantified at (**a**) 24 h, (**b**) 48 h, and (**c**) 72 h post-treatment and showed a significant increase in LDH release that was proportional to the number of dead or damaged cells compared to the control group (**** $p < 0.0001$). +ve Control represents the maximum release of LDH after 100% cell death. The asterisks denote significant results regarding the respective measurement indicated with the bar. Values are given as the mean ± SEM of three independent experiments conducted in 15 replicates.

3.4.3. Cell Proliferation by BrdU Assay

The percentage of BrdU immunoreactive cells was determined in non-neuronal cells after exposure to various concentrations of paclitaxel at 24 h, 48 h, and 72 h post-treatment. At all investigated time points, a significantly lower number of BrdU-positive cells was found in treated cultures with different paclitaxel concentrations compared to the vehicle control group ($p < 0.0001$) (Figure 7a–d). As no significant difference was detected between different paclitaxel concentrations, no concentration-dependent effect was assumed (Figure S5a). In contrast, a significant difference between different timelines for all applied concentrations of paclitaxel was found, revealing a time-dependency of anti-proliferative effects (Figure S5b).

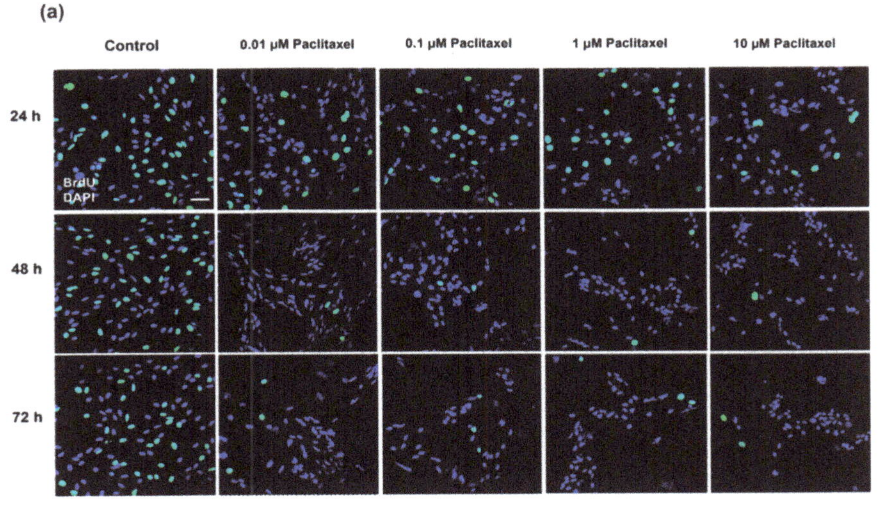

Figure 7. *Cont.*

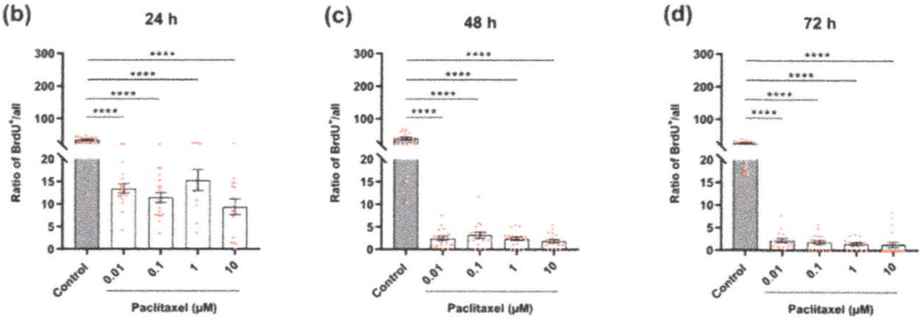

Figure 7. Effects of different concentrations of paclitaxel on cell proliferation of DRG non-neuronal cells using BrdU assay. (**a**) Representative immunofluorescence images of different non-neuronal cells treated with 0.01 µM, 0.1 µM, 1 µM, and 10 µM paclitaxel at 24 h, 48 h, and 72 h post-treatment show proliferating cells labeled with BrdU antibody (green) and all nuclei stained with DAPI (blue). 5–8 areas were recorded randomly per each coverslip; Scale bar = 75 µm. Bar charts demonstrated a significant decrease in the rate of cell proliferation after treatment compared to the control group (**** $p < 0.0001$) at (**b**) 24 h, (**c**) 48 h, and (**d**) 72 h post-treatment. The asterisks denote significant results regarding the respective measurement indicated with the bar. Values served as the mean ± SEM of three independent experiments performed in 15 replicates.

3.4.4. Cellular Morphological Changes

Except for 0.01 µM, all applied paclitaxel concentrations showed hallmarks of cell death and a variety of toxic alterations to the morphology of non-neuronal cells, including cell shrinkage, swollen cell bodies, or reductions in the length of processes. Additionally, other morphologic changes were observed in nuclei, such as nuclear fragmentation and chromatin condensation (Figure 8). The number of viable DRG non-neuronal cells was significantly reduced ($p < 0.05$) compared to the control group at all time windows (Figure 8).

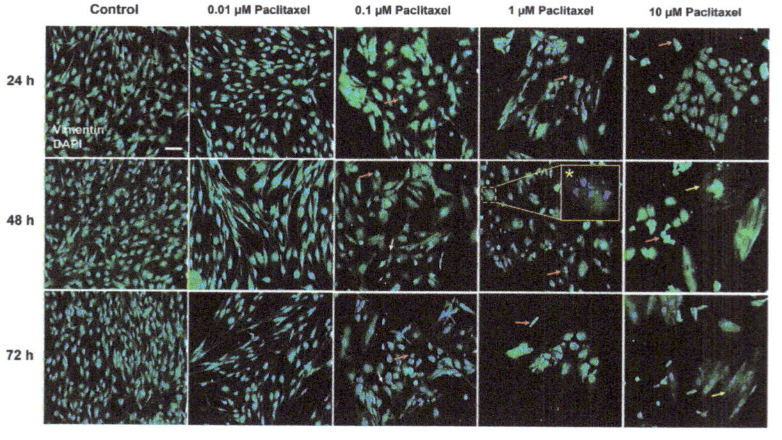

Figure 8. Effects of various paclitaxel concentrations on cellular morphology at different investigated time points using immunofluorescence staining. Representative microphotographs demonstrate cells stained with vimentin antibody (green) and nuclei counterstained with DAPI (blue). Paclitaxel (0.1 µM, 1 µM, and 10 µM) strongly affected the cell morphology of non-neuronal cells including shrinkage of cells' bodies (red arrows) and retraction of processes (white arrows). In addition, some cells treated with 10 µM paclitaxel were swelling (yellow arrows). Additionally, nuclear changes were observed, such as nuclear fragmentation (indicated by an asterisk in the inlet) and condensation. Five to eight regions were recorded randomly per coverslip by fluorescence microscopy. Scale bar = 75 µm.

3.4.5. Analysis of Changes in Nuclear Morphology

The effects of paclitaxel on nuclear morphology were investigated 24 h, 48 h, and 72 h after treatment. Paclitaxel induces nuclear fragmentation and condensation, which are hallmarks of apoptosis (Figure 9a). Different paclitaxel concentrations revealed a substantially increased number of apoptotic cells when compared to the control group ($p < 0.05$) at all investigated time points (Figure 9b–d). There was a significant difference between paclitaxel concentrations, indicating concentration dependence at the various time points studied (Figure S6a). Moreover, there was a significant difference between different investigated time windows, particularly between 24 h and 48 h for all paclitaxel concentrations, indicating a time dependence for the effects of different paclitaxel concentrations (Figure S6b).

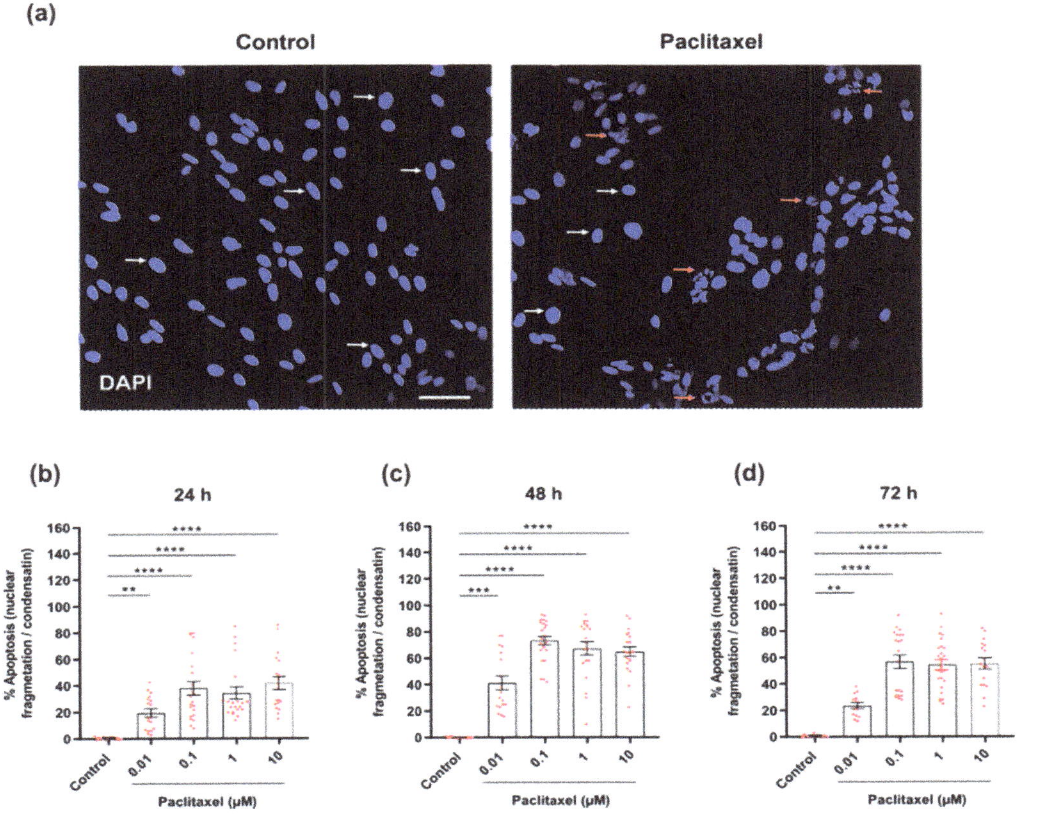

Figure 9. Effects of different concentrations of paclitaxel on nuclear morphology of DRG non-neuronal cells were analyzed by DAPI staining to detect % of apoptosis. (**a**) Representative images show DAPI-stained nuclei of non-neuronal cells of the control group (left) or 1 µM paclitaxel group (right) at 48 h post-treatment, Scale bar = 75 µm. White arrows indicate healthy and uniformly stained nuclei, whereas red arrows identify apoptotic nuclei. (**b–d**) A significant increase in % of apoptotic cells with fragmented or condensed nuclei was observed in different cultures treated with various paclitaxel concentrations (0.01 µM, 0.1 µM, 1 µM, and 10 µM) in comparison with the control group (** $p < 0.01$, *** $p < 0.001$, **** $p < 0.0001$). Data represented as mean ± SEM. The experiments were performed at least three independent times with n = 15 replicas. The asterisk denotes significant results regarding the respective measurement indicated with the bar graphs.

3.4.6. Detection of Cell Death by Propidium Iodide Staining

To detect degenerating non-neuronal cells with late apoptosis and necrosis, combined staining with PI and DAPI was performed. Dead cells showed pycnotic highly condensed or fragmented nuclei in bright pink, while live cells showed normal nuclei with homogenously distributed chromatin and regular morphology (Figure 10a). Except for 0.01 µM at 48 h post-treatment, all treated groups at all time points showed an apparent increase in the ratio of positive PI cells when compared to their corresponding untreated control group ($p < 0.05$) (Figure 10b–d). The presence of dead cells also increased with increasing paclitaxel concentrations when compared to the control, confirming a concentration dependency at different investigated time points (Figure S7a). Furthermore, a time-dependent increase in the ratio of cell death to DRG non-neuronal cells was observed except for 0.01 µM paclitaxel at 48 h (Figure S7a).

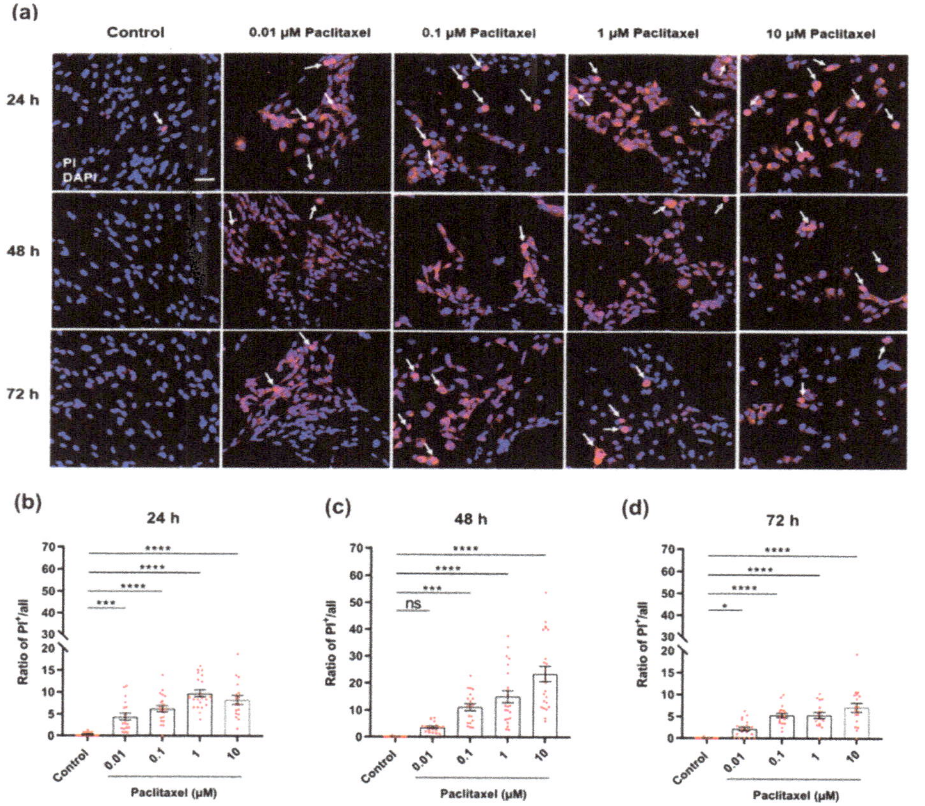

Figure 10. Effects of different paclitaxel concentrations on cell death of DRG non-neuronal cells by using PI assay. (**a**) Representative immunofluorescent fields show the amount of damaged non-neuronal cells (PI-positive) in treated groups compared to control fields. The white arrows represent degenerating cells (bright pink nuclei), Scale bar = 75 µm. At 24 h (**b**), 48 h (**c**), and 72 h (**d**) post-treatment, all concentrations of paclitaxel led to a massive increase in the number of dead cells compared to the control group (* $p < 0.05$, *** $p < 0.001$, **** $p < 0.0001$), except for 0.01 µM paclitaxel concentration at 48 h ($p > 0.05$). Values served as mean ± SEM, and the experiments were carried out three times independently with n = 15 replicas. The asterisk denotes significant results regarding the respective measurement indicated with the bar graphs, ns: non-significant.

4. Discussion

Primary DRG non-neuronal cells play a crucial role in supporting DRG neurons [22,31]. Previous studies investigated the toxic effects of paclitaxel on Primary DRG neurons, but little is known about the toxicity of paclitaxel on non-neuronal cells. Furthermore, the time course and concentration-dependency of paclitaxel's toxic effects on non-neuronal cells attracted little attention. To address these aspects, a more comprehensive approach using a variety of techniques, time points, and concentrations was chosen to demonstrate the effects of paclitaxel on DRG non-neuronal cell culture in vitro.

Paclitaxel exhibited several toxicological effects on primary DRG non-neuronal cells, including a decrease in cell viability, an increase in cell death, inhibition of cell proliferation, and cellular and nuclear changes, all of which were concentration- and time-dependent. Our findings on DRG SCs are consistent with previous research that studied paclitaxel effects on viability in a model of isolated SCs from the sciatic nerve [31]. These effects were attributed to paclitaxel's fast and strong mechanism of action on primary DRG non-neuronal cells, as these cells are non-transformed and proliferating cells. Therefore, paclitaxel selectively induces the death of transformed cells, possibly by arresting the cell cycle at G1 as well as G2/M phases [54–58].

Our results also revealed that paclitaxel significantly reduced the proliferation rate of DRG non-neuronal cells at various investigated timelines regardless of the applied concentration, but this suppression increased in a time-dependent manner. These findings expand the data of previous research, which reported a decrease in cell proliferation of SGCs of DRG after 24 h of treatment with 1 μM and 5 μM paclitaxel [14]. The authors postulated a paclitaxel stabilizing effect on microtubules by binding to beta-tubulin units, which disrupts microtubule dynamics [58]. As a result, mitosis was arrested between metaphase and anaphase (G2/M phase), suggesting a mitotic block and proliferation inhibition [57–59].

The majority of anticancer drugs have been shown to induce apoptosis in vulnerable cells [60–62]. Cellular and nuclear changes induced by anticancer drugs are very common and involve shrinkage of cell bodies, nuclear condensation, and chromatin fragmentation [54–56]. As shown here, the response to paclitaxel seems similar in primary DRG non-neuronal cells and affects all cellular subtypes.

Interestingly, we found that the percentage of apoptotic cells in DRG non-neuronal cell culture detected by DAPI staining at different investigated time points was higher when compared to the proportion of dead cells determined by the PI assay. This seeming discrepancy can be explained as DAPI staining detects cells in the early and late stages of apoptosis based on their nuclear morphology [63], but PI labels late apoptotic and dead cells with damaged cell membranes [64].

Furthermore, retraction and loss or shortening of processes increased strongly with the duration of treatment. These results add to the time- and concentration-dependency of paclitaxel effects and support previous research that reported a loss or shortening of processes in non-neuronal cells, however, in primary DRG co-culture after 24 h of exposure to paclitaxel [31,39]. These effects are comparable to those found in sensory neurons [17,37,65], implying the strong toxicity of paclitaxel on DRG non-neuronal cells, which might have adverse functional consequences for DRG sensory neurons.

Dose- and time-dependent pharmacokinetics have been reported more frequently for anticancer drugs than for other medications [66–70]. Our findings revealed that the effects of paclitaxel on Primary DRG non-neuronal cell culture are concentration- and time-dependent. Previous studies also reported similar findings on primary DRG neuronal and non-neuronal cells [14,17,39]. Low concentrations of paclitaxel (0.01–0.1 μM) were reported to suppress microtubule dynamics and inhibit mitotic spindle formation, resulting in a cell cycle arrest at the G2/M phase [55]. Considerably, low concentrations of paclitaxel showed no effect on the overall architecture of the microtubule cytoskeleton (Jordan et al., 1993), as noticed with 0.01 μM paclitaxel in the current study. In contrast, higher doses of paclitaxel were found to cause massive microtubule damage [59,71,72] and activate kinase pathways such as JNK/SAPK and p34 (cdc2) pathways [73–76], all of which are associated with paclitaxel-induced apoptosis [57]. It is important to note that apoptosis induced by these pathways is not dependent on mitotic arrest at higher concentrations, suggesting that it may occur in cells at any phase of the cell cycle [55]. This interpretation is consistent with our data that 0.01 μM paclitaxel did not exhibit a significant toxic effect on the morphology of primary DRG non-neuronal cells, whereas the higher concentrations (1 and 10 μM) did.

In the current study, the effect of paclitaxel on the cell viability of primary DRG co-culture by MTT assay was time-dependent and modulated by the presence of neuronal and non-neuronal cells in primary DRG culture. For example, the toxic effects of paclitaxel on the viability of non-neuronal cells alone were apparent earlier, at 24 h post-treatment, while a significant reduction appeared at 72 h after treatment in DRG neuronal cells [17]. However, in primary DRG co-cultures containing neuronal and non-neuronal cells, the effect was present 48 h post-treatment. A possible explanation might be that non-neuronal cells are more susceptible and sensitive to paclitaxel treatment when compared to neurons. As a result, paclitaxel's effects on non-neuronal cells become more apparent because they are actively growing, whereas post mitotic neurons need longer to respond to cell death [77,78]. Importantly, the effects of paclitaxel on the viability of primary DRG co-culture appeared at 48 h, not 24 h post-treatment, implying that there are cell-cell interactions between neurons and non-neuronal cells and modulating signaling pathways that impact the paclitaxel toxicity in the co-culture. Furthermore, the fate of cells after paclitaxel treatment might be affected by both paclitaxel concentrations and exposure time [51,59].

Neuronal function studies showed that neurons are not the only cell type that contributes to neuronal signaling. In the CNS, non-neuronal cells such as astrocytes, oligodendrocytes, and microglia all play important roles in influencing neuronal activity via interactions between neuronal cells and both glial cells and SGCs [79–82]. Non-neuronal glial cells and macrophages were shown to play critical roles in neuronal excitability modulation as well as in nutrition, structural, and maintenance functions [83,84]. In addition, they become activated following peripheral nerve injury or chronic inflammation and are involved in controlling neuronal excitability [85]. An interesting structural feature of the sensory ganglia is that the somata of sensory neurons do not form synaptic contacts with one another [86]. Additionally, neuronal cell bodies are enwrapped by SGCs inside the ganglia to form a structural and functional unit [27]. This specific structural arrangement stands for the communication between neurons and SGCs and is a determinant of somatic activity, as recently reported [82]. Changes in communication after injury are critical for understanding the development of abnormal ectopic discharges in somata that influence afferent signaling [28]. As a result, interactions between DRG neurons and glia and the activation of signaling pathways are believed to play an important role in the management of peripheral neuropathy [82].

5. Conclusions

Paclitaxel showed a set of toxicological effects on primary DRG non-neuronal cells that included a reduction in cell viability, an increase in cell death, inhibition of cell proliferation, and morphological changes. The effects of paclitaxel on primary DRG non-neuronal cells are concentration- and time- dependent. Given the crucial role of primary DRG non-neuronal cells in supporting DRG neurons and in the development and maintenance of neuropathic pain, the described adverse effects of paclitaxel on DRG non-neuronal cells might have functional consequences for sensory neurons in the DRG and should be considered in the management of peripheral neuropathy. Future research should investigate the potential negative effects of paclitaxel on signaling pathways and interactions between DRG neuronal and non-neuronal cells.

Supplementary Materials: The following supporting information can be downloaded at: https://www.mdpi.com/article/10.3390/toxics11070581/s1, Figure S1: (a) DRG isolation from 6–8 weeks old Wister rats, and (b) extraction and purification of DRG non-neuronal cells by using the density gradient centrifugation method; Figure S2: Representative example of automatic counting of nuclei of non-neuronal cells by the FIJI program for the control group; Figure S3: Effects of different concentrations of paclitaxel on the viability of DRG non-neuronal cells at different investigated time points using the MTT assay; Figure S4: Effects of different concentrations of paclitaxel on the percentage of cytotoxicity of DRG non-neuronal cells at different investigated time windows using the LDH assay; Figure S5: Effects of different concentrations of paclitaxel on the rate of cell proliferation of DRG non-neuronal cells at 24 h, 48 h, and 72 h post-treatment using BrdU assay; Figure S6: Effects of different concentrations of paclitaxel on nuclear morphology (% apoptosis) of DRG non-neuronal cells at 24 h, 48 h, and 72 h post-treatment by DAPI staining; Figure S7: Effects of different concentrations of paclitaxel on the ratio of PI^+ of DRG non-neuronal cells at 24 h, 48 h, and 72 h post-treatment by propidium iodide assay.

Author Contributions: Conceptualization, F.D., and A.E.; methodology, A.E.; formal analysis, A.E.; investigation, A.E.; resources, F.D.; data curation, A.E.; writing—original draft preparation, A.E. and F.D.; writing—review and editing, F.D. and A.E.; visualization, A.E.; supervision, F.D.; project administration, F.D. All authors have read and agreed to the published version of the manuscript.

Funding: This research received no external funding.

Institutional Review Board Statement: All animal experiments were carried out in accordance with the policy on ethics and the policy on the use of animals in neuroscience research, as specified in Directive 2010/63/EU of the European Parliament and of the Council of the European Union on the protection of animals used for scientific purposes, and were approved by local authorities for laboratory animal care and use (State of Saxony-Anhalt, Germany, permission number: I11M27).

Informed Consent Statement: Not applicable.

Data Availability Statement: All datasets generated for this study are included in the article/supplementary material.

Acknowledgments: We acknowledge the Katholischer Akademischer Ausländer-Dienst (KAAD) for the Ph.D. scholarship to A.E. The authors would like to thank Chalid Ghadban and Candy Rothgänger-Strube for excellent technical assistance. We acknowledge the financial support of the Open Access Publication Fund of Martin Luther University Halle-Wittenberg.

Conflicts of Interest: The authors declare no conflict of interest.

References

1. Kim, E. Chemotherapy-Induced Peripheral Neuropathy: Bench to Clinical Practice. *Korean J. Pain* **2020**, *33*, 291–293. [CrossRef] [PubMed]
2. Seretny, M.; Currie, G.L.; Sena, E.S.; Ramnarine, S.; Grant, R.; MacLeod, M.R.; Colvin, L.A.; Fallon, M. Incidence, Prevalence, and Predictors of Chemotherapy-Induced Peripheral Neuropathy: A Systematic Review and Meta-Analysis. *Pain* **2014**, *155*, 2461–2470. [CrossRef] [PubMed]
3. Han, Y.; Smith, M.T. Pathobiology of Cancer Chemotherapy-Induced Peripheral Neuropathy (CIPN). *Front. Pharmacol.* **2013**, *4*, 156. [CrossRef] [PubMed]
4. Miltenburg, N.C.; Boogerd, W. Chemotherapy-Induced Neuropathy: A Comprehensive Survey. *Cancer Treat. Rev.* **2014**, *40*, 872–882. [CrossRef]
5. Argyriou, A.; Kyritsis, A.; Makatsoris, T.; Kalofonos, H. Chemotherapy-Induced Peripheral Neuropathy in Adults: A Comprehensive Update of the Literature. *Cancer Manag. Res.* **2014**, *6*, 135–147. [CrossRef] [PubMed]
6. Boyette-Davis, J.A.; Hou, S.; Abdi, S.; Dougherty, P.M. An Updated Understanding of the Mechanisms Involved in Chemotherapy-Induced Neuropathy. *Pain Manag.* **2018**, *8*, 363–375. [CrossRef]
7. Zajączkowska, R.; Kocot-Kępska, M.; Leppert, W.; Wrzosek, A.; Mika, J.; Wordliczek, J. Mechanisms of Chemotherapy-Induced Peripheral Neuropathy. *Int. J. Mol. Sci.* **2019**, *20*, 1451. [CrossRef]
8. Gießen-Jung, C.; von Baumgarten, L. Chemotherapie-Induzierte Periphere Neuropathie. *DMW—Dtsch. Med. Wochenschr.* **2018**, *113*, 970–978. [CrossRef]
9. Windebank, A.J.; Smith, A.G.; Russell, J.W. The Effect of Nerve Growth Factor, Ciliary Neurotrophic Factor, and ACTH Analogs on Cisplatin Neurotoxicity in Vitro. *Neurology* **1994**, *44 Pt 1*, 488–494. [CrossRef]
10. Melli, G.; Höke, A. Dorsal Root Ganglia Sensory Neuronal Cultures: A Tool for Drug Discovery for Peripheral Neuropathies. *Expert Opin. Drug Discov.* **2009**, *4*, 1035–1045. [CrossRef]
11. Li, Y.; Marri, T.; North, R.Y.; Rhodes, H.R.; Uhelski, M.L.; Tatsui, C.E.; Rhines, L.D.; Rao, G.; Corrales, G.; Abercrombie, T.J.; et al. Chemotherapy-Induced Peripheral Neuropathy in a Dish: Dorsal Root Ganglion Cells Treated in Vitro with Paclitaxel Show Biochemical and Physiological Responses Parallel to That Seen in Vivo. *Pain* **2021**, *162*, 84–96. [CrossRef]

12. Jimenez-Andrade, J.M.; Herrera, M.B.; Ghilardi, J.R.; Vardanyan, M.; Melemedjian, O.K.; Mantyh, P.W. Vascularization of the Dorsal Root Ganglia and Peripheral Nerve of the Mouse: Implications for Chemical-Induced Peripheral Sensory Neuropathies. *Mol. Pain* **2008**, *4*, 10. [CrossRef] [PubMed]
13. Cavaletti, G.; Cavalletti, E.; Oggioni, N.; Sottani, C.; Minoia, C.; D'Incalci, M.; Zucchetti, M.; Marmiroli, P.; Tredici, G. Distribution of Paclitaxel within the Nervous System of the Rat after Repeated Intravenous Administration. *Neurotoxicology* **2000**, *21*, 389–393. [CrossRef] [PubMed]
14. Klein, I.; Boenert, J.; Lange, F.; Christensen, B.; Wassermann, M.K.; Wiesen, M.H.J.; Olschewski, D.N.; Rabenstein, M.; Müller, C.; Lehmann, H.C.; et al. Glia from the Central and Peripheral Nervous System Are Differentially Affected by Paclitaxel Chemotherapy via Modulating Their Neuroinflammatory and Neuroregenerative Properties. *Front. Pharmacol.* **2022**, *13*, 1038285. [CrossRef] [PubMed]
15. Gill, J.S.; Windebank, A.J. Cisplatin-Induced Apoptosis in Rat Dorsal Root Ganglion Neurons Is Associated with Attempted Entry into the Cell Cycle. *J. Clin. Investig.* **1998**, *101*, 2842–2850. [CrossRef]
16. Fischer, S.J.; McDonald, E.S.; Gross, L.; Windebank, A.J. Alterations in Cell Cycle Regulation Underlie Cisplatin Induced Apoptosis of Dorsal Root Ganglion Neurons in Vivo. *Neurobiol. Dis.* **2001**, *8*, 1027–1035. [CrossRef] [PubMed]
17. Elfarnawany, A.; Dehghani, F. Palmitoylethanolamide Mitigates Paclitaxel Toxicity in Primary Dorsal Root Ganglion Neurons. *Biomolecules* **2022**, *12*, 1873. [CrossRef]
18. Delree, P.; Leprince, P.; Schoenen, J.; Moonen, G. Purification and Culture of Adult Rat Dorsal Root Ganglia Neurons. *J. Neurosci. Res.* **1989**, *23*, 198–206. [CrossRef] [PubMed]
19. Grothe, C.; Unsicker, K. Neuron-Enriched Cultures of Adult Rat Dorsal Root Ganglia: Establishment, Characterization, Survival, and Neuropeptide Expression in Response to Trophic Factors. *J. Neurosci. Res.* **1987**, *18*, 539–550. [CrossRef] [PubMed]
20. Li, R.; Sliwkowski, M.X.; Lo, J.; Mather, J.P. Establishment of Schwann Cell Lines from Normal Adult and Embryonic Rat Dorsal Root Ganglia. *J. Neurosci. Methods* **1996**, *67*, 57–69. [CrossRef]
21. Wrathall, J.R.; Rigamonti, D.D.; Braford, M.R.; Kao, C.C. Non-Neuronal Cell Cultures from Dorsal Root Ganglia of the Adult Cat: Production of Schwann-like Cell Lines. *Brain Res.* **1981**, *229*, 163–181. [CrossRef] [PubMed]
22. Ji, R.-R.; Berta, T.; Nedergaard, M. Glia and Pain: Is Chronic Pain a Gliopathy? *Pain* **2013**, *154* (Suppl. S1), S10–S28. [CrossRef] [PubMed]
23. Ji, R.-R.; Chamessian, A.; Zhang, Y.-Q. Pain Regulation by Non-Neuronal Cells and Inflammation. *Science* **2016**, *354*, 572–577. [CrossRef]
24. Nascimento, R.S.; Santiago, M.F.; Marques, S.A.; Allodi, S.; Martinez, A.M.B. Diversity among Satellite Glial Cells in Dorsal Root Ganglia of the Rat. *Brazilian J. Med. Biol. Res.* **2008**, *41*, 1011–1017. [CrossRef]
25. Hanani, M.; Spray, D.C. Emerging Importance of Satellite Glia in Nervous System Function and Dysfunction. *Nat. Rev. Neurosci.* **2020**, *21*, 485–498. [CrossRef] [PubMed]
26. Avraham, O.; Deng, P.-Y.; Jones, S.; Kuruvilla, R.; Semenkovich, C.F.; Klyachko, V.A.; Cavalli, V. Satellite Glial Cells Promote Regenerative Growth in Sensory Neurons. *Nat. Commun.* **2020**, *11*, 4891. [CrossRef] [PubMed]
27. Pannese, E. The Structure of the Perineuronal Sheath of Satellite Glial Cells (SGCs) in Sensory Ganglia. *Neuron Glia Biol.* **2010**, *6*, 3–10. [CrossRef] [PubMed]
28. Huang, L.-Y.M.; Gu, Y.; Chen, Y. Communication between Neuronal Somata and Satellite Glial Cells in Sensory Ganglia. *Glia* **2013**, *61*, 1571–1581. [CrossRef]
29. Jasmin, L.; Vit, J.-P.; Bhargava, A.; Ohara, P.T. Can Satellite Glial Cells Be Therapeutic Targets for Pain Control? *Neuron Glia Biol.* **2010**, *6*, 63–71. [CrossRef]
30. Bhatheja, K.; Field, J. Schwann Cells: Origins and Role in Axonal Maintenance and Regeneration. *Int. J. Biochem. Cell Biol.* **2006**, *38*, 1995–1999. [CrossRef]
31. Imai, S.; Koyanagi, M.; Azimi, Z.; Nakazato, Y.; Matsumoto, M.; Ogihara, T.; Yonezawa, A.; Omura, T.; Nakagawa, S.; Wakatsuki, S.; et al. Taxanes and Platinum Derivatives Impair Schwann Cells via Distinct Mechanisms. *Sci. Rep.* **2017**, *7*, 5947. [CrossRef] [PubMed]
32. Chen, G.; Zhang, Z.; Wei, Z.; Cheng, Q.; Li, X.; Li, W.; Duan, S.; Gu, X. Lysosomal Exocytosis in Schwann Cells Contributes to Axon Remyelination. *Glia* **2012**, *60*, 295–305. [CrossRef] [PubMed]
33. Kidd, G.J.; Ohno, N.; Trapp, B.D. Biology of Schwann Cells. In *Handbook of Clinical Neurology*; Elsevier: Amsterdam, The Netherlands, 2013; Volume 115, pp. 55–79. [CrossRef]
34. Su, W.; Gu, Y.; Wei, Z.; Shen, Y.; Jin, Z.; Yuan, Y.; Gu, X.; Chen, G. Rab27a/Slp2-a Complex Is Involved in Schwann Cell Myelination. *Neural Regen. Res.* **2016**, *11*, 1830. [CrossRef]
35. Flatters, S.J.L.; Dougherty, P.M.; Colvin, L.A. Clinical and Preclinical Perspectives on Chemotherapy-Induced Peripheral Neuropathy (CIPN): A Narrative Review. *Br. J. Anaesth.* **2017**, *119*, 737–749. [CrossRef] [PubMed]
36. Cirrincione, A.M.; Pellegrini, A.D.; Dominy, J.R.; Benjamin, M.E.; Utkina-Sosunova, I.; Lotti, F.; Jergova, S.; Sagen, J.; Rieger, S. Paclitaxel-Induced Peripheral Neuropathy Is Caused by Epidermal ROS and Mitochondrial Damage through Conserved MMP-13 Activation. *Sci. Rep.* **2020**, *10*, 3970. [CrossRef] [PubMed]
37. Park, S.H.; Eber, M.R.; Fonseca, M.M.; Patel, C.M.; Cunnane, K.A.; Ding, H.; Hsu, F.-C.; Peters, C.M.; Ko, M.-C.; Strowd, R.E.; et al. Usefulness of the Measurement of Neurite Outgrowth of Primary Sensory Neurons to Study Cancer-Related Painful Complications. *Biochem. Pharmacol.* **2021**, *188*, 114520. [CrossRef] [PubMed]

38. Scuteri, A.; Nicolini, G.; Miloso, M.; Bossi, M.; Cavaletti, G.; Windebank, A.J.; Tredici, G. Paclitaxel Toxicity in Post-Mitotic Dorsal Root Ganglion (DRG) Cells. *Anticancer Res.* **2006**, *26*, 1065–1070.
39. Guo, L.; Hamre, J.; Eldridge, S.; Behrsing, H.P.; Cutuli, F.M.; Mussio, J.; Davis, M. Multiparametric Image Analysis of Rat Dorsal Root Ganglion Cultures to Evaluate Peripheral Neuropathy-Inducing Chemotherapeutics. *Toxicol. Sci.* **2017**, *156*, 275–288. [CrossRef]
40. Hansen, J.; Bross, P. A Cellular Viability Assay to Monitor Drug Toxicity. In *Protein Misfolding and Cellular Stress in Disease and Aging: Concepts and Protocols*; Springer: Berlin/Heidelberg, Germany, 2010; pp. 303–311. [CrossRef]
41. Kumar, P.; Nagarajan, A.; Uchil, P.D. Analysis of Cell Viability by the Lactate Dehydrogenase Assay. *Cold Spring Harb. Protoc.* **2018**, *2018*, pdb.prot095497. [CrossRef]
42. Brana, C.; Benham, C.; Sundstrom, L. A Method for Characterising Cell Death in Vitro by Combining Propidium Iodide Staining with Immunohistochemistry. *Brain Res. Protoc.* **2002**, *10*, 109–114. [CrossRef]
43. Liu, R.; Lin, G.; Xu, H. An Efficient Method for Dorsal Root Ganglia Neurons Purification with a One-Time Anti-Mitotic Reagent Treatment. *PLoS ONE* **2013**, *8*, e60558. [CrossRef] [PubMed]
44. Gao, W.; Zan, Y.; Wang, Z.J.; Hu, X.; Huang, F. Quercetin Ameliorates Paclitaxel-Induced Neuropathic Pain by Stabilizing Mast Cells, and Subsequently Blocking PKCε-Dependent Activation of TRPV1. *Acta Pharmacol. Sin.* **2016**, *37*, 1166–1177. [CrossRef]
45. Jang, H.J.; Hwang, S.; Cho, K.Y.; Kim, D.K.; Chay, K.-O.; Kim, J.-K. Taxol Induces Oxidative Neuronal Cell Death by Enhancing the Activity of NADPH Oxidase in Mouse Cortical Cultures. *Neurosci. Lett.* **2008**, *443*, 17–22. [CrossRef] [PubMed]
46. Pittman, S.K.; Gracias, N.G.; Vasko, M.R.; Fehrenbacher, J.C. Paclitaxel Alters the Evoked Release of Calcitonin Gene-Related Peptide from Rat Sensory Neurons in Culture. *Exp. Neurol.* **2014**, *253*, 146–153. [CrossRef] [PubMed]
47. Malin, S.A.; Davis, B.M.; Molliver, D.C. Production of Dissociated Sensory Neuron Cultures and Considerations for Their Use in Studying Neuronal Function and Plasticity. *Nat. Protoc.* **2007**, *2*, 152–160. [CrossRef] [PubMed]
48. Akin, E.J.; Alsaloum, M.; Higerd, G.P.; Liu, S.; Zhao, P.; Dib-Hajj, F.B.; Waxman, S.G.; Dib-Hajj, S.D. Paclitaxel Increases Axonal Localization and Vesicular Trafficking of Nav1.7. *Brain* **2021**, *144*, 1727–1737. [CrossRef]
49. Shin, G.J.; Pero, M.E.; Hammond, L.A.; Burgos, A.; Kumar, A.; Galindo, S.E.; Lucas, T.; Bartolini, F.; Grueber, W.B. Integrins Protect Sensory Neurons in Models of Paclitaxel-Induced Peripheral Sensory Neuropathy. *Proc. Natl. Acad. Sci. USA* **2021**, *118*, e2006050118. [CrossRef]
50. Liebmann, J.; Cook, J.; Lipschultz, C.; Teague, D.; Fisher, J.; Mitchell, J. Cytotoxic Studies of Paclitaxel (Taxol®) in Human Tumour Cell Lines. *Br. J. Cancer* **1993**, *68*, 1104–1109. [CrossRef]
51. Zasadil, L.M.; Andersen, K.A.; Yeum, D.; Rocque, G.B.; Wilke, L.G.; Tevaarwerk, A.J.; Raines, R.T.; Burkard, M.E.; Weaver, B.A. Cytotoxicity of Paclitaxel in Breast Cancer Is Due to Chromosome Missegregation on Multipolar Spindles. *Sci. Transl. Med.* **2014**, *6*, 229ra43. [CrossRef]
52. Tolkovsky, A.M.; Brelstaff, J. Sensory Neurons from Tau Transgenic Mice and Their Utility in Drug Screening. *Methods Mol. Biol.* **2018**, *1727*, 93–105. [CrossRef]
53. Aras, M.A.; Hartnett, K.A.; Aizenman, E. Assessment of Cell Viability in Primary Neuronal Cultures. *Curr. Protoc. Neurosci.* **2008**, *44*, 7.18.1–7.18.15. [CrossRef] [PubMed]
54. Trielli, M.O.; Andreassen, P.R.; Lacroix, F.B.; Margolis, R.L. Differential Taxol-Dependent Arrest of Transformed and Nontransformed Cells in the G1 Phase of the Cell Cycle, and Specific-Related Mortality of Transformed Cells. *J. Cell Biol.* **1996**, *135*, 689–700. [CrossRef]
55. Wang, T.-H.; Wang, H.-S.; Soong, Y.-K. Paclitaxel-Induced Cell Death. *Cancer* **2000**, *88*, 2619–2628. [CrossRef] [PubMed]
56. Weaver, B.A. How Taxol/Paclitaxel Kills Cancer Cells. *Mol. Biol. Cell* **2014**, *25*, 2677–2681. [CrossRef] [PubMed]
57. Hammad, A.; Mohamed M, S.A.; Khalifa, M.; El-Daly, M. Mechanisms of Paclitaxel-Induced Peripheral Neuropathy. *J. Adv. Biomed. Pharm. Sci.* **2023**, *6*, 25–35. [CrossRef]
58. Klein, I.; Lehmann, H. Pathomechanisms of Paclitaxel-Induced Peripheral Neuropathy. *Toxics* **2021**, *9*, 229. [CrossRef]
59. Jordan, M.A.; Toso, R.J.; Thrower, D.; Wilson, L. Mechanism of Mitotic Block and Inhibition of Cell Proliferation by Taxol at Low Concentrations. *Proc. Natl. Acad. Sci. USA* **1993**, *90*, 9552–9556. [CrossRef]
60. Hickman, J.A. Apoptosis Induced by Anticancer Drugs. *Cancer Metastasis Rev.* **1992**, *11*, 121–139. [CrossRef]
61. Pistritto, G.; Trisciuoglio, D.; Ceci, C.; Garufi, A.; D'Orazi, G. Apoptosis as Anticancer Mechanism: Function and Dysfunction of Its Modulators and Targeted Therapeutic Strategies. *Aging* **2016**, *8*, 603–619. [CrossRef]
62. Fischer, U.; Schulze-Osthoff, K. Apoptosis-Based Therapies and Drug Targets. *Cell Death Differ.* **2005**, *12*, 942–961. [CrossRef]
63. Figueroa-Masot, X.A.; Hetman, M.; Higgins, M.J.; Kokot, N.; Xia, Z. Taxol Induces Apoptosis in Cortical Neurons by a Mechanism Independent of Bcl-2 Phosphorylation. *J. Neurosci.* **2001**, *21*, 4657–4667. [CrossRef] [PubMed]
64. Rieger, A.M.; Nelson, K.L.; Konowalchuk, J.D.; Barreda, D.R. Modified Annexin V/Propidium Iodide Apoptosis Assay for Accurate Assessment of Cell Death. *J. Vis. Exp.* **2011**, *50*, e2597. [CrossRef]
65. Yang, I.H.; Siddique, R.; Hosmane, S.; Thakor, N.; Höke, A. Compartmentalized Microfluidic Culture Platform to Study Mechanism of Paclitaxel-Induced Axonal Degeneration. *Exp. Neurol.* **2009**, *218*, 124–128. [CrossRef] [PubMed]
66. Li, J.; Chen, R.; Yao, Q.; Liu, S.; Tian, X.; Hao, C.; Lu, W.; Zhou, T. Time-Dependent Pharmacokinetics of Dexamethasone and Its Efficacy in Human Breast Cancer Xenograft Mice: A Semi-Mechanism-Based Pharmacokinetic/Pharmacodynamic Model. *Acta Pharmacol. Sin.* **2018**, *39*, 472–481. [CrossRef] [PubMed]

67. Madeddu, C.; Deidda, M.; Piras, A.; Cadeddu, C.; Demurtas, L.; Puzzoni, M.; Piscopo, G.; Scartozzi, M.; Mercuro, G. Pathophysiology of Cardiotoxicity Induced by Nonanthracycline Chemotherapy. *J. Cardiovasc. Med.* **2016**, *17*, e12–e18. [CrossRef]
68. Lennernäs, B.; Albertsson, P.; Lennernäs, H.; Norrby, K. Chemotherapy and Antiangiogenesis. *Acta Oncol.* **2003**, *42*, 294–303. [CrossRef]
69. Powis, G.; Ames, M.; Kovach, J. Dose-Dependent Pharmacokinetics and Cancer Chemotherapy. *Cancer Chemother. Pharmacol.* **1981**, *6*, 1–9. [CrossRef]
70. WANG, P.; SONG, J.; SONG, D.; ZHANG, J.; HAO, C. Role of Death Receptor and Mitochondrial Pathways in Conventional Chemotherapy Drug Induction of Apoptosis. *Cell. Signal.* **2006**, *18*, 1528–1535. [CrossRef]
71. Jordan, M.A.; Wilson, L. Microtubules as a Target for Anticancer Drugs. *Nat. Rev. Cancer* **2004**, *4*, 253–265. [CrossRef]
72. Jordan, M.A.; Wendell, K.; Gardiner, S.; Derry, W.B.; Copp, H.; Wilson, L. Mitotic Block Induced in HeLa Cells by Low Concentrations of Paclitaxel (Taxol) Results in Abnormal Mitotic Exit and Apoptotic Cell Death. *Cancer Res.* **1996**, *56*, 816–825.
73. Wang, T.-H.; Popp, D.M.; Wang, H.-S.; Saitoh, M.; Mural, J.G.; Henley, D.C.; Ichijo, H.; Wimalasena, J. Microtubule Dysfunction Induced by Paclitaxel Initiates Apoptosis through Both C-Jun N-Terminal Kinase (JNK)-Dependent and -Independent Pathways in Ovarian Cancer Cells. *J. Biol. Chem.* **1999**, *274*, 8208–8216. [CrossRef]
74. Lee, L.-F.; Li, G.; Templeton, D.J.; Ting, J.P.-Y. Paclitaxel (Taxol)-Induced Gene Expression and Cell Death Are Both Mediated by the Activation of c-Jun NH2-Terminal Kinase (JNK/SAPK). *J. Biol. Chem.* **1998**, *273*, 28253–28260. [CrossRef] [PubMed]
75. Wang, T.-H.; Wang, H.-S.; Ichijo, H.; Giannakakou, P.; Foster, J.S.; Fojo, T.; Wimalasena, J. Microtubule-Interfering Agents Activate c-Jun N-Terminal Kinase/Stress-Activated Protein Kinase through Both Ras and Apoptosis Signal-Regulating Kinase Pathways. *J. Biol. Chem.* **1998**, *273*, 4928–4936. [CrossRef] [PubMed]
76. Scatena, C.D.; Stewart, Z.A.; Mays, D.; Tang, L.J.; Keefer, C.J.; Leach, S.D.; Pietenpol, J.A. Mitotic Phosphorylation of Bcl-2 during Normal Cell Cycle Progression and Taxol-Induced Growth Arrest. *J. Biol. Chem.* **1998**, *273*, 30777–30784. [CrossRef] [PubMed]
77. Leung, J.C.; Cassimeris, L. Reorganization of Paclitaxel-Stabilized Microtubule Arrays at Mitotic Entry: Roles of Depolymerizing Kinesins and Severing Proteins. *Cancer Biol. Ther.* **2019**, *20*, 1337–1347. [CrossRef]
78. Lieu, C.-H.; Chang, Y.-N.; Lai, Y.-K. Dual Cytotoxic Mechanisms of Submicromolar Taxol on Human Leukemia HL-60 Cells. *Biochem. Pharmacol.* **1997**, *53*, 1587–1596. [CrossRef] [PubMed]
79. Chiang, C.-Y.; Dostrovsky, J.O.; Iwata, K.; Sessle, B.J. Role of Glia in Orofacial Pain. *Neurosci.* **2011**, *17*, 303–320. [CrossRef]
80. Hanani, M.; Huang, T.Y.; Cherkas, P.S.; Ledda, M.; Pannese, E. Glial Cell Plasticity in Sensory Ganglia Induced by Nerve Damage. *Neuroscience* **2002**, *114*, 279–283. [CrossRef]
81. Kettenmann, H.; Hanisch, U.-K.; Noda, M.; Verkhratsky, A. Physiology of Microglia. *Physiol. Rev.* **2011**, *91*, 461–553. [CrossRef]
82. Pozzi, E.; Ballarini, E.; Rodriguez-Menendez, V.; Canta, A.; Chiorazzi, A.; Monza, L.; Bossi, M.; Alberti, P.; Malacrida, A.; Meregalli, C.; et al. Paclitaxel, but Not Cisplatin, Affects Satellite Glial Cells in Dorsal Root Ganglia of Rats with Chemotherapy-Induced Peripheral Neurotoxicity. *Toxics* **2023**, *11*, 93. [CrossRef]
83. Zhang, Z.-J.; Jiang, B.-C.; Gao, Y.-J. Chemokines in Neuron–Glial Cell Interaction and Pathogenesis of Neuropathic Pain. *Cell. Mol. Life Sci.* **2017**, *74*, 3275–3291. [CrossRef] [PubMed]
84. Herculano-Houzel, S. The Glia/Neuron Ratio: How It Varies Uniformly across Brain Structures and Species and What That Means for Brain Physiology and Evolution. *Glia* **2014**, *62*, 1377–1391. [CrossRef] [PubMed]
85. Tsuboi, Y.; Takeda, M.; Tanimoto, T.; Ikeda, M.; Matsumoto, S.; Kitagawa, J.; Teramoto, K.; Simizu, K.; Yamazaki, Y.; Shima, A.; et al. Alteration of the Second Branch of the Trigeminal Nerve Activity Following Inferior Alveolar Nerve Transection in Rats. *Pain* **2004**, *111*, 323–334. [CrossRef] [PubMed]
86. Pannese, E. Advances in Anatomy Embryology and Cell Biology. In *The Satellite Cells of the Sensory Ganglia*; Springer: Berlin/Heidelberg, Germany, 1981; Volume 65. [CrossRef]

toxics

MDPI

Article

Exploration of Maternal and Fetal Toxicity Risks for Metronidazole-Related Teratogenicity and Hepatotoxicity through an Assessment in Albino Rats

Mervat A. AbdRabou [1,*], Barakat M. Alrashdi [1], Hadeel K. Alruwaili [1], Reda H. Elmazoudy [2,*], Maha A. Alwaili [3], Sarah I. Othman [3], Fawzyah A. Alghamdi [4] and Gehan H. Fahmy [5]

1 Biology Department, College of Science, Jouf University, P.O. Box 2014, Sakaka 72388, Saudi Arabia
2 Biology Department, College of Science, Imam Abdulrahman Bin Faisal University, P.O. Box 1982, Dammam 31441, Saudi Arabia
3 Biology Department, College of Science, Princess Nourah Bint Abdulrahman University, Riyadh 11564, Saudi Arabia
4 Biology Department, College of Science, University of Jeddah, Jeddah 23218, Saudi Arabia
5 Biology Department, College of Science, Taibah University, Al-Madinah Al-Munawwarah 30001, Saudi Arabia
* Correspondence: mababdraboh@ju.edu.sa (M.A.A.); rhelmazoudy@alexu.edu.eg (R.H.E.); Tel.: +966-537242262 (M.A.A.)

Abstract: Metronidazole is the primary antimicrobial drug for treating acute and chronic vaginal pathogens during pregnancy; however, there has been insufficient research on placental disorders, early pregnancy loss, and preterm birth. Here, the potential activity of metronidazole on pregnancy outcomes was investigated. 130 mg/kg body weight of metronidazole was orally given individually to pregnant rats on gestation days 0–7, 7–14, and 0–20. Pregnancy outcome evaluations were carried out on gestation day 20. It was demonstrated that metronidazole could induce maternal and fetal hepatotoxicity. There is a significant increase in the activities of maternal hepatic enzymes (ALT, AST, and ALP), total cholesterol, and triglycerides compared with the control. These biochemical findings were evidenced by maternal and fetal liver histopathological alterations. Furthermore, metronidazole caused a significant decrease in the number of implantation sites and fetal viability, whereas it caused an increase in fetal lethality and the number of fetal resorptions. In addition, a significant decrease in fetal weight, placental weight, and placental diameter was estimated. Macroscopical examination revealed placental discoloration and hypotrophy in the labyrinth zone and the degeneration of the basal zone. The fetal defects are related to exencephaly, visceral hernias, and tail defects. These findings suggest that the administration of metroniazole during gestation interferes with embryonic implantation and fetal organogenesis and enhances placental pathology. We can also conclude that metronidazole has potential maternal and fetal risks and is unsafe during pregnancy. Additionally, it should be strictly advised and prescribed, and further consideration should be given to the associated health risks.

Keywords: metronidazole; pregnancy; embryos; preterm birth; birth defects; placenta

Citation: AbdRabou, M.A.; Alrashdi, B.M.; Alruwaili, H.K.; Elmazoudy, R.H.; Alwaili, M.A.; Othman, S.I.; Alghamdi, F.A.; Fahmy, G.H. Exploration of Maternal and Fetal Toxicity Risks for Metronidazole-Related Teratogenicity and Hepatotoxicity through an Assessment in Albino Rats. *Toxics* **2023**, *11*, 303. https://doi.org/10.3390/toxics11040303

Academic Editor: Youssef Sari

Received: 25 February 2023
Revised: 17 March 2023
Accepted: 22 March 2023
Published: 25 March 2023

1. Introduction

Females are among the most vulnerable to being infected with vaginitis caused by parasites or pathologies related to childbirth and the care of women giving birth [1]. Intentionally or unintentionally, women rely broadly on medication or pharmaceutical therapy to treat protozoal and/or bacterial infections during pregnancy or when not pregnant [2].

Pregnancy evaluation focuses on the maternal and/or fetal implications of drug use. Furthermore, embryos/fetuses are more susceptible in pregnant females treated with medications [3]. Perinatal exposure to drugs leads to abnormal intrauterine embryo/fetus development manifested by growth delay, organ deterioration, and fetal resorption or death [4]. These drugs can penetrate the placental membrane's maternal-fetal barrier and disrupt normal fetal development [5].

Metronidazole is an antibiotic drug synthesized by actinobacteria and proteobacterial genera and is used to cure Bacteroides infections and certain parasitic illnesses [6]. It has been an effective prescribed medication against human vaginitis infections in gynecology and obstetrics such as *Trichomonas vaginalis*, *Entamoeba histolytica*, and *Giardia lamblia* for many years. The metronidazole-available doses are injectable, intravenous, vaginally, and rectally [7].

Various literature reports that metronidazole has therapeutic effects; however, its safety during pregnancy has not been fully confirmed [8,9]. Concerns have been raised about the potential side effects of treating pregnant women with metronidazole [10]. According to the Food and Drug Administration (FDA), metronidazole is classified as having a category B risk for damaging fetuses, but its use still provokes divided opinion among physicians. [11]. In this classification, harmful action is evident in the first trimester of the gestation period [12]. The majority do not advocate it during the first trimester, while in the second and third trimesters it is justified only in cases where alternative therapy is unsuccessful [13].

Toxicological studies demonstrated that metronidazole is bioavailable and can be distributed in body fluids [14] and extend across the maternal-fetal barrier into the embryo/fetus circulation and amniotic fluid [15]. Through this potential effect, developmental retardation, deformed organs, and fetal death can be observed. Furthermore, metronidazole administration can directly influence fetogenesis independent of maternal toxicity [16].

Metronidazole is considered to have broad toxicological prospects compared to most xenobiotics due to its biotransformation in the liver through oxidation, hydroxylation, and conjugation of metronidazole glucuronide [17]. Moreover, a cumulative number of studies on animals and humans indicated an association of metronidazole with the disturbance of alanine aminotransaminase (ALT), aspartate aminotransaminase (AST), alkaline phosphatase (ALP), total cholesterol (TC), and triglyceride (TG), which are factors involved in hepatotoxicity [18]. Metronidazole also enhances steatosis-related early-stage hepatocarcinogenesis and induces liver tumors through increased hepatic neoplasms [19].

Experimental research showed that metronidazole can induce post-implantation embryo lethality in rats [20]. Others declared a probable relationship between the administration of vaginal metronidazole and hydrocephaly during pregnancy [21]. In addition, Shennan et al. [22] reported that metronidazole therapy might increase the chance of preterm birth and/or abortions during pregnancy.

Previous published meta-analyses answered whether metronidazole exposure during the first trimester of pregnancy is associated with an increased teratogenic risk in humans. The outcome under consideration was the occurrence of birth defects in live-born infants. It was concluded that metronidazole does not appear to be associated with an increased teratogenic risk [23]. On the other hand, other meta-analysis data do not confirm the efficacy of metronidazole in reducing the risk of preterm birth and associated delivery outcomes [24]. Hence, further research is required to confirm the effect of high doses and a short duration of metronidazole treatment on preterm birth among the high-risk group.

Consequently, in addition to maternal toxicity, it is evident that metronidazole has the potential to disrupt the normal intrauterine development of the embryo/fetus. So, the present study and the body of literature were planned to investigate methodically the effect of metronidazole during different trimesters and the whole pregnancy on fetal development in pregnant rats.

2. Materials and Methods

2.1. Metronidazole Drug

Metronidazole (Flagyl®, 500 mg per capsule) was purchased from Sanofi global biopharmaceutical company, a Saudi Arabian branch (Jouf, Saudi Arabian). The dose prepared for the current study was calculated (130 mg/kg) according to [2] and was dissolved in a physiological saline solution.

2.2. Animal Care and Use

All procedures of the present study were carried out in compliance with standards, rules of Institutional Ethics Use and the Care Committee of Laboratory Animals at Jouf University. This study was conducted on sexually mature and experimentally naïve female rats weighing about 160–180 g at the beginning of the experiment. Female rats were kept in cages with free access to conventional drinking water and food at 25 ± 2 °C. The relative humidity levels were maintained at 50 ± 5%.

2.3. Determination of the Estrous Cycle

Every morning, vaginal smears were obtained for vaginal cytology to identify the phases of the oestrous cycle. Proestrus is characterized by a predominance of nucleated epithelial cells and a small proportion of cornified epithelial cells. During the estrus phase, cornified squamous epithelial cells will predominate. Leukocytes and cornified squamous epithelial cells are present during the metestrus phase. In the diestrus phase, leukocytes will be the predominant cell type [25].

2.4. Mating and Pregnancy Snippets

After two weeks of adaptation, females with three consecutive regular oestrous cycles were caged with a proven fertile male (2 females: 1 male). The next morning, to check if sperm were present, vaginal smears were taken. The appearance of spermatozoa in vaginal smears or the existence of a vaginal plug was regarded as the onset of gestation [26].

2.5. Experimental Design

The timed pregnant females were arbitrarily divided into four groups, (10 females/each). The control group (GC): pregnant rats received 0.5 mL of physiological saline solution from day zero through the twentieth day of gestation. The first experimental group (GMI): pregnant rats treated with metronidazole from day zero through the seventh day of gestation. The second experimental group (GMII): pregnant rats received metronidazole from day zero through the fourteenth day of pregnancy. The third experimental group (GMIII): pregnant females received metronidazole from day zero through the twentieth day of gestation. Metronidazole was given orally by gavage at a daily dosage of 130 mg/kg body weight in 0.5 mL of physiological saline solution after confirmation of mating.

2.6. Narrative Toxicological Signs of Dams

The perinatal screening of pregnant female rats was reviewed to determine whether metronidazole treatment induces vaginal bleeding. In addition, a gross analysis of the maternal toxicity of pregnant females from all groups was monitored throughout the pregnancy trimesters. Preterm births, spontaneous abortions, and maternal deaths before the 20th day of gestation (via cesarean section) were also screened during the period of pregnancy.

3. Pregnancy Outcomes Evaluation

3.1. Dissection Procedure of Pregnant Rats

On the 20th day of gestation (9:00 a.m.), all pregnant rats from each group were intramuscularly anesthetized with (80 mg/kg ketamine hydrochloride and 6 mg/kg xylazine intramuscularly) and then euthanized and dissected under a dissecting stereomicroscope. Dissection procedures are performed on a black background and performed as quickly as possible to maintain the viability of the samples. After a transverse incision of the abdomen with scissors, the gravid uterine horns were carefully exposed and excised with forceps by cutting at the cervix and along the mesometrium. At necropsy, the fetuses were pulled out, aseptically detached, and removed from the uterus with the attached and intact placenta and yolk sac. The placenta discs were secluded from the uterine wall. After that, the isolated placenta and fetuses were externally washed with 70% ethanol, air-dried, and individually weighed. The uterine horns were inspected to estimate the number and location of implantation sites, fetal resorptions (early or late), fetal death, and viable fetuses. Fetal viability was evaluated by the presence or absence of fetal size, fetal movement, skin coloration, and fetal developmental stage. The uteri of females that cannot be easily seen or noticed to be pregnant can be inspected by staining the uterine horns with 10% ammonium sulfide [27].

3.2. *Macroscopical Evaluation*

3.2.1. Fetus and Placenta

The crown-rump length of each fetus was measured. Thereafter, the collected fetuses were inspected for the skull, eyes, ears, limbs, and tail to carry out a possible systematic inquiry and identify the existence of the malformations of external morphology. Macroscopically, at necropsy, the placenta was also individually evaluated for abnormal or distinctive attributes or aspects in anatomy and pathology, and an image was acquired using a dissecting stereomicroscope.

3.2.2. Blood and Serum Biochemistry

The blood samples were placed in a 1.5 mL anticoagulation centrifuge tube for one hour and then centrifuged at 4 °C and 10,000× *g* rpm for 10 min to obtain the serum. Alanine aminotransferase (ALT), aspartate aminotransferase (AST), alkaline phosphatase (ALP), total protein (TP), total cholesterol (TC), triglyceride (TG), high-density lipoprotein cholesterol (HDL-C), and low-density lipoprotein cholesterol (LDL-C) assay kits (Mindray, Shenzhen, China) were removed from the refrigerator at 4 °C and tested by a fully automated biochemical analyzer.

3.2.3. Histopathological Analysis

The largest lobe of the weighed liver was soaked in 4% paraformaldehyde, dehydrated with a certain gradient of ethanol, embedded in paraffin, and stained with hematoxylin-eosin (HE) at a thickness of 4 μm. Microscopy examined histopathology to assess the extent of liver injury in all sections [28].

4. Statistical Analysis

Statistical data analysis package for science (Origin 2019b SPSS—version 23) for Windows was applied. To analyze differences among all treatments, the one-way analysis of variance (ANOVA) is followed by Duncan's test for the difference between groups. The data were expressed as the mean and standard deviation (mean ± standard deviations (SDs)) at $p \leq 0.05$.

5. Results

5.1. *Maternal Toxicity*

Observational findings evaluated the safety of metronidazole administration, and the statistical analysis showed no significant results during the three trimesters. Pregnant rats of the three groups (GC, GMI, GMII, and GMIII) did not display any evident symptoms of maternal toxicity or undesirable behavior. No maternal mortality or morbidity was noticed among the pregnant rats during the pregnancy period. Regarding vaginal bleeding, most pregnant rats did not exhibit it, starting around GD 0–7th, 7–14th, or 0–20th days. Furthermore, no clinical signs of hemorrhage were observed in gravid uteri in treated females at necropsy compared with controls. No changes in skin and fur, eyes, and mucous membranes, respiratory and digestive distress, behavior patterns, or coma were observed in the pregnant females. Furthermore, no gross pathological changes were monitored at necropsy in the tissues and organs of surviving rats.

5.2. Preterm Birth and Abortion

Statistically, there were no preterm births among all pregnant female rats treated with metronidazole in any trimester compared to pregnant female rats without metronidazole treatment. Furthermore, no significant abortion was observed among metronidazole-treated pregnant female rats.

5.3. Maternal Body Weight

Excluding the dead rats, the females in the experimental and control groups showed a steady body weight during the days of the experimental period.

5.4. Liver Coefficients

Compared with the control group, the differences in liver coefficients in the GI and GII experimental groups of rats were not statistically significant, while the difference in liver coefficients in the GIII experimental group was significantly decreased ($p \leq 0.05$, Figure 1).

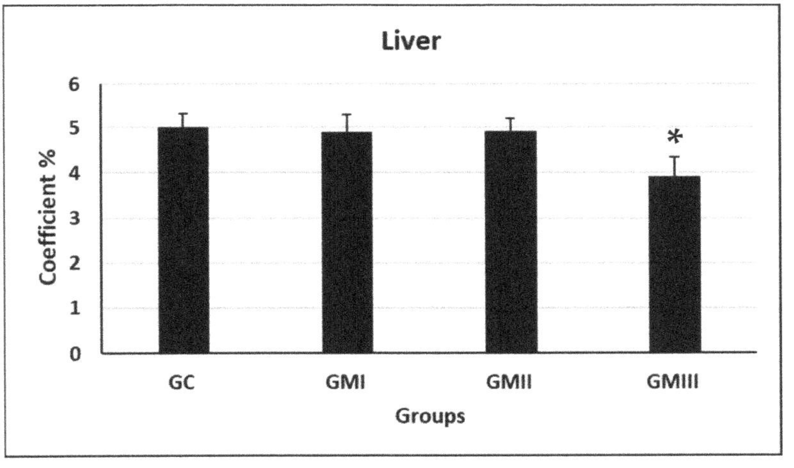

Figure 1. Liver coefficient of pregnant rats (n = 10.) * Significant differences with respect to the control group (GC) $p \leq 0.05$.

6. Indices of Pregnancy Outcomes

6.1. Gross Gravid Uterine Horns

Each gravid uterine horn contains multiple healthy fetuses, each within its own separate yolk sac and attached to the uterus via the umbilical cord and a discoid placenta (Figure 2a,b). The number of implanted fetuses per gravid uterine horn decreased compared with the control group. On the other hand, there was an increase in resorbed fetuses in the uterine horns of females in groups I and II (Arrow, Figure 2c,d) and group III (Arrow, Figure 2e). Furthermore, group III uterine horns suffered from complete fetal resorption and appeared as resorbed implantation sites (Arrowhead, Figure 2f).

6.2. Day 0–7 Experiment Findings

In this group, there was a significant decrease in both the number of implantation sites and the number of viable fetuses compared to the control group ($p \leq 0.05$, Table 1). On the other hand, the number of resorption sites was significantly increased compared with the control group. The fetal body weight and crown-rump length were not significantly affected in this treated group compared to the ad libitum control group ($p \leq 0.05$, Table 1). No significant difference was observed in the number of dead fetuses in metronidazole-treated rats compared to the control group ($p \leq 0.05$, Table 1).

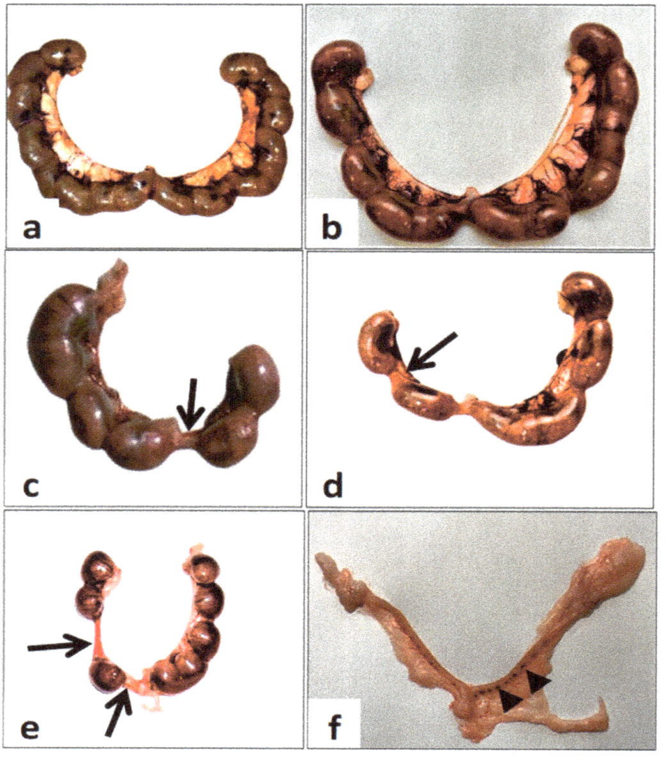

Figure 2. The whole mount of gravid uterine horns from pregnant rats was sacrificed on the 20th day of gestation. (**a**,**b**) The control group with full-term fetuses. (**c**) group I (day 0–7) show missed and resorbed fetuses (arrows). (**d**) Group II (day 7–14) shows resorbed fetuses (arrows) and intrauterine growth retardation of fetuses (arrows). (**e**) Group III (day 0–20) show resorbed fetuses (arrows) and dead and retarded fetuses. (**f**) Postimplantation loss (arrowheads).

Table 1. Pregnancy outcomes of female rats treated with metronidazole at different gestation periods and sacrificed on the 20th day of gestation.

Parameters	Experimental Groups			
	GC	GMI	GMII	GMIII
Number of litters	10	10	10	10
Number of fetuses	132.3	77.80	68.20	60.80
Number of implantation sites/litter	13.50 ± 1.25	8.75 ± 0.95 *	8.69 ± 0.85 *	7.28 ± 0.95 *
Number of resorption sites/litter	0.27 ± 0.015	0.97 ± 0.013 *	1.21 ± 0.002 *	1.42 ± 0.002 *
Number of dead fetuses/litter	0.00	0.00	0.66 ± 0.04 *	0.78 ± 0.04 *
Number of live fetuses/litter	13.23 ± 0.57	7.78 ± 1.30 *	6.82 ± 1.30 *	6.08 ± 1.30 *
Fetal body weight (g)	5.02 ± 0.130	4.89 ± 0.300	3.52 ± 0.167 *	2.99 ± 0.090 *
Crown-Rump Length (mm)	40.01 ± 1.42	39.20 ± 2.21	26.32 ± 1.12 *	21.50 ± 1.23 *

Values are expressed as mean ± standard deviation (mean ±SD) n= 10/group. * The values are significantly different at $p \leq 0.05$ (ANOVA) with Duncan's multiple range test.

6.3. Day 7–14 Experiment Findings

The statistical analysis of pregnancy outcomes showed a significant decrease in both the number of implantation sites and the number of viable fetuses compared to the control group ($p \leq 0.05$, Table 1). While the number of resorptions and dead fetuses significantly increased compared to the control group ($p \leq 0.05$, Table 1). Concerning the fetal growth parameters, metronidazole produced a significant reduction in fetal body weight, and crown-rump length ($p \leq 0.05$, Table 1).

6.4. Day 0–20 Experiment Findings

The results of the whole pregnancy period (day 0–20) treatment with metronidazole showed a significant decrease in the mean number of implantation sites and the number of live fetuses ($p \leq 0.05$, Table 1) compared to the control group. A significant decrease in fetal body weight and fetal crown-rump length was observed compared to the control group ($p \leq 0.05$, Table 1). The total number of fetuses per group was significantly decreased in all treated groups during the pregnancy trimesters compared to the corresponding control group ($p \leq 0.05$, Table 1).

6.5. Placenta Weight and Diameter

Compared with the control group, the differences in placenta weight and diameter in the GII and GIII experimental groups of rats were significantly decreased compared with the control group ($p \leq 0.05$, Figure 3). While no significant difference was observed in the GI compared with the control group ($p \leq 0.05$, Figure 3).

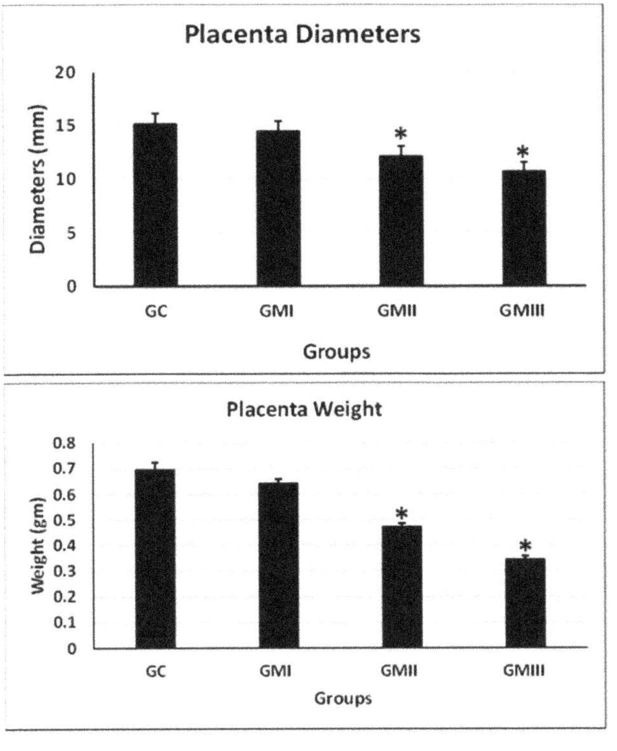

Figure 3. Evaluation of placental parameters (weights and diameters) in the treated pregnant rats with metronidazole. Values are expressed as mean ± standard deviation (M ± SD) $n = 10$/group. * The values are significantly different at $p \leq 0.05$ (ANOVA) with Duncan's multiple range test.

6.6. Placental Morphology and Anatomy

Normally, in the frontal view, the fetal surface of the placenta facing the fetus wherein enters the umbilicus appears as a dark red zone due to the high vascularization of the labyrinth (L, Figure 4a), called the chorionic plate (H, Figure 4a). The adjacent zone appears yellow, representing the junctional zone. The fetal surface placenta comprises the fetus and is composed of three compartments, encompassing the yolk sac, chorionic plate, labyrinth, and junctional zone. While the maternal surface includes the decidua and is called the basal plate (Arrow, Figure 4a). As shown in Figure 4a and compared with the normal rat placentas, the GI placenta showed normal architecture with three compartments: the chorionic zone (H, Figure 4b), the labyrinth zone (L, Figure 4b), and the maternal decidua basalis (Arrow, Figure 4b). However, the labyrinth zone shows discoloration and adherence of the umbilical cord to the chorionic surface (L, Figure 4b). The GII placentas exhibited hypotrophy with disrupted structures such as the labyrinth (L, Figure 4c) and decreased thickness of the decidual basal layer that lacked differentiation (Arrow, Figure 4c). On the other hand, the placentas in the GIII appeared more hypotrophic with decreased diameter and thickness compared with control placentas (Figure 4d). In addition, the degeneration of the decidual basal layer (Arrow, Figure 4d).

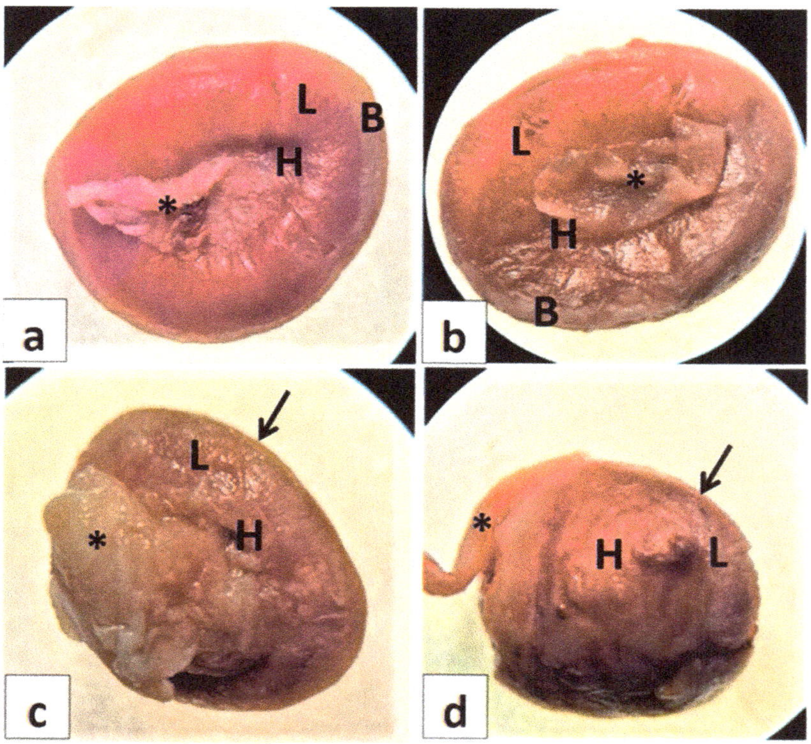

Figure 4. Photomacrographs of the frontal view of placental morphologies in control and pregnant rats treated with metronidazole at different gestation periods. (**a**) Normal control, red-colored labyrinth (L), thick basal decidua (B). (**b**) Group I indicate discoloration of the labyrinth (L) and adherence of the umbilical cord (*) to the chorionic surface (H). (**c**) Group II indicates placental hypotrophy, and thin basal decidua (arrow). (**d**) Group III indicates reduced placenta and placental hypotrophy, necrosis, and the disappearance of basal decidua (arrow).

6.7. Gross Morphology of Fetal Abnormalities

6.7.1. Fetal Growth

Figure 5 demonstrates the gross morphological abnormalities of fetuses. These results showed that metronidazole resulted in intrauterine growth retardation or restriction at 20 days of gestation, indicating that metronidazole can be implicated in fetal development. The hematoma was recorded to occur more frequently among the treated groups compared with the control groups.

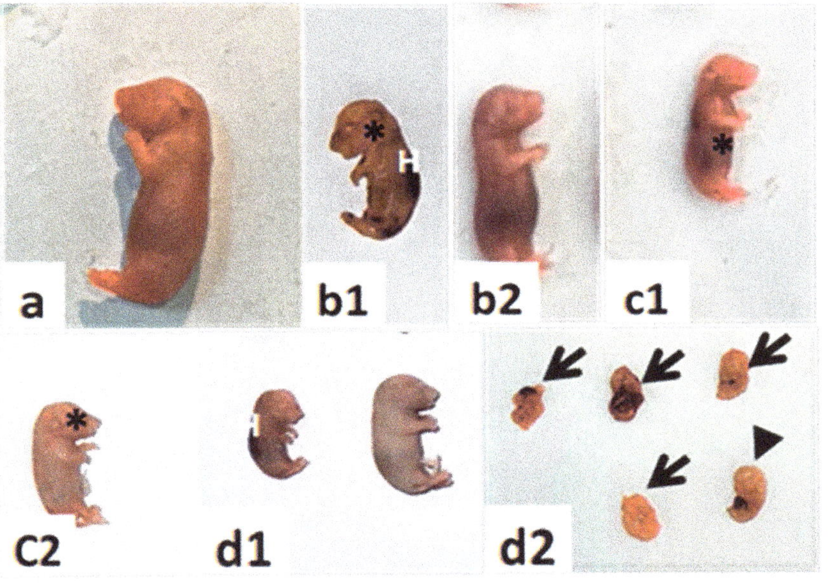

Figure 5. Photomacrographs of the whole mount of twentieth-day-old rat fetuses of pregnant rats treated with metronidazole at different gestation periods showing various morphological intrauterine growth restrictions (IUGR). (**a**) In the control group, the fetuses appeared healthy and normal. (**b1**, **b2**) Group I, (**c1**, **c2**) Group II, and (**d1**, **d2**) Group III. Notice growth retardation (*), dead fetus (arrowhead), partially resorbed fetuses (arrows), and hematomas (H).

6.7.2. Major Congenital Anomalies

Congenital anomalies were observed more frequently in GMIII compared with the GMI, GMII, and GC control groups (Figure 6). Such that, treatment with metronidazole during the three trimesters (day 0–20) produced morphological anomalies in 20-day-old fetuses compared with control fetuses (Figure 6a). The major congenital malformations were exencephaly anomalies, visceral hernias, and tail defects (Figure 6b–d).

6.7.3. Blood Chemistry

Compared with the control group, the GIII experimental group caused significant alterations in liver functions ($p \leq 0.05$, Figure 7). The GIII group caused a significant increase in levels of ALT, AST, ALP, and total protein compared with the control group ($p \leq 0.05$, Figure 7). Furthermore, compared with the control group, triglycerides (TG, lipid index), total cholesterol, and HDL-C only increased in the GIII group ($p \leq 0.05$, Table 2). There is no significant alteration in the activity of LDL-C compared with the control group ($p \leq 0.05$, Table 2).

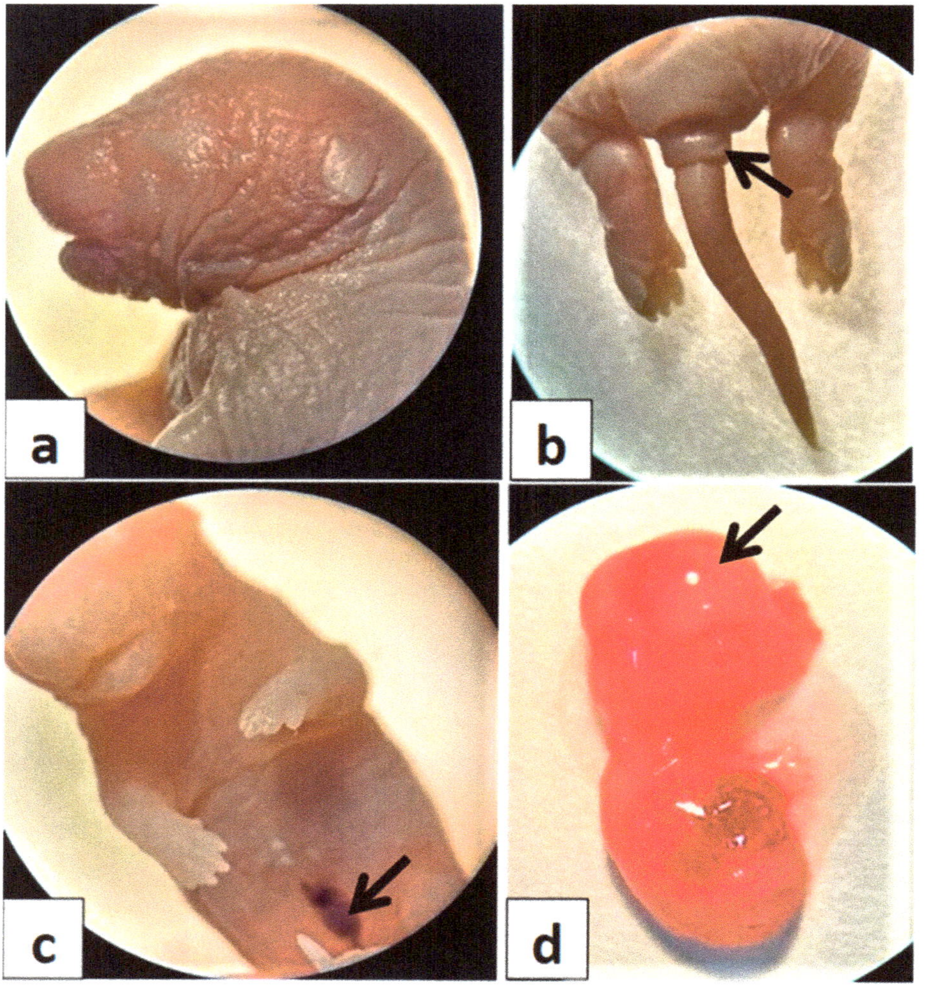

Figure 6. The major external and visceral morphological anomalies of pregnant rats treated with metronidazole during the 0–20th days of the gestational period. (**a**) control group, (**b**) show tail defects (arrow), (**c**) show visceral hernia (arrow), and (**d**) indicate exencephaly (arrow).

Table 2. Effect of metronidazole on lipid profile in pregnant albino rats.

Lipid Index	Groups			
	GC	GMI	GMII	GMIII
TC (mg/dL)	81.51 ± 5.11	82.61 ± 6.12	85.43 ± 4.22	131.42 ± 7.12 *
TG (mg/dL)	89.14 ± 6.31	88.33 ± 5.67	87.86 ± 5.17	166.8 ± 10.31 *
HDL (mg/dL)	37.33 ± 3.22	40.21 ± 4.01	42.62 ± 3.15	35.04 ± 2.09
LDL (mg/dL)	16.24 ± 3.46	17.88 ± 1.96	18.21 ± 1.77	31.22 ± 2.41 *

Values are expressed as the mean ± standard deviation (mean ± SD). (n = 10). * Significant difference with respect to the control group at $p \leq 0.05$ (ANOVA) with Duncan's multiple range test.

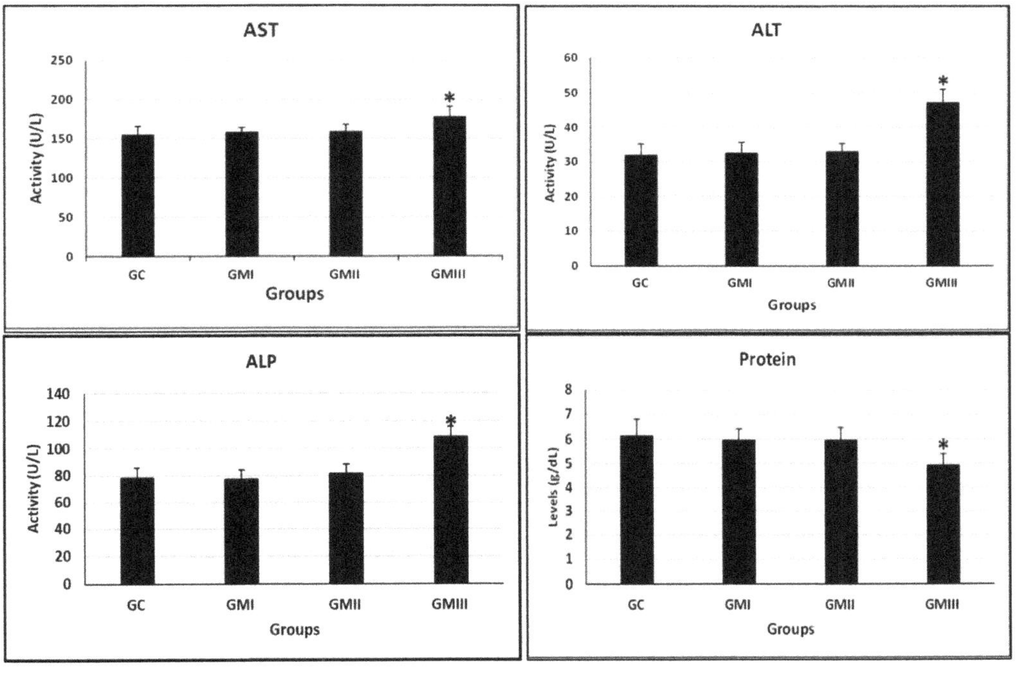

Figure 7. Effect of metronidazole on the serum liver function index of pregnant albino rats. Values are expressed as mean ± standard deviation (mean ± SD) $n = 10$/group. * The values are significantly different at $p \leq 0.05$ (ANOVA) with Duncan's multiple range test.

6.8. Hepatic Histopathology

6.8.1. Maternal Hepatotoxicity

Histological examination illustrates that the maternal liver displays a normal architecture of hepatocytes with a distinct mitotic index in the liver (Figure 8a). The histological analysis of the maternal liver showed different degrees of histopathological alterations in all experimental groups compared with the control group (Figure 8). In contrast to the control group, increased severity of ground parenchyma was observed in the GII and GIII groups (Figure 8e–h). The liver in the GII group showed mild inflammatory cell infiltrates and blood vessel congestion (Figure 8e,f). The liver tissue of rats in the GIII group showed the disappearance of the hepatic cord, swollen hepatocytes, broken cells, and pyknosis of hepatocytic nuclei (Figure 8g,h). Compared with the control group, the GI group had less liver damage (Figure 8c,d).

6.8.2. Fetal Hepatotoxicity

Histological examination illustrates that the normal fetal liver is mostly comprised of hepatic cords and sinusoids at this stage of development. The hepatic cords are composed of primarily and largely undifferentiated hepatoblasts (Figure 9a,b). RBCs are found within the vessels. Most of the hematopoietic cell population is of the erythroid lineage and can be identified by the intense, hyperchromatic nuclei (Figure 9a,b). The architecture of the liver in the GI tract is nearly identical to that seen in the normal fetal liver (Figure 9c,d). Whereas GII showed disruptions in the hepatic organization of hepatocytes, defective hepatocyte maturation, and abnormal hepatic cord arrangements (Figure 9e,f). In the liver tissue sections of GIII, hepatocytes are small, round, and loose and associated with disruptions in hepatic architecture and cell morphology observable on stained liver tissue sections (Figure 9g,h). The liver parenchyma appears looser and less organized. In addition, megakaryocytes are present (Figure 9g,h).

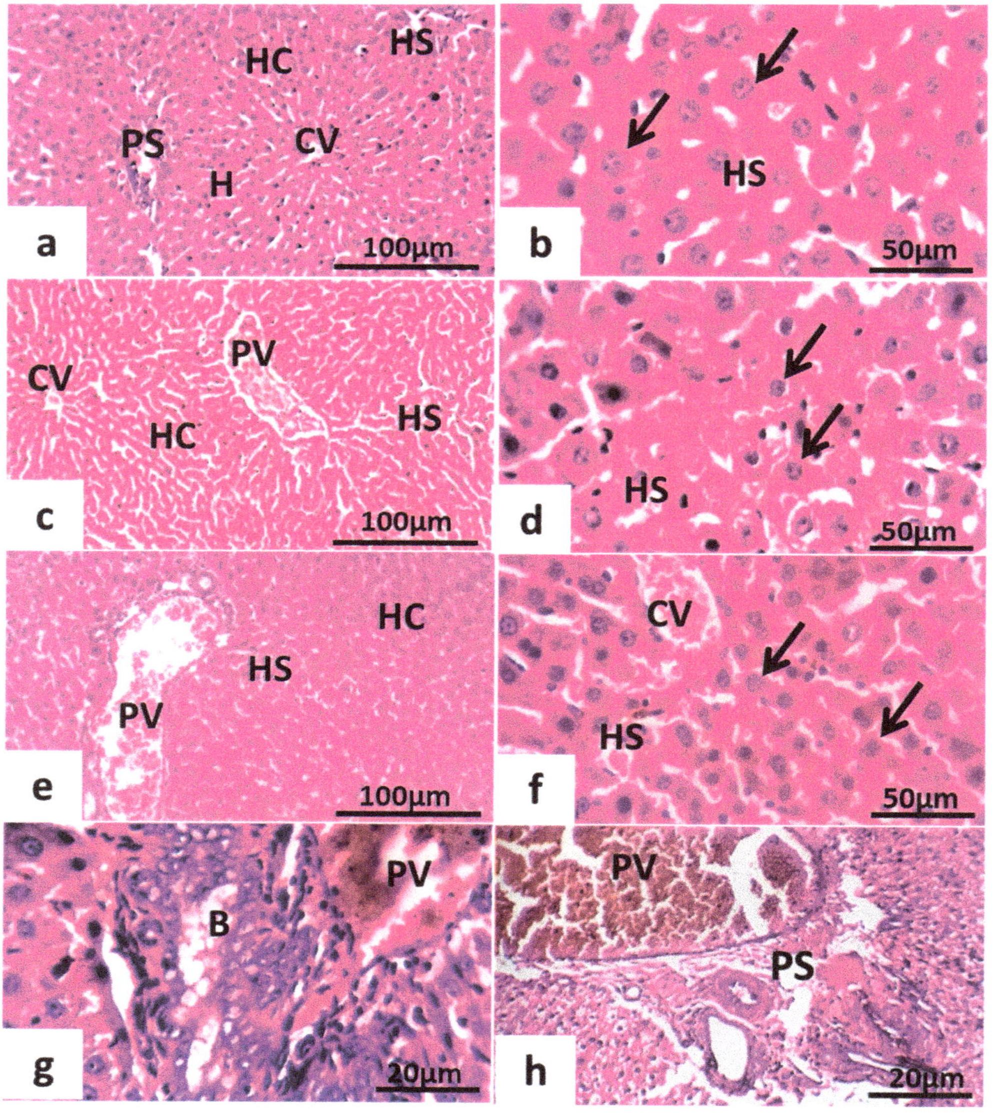

Figure 8. Photomicrograph of maternal rats' liver. (**a**,**b**) Normal control group. (**c**,**d**) treated females at (0–7th dpc). (**e**,**f**) treated female at (7–14th dpc). (**g**,**h**) treated female at (0–20th dpc). Notice the portal space (PS), portal vein (PV), central vein (CV), hepatocytes (HC, arrows) of hepatic strands (H), and hepatic sinusoids (HS). H&E.

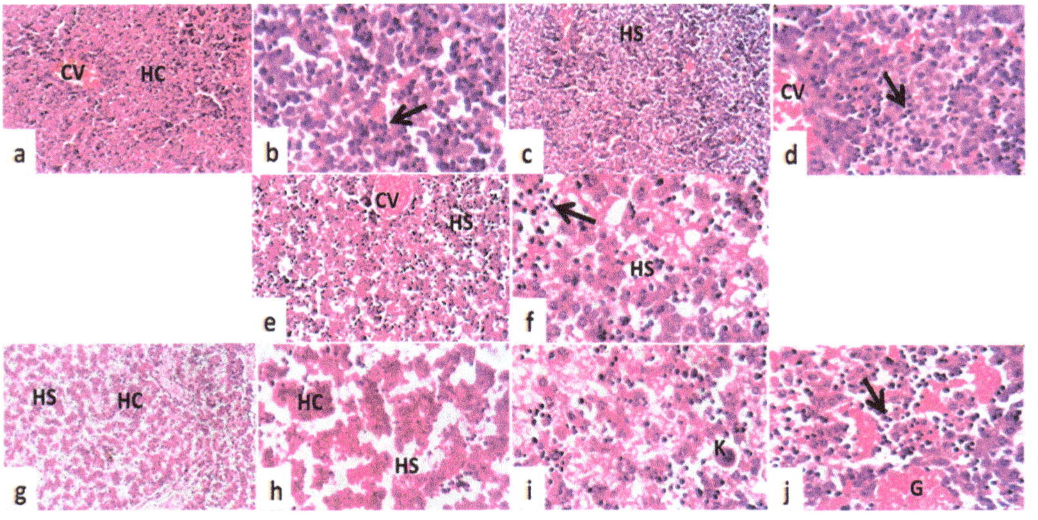

Figure 9. Photomicrographs of fetal liver tissue sections of metronidazole-maternally treated mice. (**a**,**b**) Normal control group, (**c**,**d**) treated female at (0–7th dpc), (**e**,**f**) treated female at (7–14th dpc), and (**g**–**j**) treated female at (0–20th dpc). Notice the central vein (CV), hepatic cords (HC), hepatic sinusoids (HS), circulate nucleated erythrocytes, mature hemopoietic cells (arrows), megakaryocytes (K), and hemorrhage (G). H&E. ((**a**,**c**,**e**,g) = Scale 100 μm) ((**b**,**d**,**f**,**h**–**j**) = Scale 50 μm).

7. Discussion

Metronidazole is the primary antimicrobial drug for treating acute and chronic vaginal pathogens during the gestation period; however, the limited literature on placental disorders and pregnancy outcomes has not been as sufficient as required. Furthermore, the placenta is crucial for fetal development and pregnancy success. So, in this study, we have attempted to confirm the in-utero effects of metronidazole given to pregnant rats on the observable disorders of the placenta along with pregnancy outcomes.

The findings showed that metronidazole induced discernible lesions in the placenta and had detrimental effects on the consequences of conception disturbances in the normal gestational consistency. The reduction in the number of live fetuses observed in the present results was consistent with several reports. Consequently, there is a relationship between metronidazole administration during pregnancy and low fetal weight, the number of implantation sites, fetal viability, and congenital anomalies among the outcomes of pregnant female rats.

The crown-rump length, embryonic resorption or death, number of implantation sites, and embryonic morphology are indicators of great significance in reproductive toxicology during the embryonic development [29]. In pregnant rats of GMI, GMII, and GMIII, a decrease in the number of live fetuses and a reduction in the implantation sites may indicate that metronidazole is implicated in the pre-implantation and/or post-implantation processes [2]. In addition, a decrease in fetal viability may be attributed to the expelling effect of metronidazole on the blastocyst after fertilization or its potential cytotoxicity on oocyte liberation [30]. The evidence indicates that metronidazole can interfere with morphogenic pathways, inducing malformations and developmental toxicity such as adduction and transversions of GC-CG DNA [31]. It is also worth noting that metronidazole disrupts apoptosis and the proliferation of cell migration and maturation, causing embryonic defects [21]. Moreover, several studies point out that perinatal exposure to metronidazole increases intrauterine fetal growth restriction and malformation occurrence, referring to its mutagenic and teratogenic potency [32]. In this context, fetal resorption, death, and teratogenicity malformations could be strictly explained by the cytotoxicity and/or genotoxicity of metronidazole, which is in agreement with its suggestion of its ability to induce genotoxic effects on embryonic cells [33].

According to Talapatra et al. [34], metronidazole induces micronucleus and binucleus formation and increases the number of chromosome aberrations due to its genotoxic, cytogenetic, and carcinogenic damage. Likewise, Roy et al. [35] found that metronidazole's genotoxicity may be due to the sensitization of bone marrow cells. In addition, Menendez et al. [36] indicated that metronidazole hydroxy metabolite, in rat hepatocytes, produced an increase in micronuclei and DNA breaks. As such, several available reports indicated that apoptosis or necrosis reflects DNA damage such as variation of bases, single-strand breaks, and crossing between DNA-protein or DNA-DNA, finally resulting in early embryonic defects [37]. It has been indicated that metronidazole also causes an increase in isochromatic and chromatid breaks [30]. Hence, we could establish that such degenerative damages and deleterious effects of metronidazole are implicated to a great extent in morphological defects and teratogenicity.

Indeed, intracellular metabolic conversion plays an important role in the cytotoxic activity of metronidazole. The reduced metronidazole binds to the DNA, enhancing the destabilization of helix strands and consequent DNA breakage [38]. Furthermore, the toxicity of metronidazole may result from its derivative, the thiamin analogue [39], or from free radical-mediated damage generated during metronidazole metabolism, which causes cell death [40]. The precise mechanism of action of metronidazole is unclear; however, the reduced form of metronidazole and free radicals can interact with DNA, leading to inhibition of DNA synthesis and DNA degradation, leading to cellular death [41].

Whatever the disruptor, an in utero placental lesion may cause prenatal growth retardation, early pregnancy loss, and increase the risk for fetal disorders through the placenta–organ axis [42]. In the present results, metronidazole produced placental hypotrophy associated with a decrease in placental weight and a reduction in the placental basal zone compared with the control. The collective findings from the present study suggest that the intrauterine fetal growth disorders and pronounced teratogenicity may be due to the disruption of normal placental morphology due to the toxic effects of metronidazole after metabolic reduction. In addition, defects in placental architecture may include the histopathology of placental zones. These also include placental blood disorders, ensuring that metronidazole implicates angiogenesis during early placental development [43].

Indeed, intrauterine embryonic lethality or viability emerges from labyrinth defects as a prominent interface of placental disorders. Many studies supported the idea that there is a direct relationship between the labyrinth and fetal development [44]. Micropathologically, the decrease in placental weight observed in the present finding refers to the apoptosis, necrosis, and degeneration of trophoblasts due to placental damage induced by metronidazole [45]. Furthermore, the placenta discoloration, adhesion of the yolk sac on the chorionic surface of the placenta, and reduction in the labyrinth zone noticed macroscopically in the present findings might be due to placental necrosis in the trophoblasts of the labyrinth zone [46]. In the present results, there is an increase in intrauterine growth restriction (IUGR), indicating placental apoptosis that may be due to the mutagenic activity of metronidazole [47]. DNA damage, arrest of the cell cycle in trophoblasts, and diminished spongiotrophoblast proliferation may interpret the reduction in diameter in the labyrinth zone and basal zone, which is consistent with metronidazole cytotoxicity [48]. Furthermore, we can explain that a reduction in placenta size and placenta weight is attributed to the growth suppression of the labyrinth zone, the retardation of the development of the basal zone, and the cystic deterioration of glycogen cells induced by metronidazole [49].

The liver coefficients of pregnant rats in the GIII experimental group were significantly reduced. It indicates that the liver is one of the target organs for metronidazole, so the practical clinical significance needs further investigation in combination with blood biochemical indices and pathological sections.

After gavage, metronidazole is digested and absorbed by the gastrointestinal tract, metabolized in the liver, and the metabolites (hydroxy metronidazole) are excreted through the kidneys. The serum biochemical findings can further detect liver damage [50]. Elevations in the levels of ALP, ALT, AST, and total protein are a sign of liver damage. The levels of total protein can reflect protein synthesis ability and immunity [51]. The liver is an important organ for metabolism and is the main site of fat and protein metabolism [52]. TC, TG, LDL-C, and HDL-C indicators are associated with dyslipidemia [53]. Therefore, biochemical parameters are determined in the present study using data obtained from the liver. Based on the present findings, the GIII group increased the levels of ALT, AST, and ALP, and the changes were more evident with an increase in the experimental period (0–20th dpc). In general, increases in serum concentration levels of ALT, AST, ALP, and TP are biological markers of hepatic damage [54]. The effect on the liver, a pivotal organ of metabolic homeostasis, is reflected in the levels of AST [55]. Total protein is one of the important indicators of biochemical detection that plays an immune role during the administration of toxic substances (xenobiotics). Therefore, the simultaneous elevation in TP levels often indicates the presence of toxicity in the body [56]. This suggests that metronidazole may cause hepatocellular damage and abnormal liver metabolic function. The GIII caused elevations of hepatic function indicators, and their elevation often indicates excessive protein intake or abnormal hepatic metabolism [57]. Studies have reported that liver dysfunction often leads to disturbances in lipid metabolism, resulting in increased serum TG [58]. The same results were observed in the GIII; thus, it was assumed that metronidazole may have an increased risk of liver dysfunction and abnormal lipid metabolism.

8. Conclusions

The developmental defects observed in present findings disclose the potency of metronidazole administration on pregnancy outcomes and have pathological effects on the placental development of pregnant female rats. Furthermore, the toxic effects of metronidazole are evidenced by a significant intrauterine fetal growth restriction and teratogenicity. In addition, metronidazole causes significant impacts on the maternal liver in pregnant rats at GIII and affects their lipid metabolism. The toxicity was also extended to the fetal liver, mainly by the maternal-fetal-placental vectors. Compared with the control group, all experimental groups showed varying degrees of histopathological alterations, including hepatocyte damage and increased inflammatory cells. Hepatic histopathology showed that the GIII group had the most severe liver tissue damage compared with the GI and GII groups. These findings were consistent with the biochemical index findings. So, from the presently established findings, we can conclude that metronidazole administration is unsafe during gestation for dams and fetuses and should be strictly advised and prescribed for its use and prescription. Additionally, further consideration should be given to the associated health risks.

Author Contributions: M.A.A. (Mervat A. AbdRabou) conceptualized and conducted the design of the study. B.M.A. participated in the design of the study. H.K.A. performed methodology, formal analysis, and data curation. R.H.E. was responsible for writing- original draft, reviewing it, and editing it. G.H.F. was responsible for methodology and resources. M.A.A. (Maha A. Alwaili) and S.I.O. performed microscopic evaluations. F.A.A. participated in the design and helped to draft the manuscript. All authors have read and agreed to the published version of the manuscript.

Funding: This research received no external funding.

Institutional Review Board Statement: Not applicable.

Informed Consent Statement: Not applicable.

Data Availability Statement: Data is contained within the article.

Conflicts of Interest: The authors declare no conflict of interest.

References

1. Telfer, S.M.; Lambin, X.N.; Birtles, R.N.; Beldomenico, P.; Burthe, S.; Paterson, S.F.; Begon, M. Species interactions in a parasite community drive infection risk in a wildlife population. *Sci. J.* **2010**, *330*, 243–246. [CrossRef]
2. da Silva, W.E.; de Melo, I.M.F.; de Albuquerque, Y.M.L.; Mariano, A.F.; Wanderley-Teixeira, V.; Teixeira, Á.A. Effect of metronidazole on placental and fetal development in albino rats. *Anim. Reprod.* **2019**, *16*, 810–818. [CrossRef] [PubMed]
3. Chisolm, M.S.; Payne, J.L. Management of psychotropic drugs during pregnancy. *BMJ* **2016**, 352. [CrossRef] [PubMed]
4. Abebe, M.; Asres, K.; Bekuretsion, Y.; Woldkidan, S.; Debebe, E.; Seyoum, G. Teratogenic Effect of High Dose of *Syzygium guineense* (Myrtaceae) Leaves on Wistar Albino Rat Embryos and Fetuses. *Evid. Based Complement. Altern. Med.* **2021**, *2021*, 6677395. [CrossRef]
5. Gupta, R.K.; Gupta, R.C. Placental Toxicity. In *Reproductive and Developmental Toxicology*; Elsevier: Amsterdam, The Netherlands, 2022; pp. 1373–1397.
6. Kupc, M.; Paunkov, A.; Strasser, D.; Soki, J.; Leitsch, D. Initial expression levels of nim A are decisive for protection against metronidazole in Bacteroides fragilis. *Anaerobe* **2022**, *77*, 102630. [CrossRef] [PubMed]
7. Adil, M.; Iqbal, W.; Adnan, F.; Wazir, S.; Khan, I.; Khayam, M.U.; Khan, I.N. Association of metronidazole with cancer: A potential risk factor or inconsistent deductions? *Curr. Drug Metab.* **2018**, *19*, 902–909. [CrossRef]
8. Muzny, C.A.; Van Gerwen, O.T.; Kissinger, P. Updates in trichomonas treatment including persistent infection and 5-nitroimidazole hypersensitivity. *Curr. Opin. Infect. Dis.* **2020**, *33*, 73–77. [CrossRef]
9. Bagga, R.; Arora, P. Genital Micro-Organisms in Pregnancy. *Front. Public Health* **2020**, *16*, 225. [CrossRef]
10. Bookstaver, P.B.; Bland, C.M.; Griffin, B.; Stover, K.R.; Eiland, L.S.; McLaughlin, M.A. Review of antibiotic use in pregnancy. *Pharmacotherapy: J. Hum. Pharmacol. Drug Ther.* **2015**, *35*, 1052–1062. [CrossRef]
11. van Schalkwyk, J.; Yudin, M.H. Vulvovaginitis: Screening for and management of trichomoniasis, vulvovaginal candidiasis, and bacterial vaginosis. *J. Obstet. Gynaecol. Can.* **2015**, *37*, 266–274. [CrossRef]
12. Waltmann, A.; McKinnish, T.R.; Duncan, J.A. Nonviral sexually transmitted infections in pregnancy: Current controversies and new challenges. *Curr. Opin. Infect. Dis.* **2021**, *34*, 40–49. [CrossRef] [PubMed]
13. McCarter-Spaulding, D.E. Medications in pregnancy and lactation. *MCN Am. J. Matern. Child Nurs.* **2005**, *30*, 10–17. [CrossRef]
14. Erdemli, M.; Turkoz, Y.; Altinoz, E.; Elibol, E.; Dogan, Z. Investigation of the effects of acrylamide applied during pregnancy on fetal brain development in rats and protective role of the vitamin E. *Hum. Exp. Toxicol.* **2016**, *35*, 1337–1344. [CrossRef]
15. Erdemli, M.E.; Aladag, M.A.; Altinoz, E.; Demirtas, S.; Turkoz, Y.; Yigitcan, B.; Bag, H.G. Acrylamide applied during pregnancy causes the neurotoxic effect by lowering BDNF levels in the fetal brain. *Neurotoxicol. Teratol.* **2018**, *67*, 37–43. [CrossRef]
16. Muanda, F.T.; Sheehy, O.; Bérard, A. Use of antibiotics during pregnancy and risk of spontaneous abortion. *CMAJ* **2017**, *189*, E625–E633. [CrossRef]
17. Zemanova, N.; Lněničkova, K.; Vavrečkova, M.; Anzenbacherova, E.; Anzenbacher, P.; Zapletalova, I.; Hermanova, P.; Hudcovic, T.; Kozakova, H.; Jourova, L. Gut microbiome affects the metabolism of metronidazole in mice through regulation of hepatic cytochromes P450 expression. *PLoS ONE* **2021**, *16*, e0259643. [CrossRef] [PubMed]
18. Chong, C.Y.L.; Orr, D.; Plank, L.D.; Vatanen, T.; O'Sullivan, J.M.; Murphy, R. Randomised Double-Blind Placebo-Controlled Trial of Inulin with Metronidazole in Non-Alcoholic Fatty Liver Disease (NAFLD). *Nutrients* **2020**, *12*, 937. [CrossRef] [PubMed]
19. Eguchi, A.; Mizukami, S.; Nakamura, M.; Masuda, S.; Murayama, H.; Kawashima, M.; Inohana, M.; Nagahara, R.; Kobayashi, M.; Yamashita, R.; et al. Metronidazole enhances steatosis-related early-stage hepatocarcinogenesis in high fat diet-fed rats through DNA double-strand breaks and modulation of autophagy. *Environ. Sci. Pollut. Res. Int.* **2022**, *29*, 779–789. [CrossRef] [PubMed]
20. Mudry, M.D.; Martínez-Flores, I.; Palermo, A.M.; Carballo, M.A.; Egozcue, J.; García Caldés, M. Embryolethality induced by metronidazole (MTZ) in Rattus norvegicus. *Teratog. Carcinog. Mutagen.* **2001**, *21*, 197–205. [CrossRef]
21. Kazy, Z.; Puhó, E.; Czeizel, A.E. Teratogenic potential of vaginal metronidazole treatment during pregnancy. *Eur. J. Obstet. Gynecol. Reprod. Biol.* **2005**, *123*, 174–178. [CrossRef]
22. Shennan, A.; Crawshaw, S.; Briley, A.; Hawken, J.; Seed, P.; Jones, G.; Poston, L.A. Randomised controlled Trial of metronidazole for the prevention of preterm birth in women positive for cervicovaginal fetal fibronectin: The PREMET Study. *BJOG* **2006**, *113*, 65–74. [CrossRef] [PubMed]
23. Burtin, P.; Taddio, A.; Ariburnu, O.; Einarson, T.R.; Koren, G. Safety of metronidazole in pregnancy: A meta-analysis. *Am. J. Obstet. Gynecol.* **1995**, *172 Pt 1*, 525–529. [CrossRef]
24. Ajiji, P.; Uzunali, A.; Ripoche, E.; Vittaz, E.; Vial, T.; Maison, P. Investigating the efficacy and safety of metronidazole during pregnancy; A systematic review and meta-analysis. *Eur. J. Obstet. Gynecol. Reprod. Biol. X* **2021**, *11*, 100128. [CrossRef]
25. Marcondes, F.K.; Bianchi, F.J.; Tanno, A.P. Determination of the oestrous cycle phases of rats: Some helpful considerations Braz. *J. Biol.* **2002**, *62*, 609–614.
26. Cora, M.C.; Kooistra, L.; Travlos, G. Vaginal cytology of the laboratory rat and mouse: Review and criteria for the staging of the oestrous cycle using stained vaginal smears. *Toxicol. Pathol.* **2005**, *43*, 776–793. [CrossRef]
27. Salewski, E. Farbemethode aum makroskopischen machweis von implantations-stellen arn uterus der ratte. *Arch. Puthol. Exp. Pharmakol.* **1964**, *247*, 367. [CrossRef]
28. ElMazoudy, R.H.; Attia, A.A. Ginger causes subfertility and abortifacient in mice by targeting both oestrous cycle and blastocyst implantation without teratogenesis. *Phytomedicine* **2018**, *50*, 300–308. [CrossRef]

29. Augustine-Rauch, K.; Zhang, C.X.; Panzica-Kelly, J.M. A developmental toxicology assay platform for screening teratogenic liability of pharmaceutical compounds. *Birth Defects Res. Part B Dev. Reprod. Toxicol.* **2016**, *107*, 4–20. [CrossRef] [PubMed]
30. Koss, C.A.; Baras, D.C.; Lane, S.D.; Aubry, R.; Marcus, M.; Markowitz, L.E.; Koumans, E.H. Investigation of Metronidazole Use during Pregnancy and Adverse Birth Outcomes. *Antimicrob. Agents Chemother.* **2012**, *56*, 4800–4805. [CrossRef]
31. Ceruelos, A.H.; Romero-Quezada, L.C.; Ledezma, R.J.C.; Contreras, L.L. Therapeutic uses of metronidazole and its side effects: An update. *Eur. Rev. Med. Pharmacol. Sci.* **2019**, *23*, 397–401.
32. Zyro, D.; Radko, L.; Sliwinska, A.; Checinska, L.; Kusz, J.; Korona-Głowniak, I.; Przekora, A.; Wójcik, M.; Posyniak, A.; Ochocki, J. Multifunctional Silver(I) Complexes with Metronidazole Drug Reveal Antimicrobial Properties and Antitumor Activity against Human Hepatoma and Colorectal Adenocarcinoma Cells. *Cancers* **2022**, *14*, 900. [CrossRef] [PubMed]
33. Buschini, A.; Ferrarini, L.; Franzoni, S.; Galati, S.; Lazzaretti, M.; Mussi, F.; de Albuquerque, C.N.; Zucchi, T.A.; Poli, P. Genotoxicity Revaluation of Three Commercial Nitroheterocyclic Drugs: Nifurtimox, Benznidazole, and Metronidazole. *J. Parasitol. Res.* **2009**, *2009*, 463575. [CrossRef] [PubMed]
34. Talapatra, S.N.; Dasgupta, S.; Guha, G.; Auddy, M.; Muka Hopadhyay, A. Therapeutic efficacies of Coriandrum sativum aqueous extract against metronidazole induced genotoxicity in Channa punctatus peripheral erythrocytes. *Food Chem. Toxicol.* **2010**, *48*, 3458–3461. [CrossRef] [PubMed]
35. Roy, L.D.; Giri, S.; Singh, S.; Giri, A. Effects of radiation and vitamin C treatment on metronidazole genotoxicity in mice. *Mutat. Res. Genet. Toxicol. Environ. Mutagen.* **2013**, *753*, 65–71. [CrossRef]
36. Menendez, D.; Bendesky, A.; Rojas, E.; Salamanca, F.; Ostrosky- Wegman, P. Role of P53 functionality in the genotoxicity of metronidazole and its hydroxy metabolite. *Mutat. Res. Genet. Toxicol. Environ. Mutagen.* **2002**, *501*, 57–67. [CrossRef]
37. El-Nahas, A.F.; El-Ashmawy, I.M. Reproductive and Cytogenetic Toxicity of Metronidazole in Male Mice. *Basic Clin. Pharmacol. Toxicol.* **2004**, *94*, 226–231. [CrossRef]
38. Kovacic, P.; Somanathan, R. Nitroaromatic compounds: Environmental toxicity, carcinogenicity, mutagenicity, therapy, and mechanism. *J. Appl. Toxicol.* **2014**, *34*, 810–824. [CrossRef]
39. Dingsdag, S.A.; Hunter, N. Metronidazole: An update on metabolism, structure-cytotoxicity and resistance mechanisms. *J. Antimicrob. Chemother.* **2017**, *73*, 265–279. [CrossRef]
40. Goolsby, T.A.; Jakeman, B.; Gaynes, R.P. Clinical relevance of metronidazole and peripheral neuropathy: A systematic review of the literature. *Int. J. Antimicrob. Agents* **2018**, *51*, 319–325. [CrossRef]
41. FDA. FDA-Approved Drug Products: Flagyl (Metronidazole) Capsules for Oral Administration. 2013. Available online: https://www.accessdata.fda.gov/drugsatfda_docs/label/2013/020334s008lbl.pdf (accessed on 12 August 2003.).
42. Rosenfeld, C.S. Placental serotonin signaling, pregnancy outcomes, and regulation of fetal brain development. *Biol. Reprod.* **2020**, *102*, 532–538. [CrossRef]
43. Barut, F.; Barut, A.; Gun, B.D.; Kandemir, N.O.; Harma, M.I.; Harma, M.; Ozdamar, S.O. Intrauterine growth restriction and placental angiogenesis. *Diagn. Pathol.* **2010**, *5*, 24. [CrossRef] [PubMed]
44. Perez-Garcia, V.; Fineberg, E.; Wilson, R.; Murray, A.; Mazzeo, C.I.; Tudor, C.; Sienerth, A.; White, J.K.; Tuck, E.; Ryder, E.J.; et al. Placentation defects are highly prevalent in embryonic lethal mouse mutants. *Nature* **2018**, *555*, 463–468. [CrossRef]
45. Rani, A.; Wadhwani, N.; Chavan-Gautam, P.; Joshi, S. Altered development and function of the placental regions in preeclampsia and its association with long-chain polyunsaturated fatty acids. *Rev. Dev. Biol.* **2016**, *5*, 582–597. [CrossRef] [PubMed]
46. Woods, L.; Perez-Garcia, V.; Hemberger, M. Regulation of Placental Development, and Its Impact on Fetal Growth-New Insights from Mouse Models. *Front. Endocrinol.* **2018**, *9*, 570. [CrossRef] [PubMed]
47. Erel, C.T.; Dane, B.; Calay, Z.; Kaleli, S.; Aydinli, K. Apoptosis in the placenta of pregnancies complicated with IUGR. *Int. J. Gynecol. Obstet.* **2001**, *73*, 229–235. [CrossRef]
48. Furukawa, S.; Hayashi, S.; Usuda, K.; Abe, M.; Hagio, S.; Ogawa, I. Toxicological Pathology in the Rat Placenta. *J. Toxicol. Pathol.* **2011**, *24*, 95–111. [CrossRef] [PubMed]
49. Onopiuk, B.; Onopiuk, P.; Dąbrowska, Z.; Dąbrowska, E.; Pietruska, M.; Car, H. Effect of Metronidazole on the Oxidoreductive Processes in the Submandibular and Parotid Glands in Experimental Research. *Oxidative Med. Cell. Longev.* **2018**, *2018*, 7083486. [CrossRef]
50. Yang, J.; Wang, T.; Lin, G.; Li, M.; Zhu, R.; Yiannikouris, A.; Zhang, Y.; Mai, K. The Assessment of Diet Contaminated with Aflatoxin B(1) in Juvenile Turbot (*Scophthalmus maximus*) and the Evaluation of the Efficacy of Mitigation of a Yeast Cell Wall Extract. *Toxins* **2020**, *12*, 597. [CrossRef]
51. Li, L.; Chen, T.; Yang, Z.; Chen, Y.; Liu, D.; Xiao, H.; Liu, M.; Liu, K.; Xu, J.; Liu, S.; et al. Nephrotoxicity Evaluation of Indium Phosphide Quantum Dots with Different Surface Modifications in BALB/c Mice. *Int. J. Mol. Sci.* **2020**, *21*, 7137. [CrossRef] [PubMed]
52. Liu, C.; Shen, W.; Hou, C.; Gao, X.; Wang, Q.; Wu, X.; Zhu, J. Low temperature-induced variation in plasma biochemical indices and aquaglyceroporin gene expression in the large yellow croaker *Larimichthys crocea*. *Sci. Rep.* **2019**, *9*, 2717. [CrossRef]
53. Arnett, D.K.; Blumenthal, R.S.; Albert, M.A.; Buroker, A.B.; Goldberger, Z.D.; Hahn, E.J.; Himmelfarb, C.D.; Khera, A.; Lloyd-Jones, D.; McEvoy, J.W.; et al. 2019 ACC/AHA Guideline on the Primary Prevention of Cardiovascular Disease: Executive Summary: A Report of the American College of Cardiology/American Heart Association Task Force on Clinical Practice Guidelines. *Circulation* **2019**, *140*, e563–e595. [CrossRef]

54. Rivadeneyra-Domínguez, E.; Pérez-Pérez, J.E.; Vázquez-Luna, A.; Díaz-Sobac, R.; Rodríguez-Landa, J.F. Effects of Cassava Juice (Manihot esculenta Crantz) on Renal and Hepatic Function and Motor Impairments in Male Rats. *Toxins* **2020**, *12*, 708. [CrossRef]
55. Kraft, R.; Herndon, D.N.; Al-Mousawi, A.M.; Williams, F.N.; Finnerty, C.C.; Jeschke, M.G. Burn size and survival probability in paediatric patients in modern burn care: Prospective. observational cohort study. *Lancet* **2012**, *379*, 1013–1021. [CrossRef] [PubMed]
56. Liu, P.; Deng, G.F.; Guo, X.Q.; Kuang, J.; Zhang, C.Y.; Cao, H.B.; Hu, G.L. Clinicopathology of Gout in Growing Layers Induced by Avian Nephrotrophic Strains of Infectious Bronchitis Virus. *Pak. Vet. J.* **2015**, *35*, 345–349.
57. Thakur, R.; Sharma, A.; Lingaraju, M.C.; Begum, J.; Kumar, D.; Mathesh, K.; Kumar, P.; Singh, T.U.; Kumar, D. Ameliorative effect of ursolic acid on renal fibrosis in adenine-induced chronic kidney disease in rats. *Biomed. Pharmacother.* **2018**, *101*, 972–980. [CrossRef] [PubMed]
58. Zhao, C.N.; Tang, G.Y.; Liu, Q.; Xu, X.Y.; Cao, S.Y.; Gan, R.Y.; Zhang, K.Y.; Meng, S.L.; Li, H.B. Five-Golden-Flowers Tea: Green Extraction and Hepatoprotective Effect against Oxidative Damage. *Molecules* **2018**, *23*, 2216. [CrossRef] [PubMed]

toxics

Review

Xylazine Poisoning in Clinical and Forensic Practice: Analysis Method, Characteristics, Mechanism and Future Challenges

Tingting Mai [1,2], Youyou Zhang [3] and Shuquan Zhao [1,2,*]

1 Faculty of Forensic Medicine, Zhongshan School of Medicine, Sun Yat-Sen University, Guangzhou 510275, China; maitt@mail2.sysu.edu.cn
2 Guangdong Province Translational Forensic Medicine Engineering Technology Research Center, Guangzhou 510275, China
3 Department of Geriatric Neurology, The Second Affiliated Hospital of Xi'an Jiaotong University, Xi'an 710004, China; tjf1289@126.com
* Correspondence: zhaoshq27@mail.sysu.edu.cn

Abstract: Xylazine abuse is emerging globally, while the identification of xylazine lethal cases poses a great challenge in clinical and forensic practice. The non-specific symptoms delay the diagnosis and treatment of xylazine poisoning, the pathological changes and lethal concentration of xylazine in body fluid and organs of fatal xylazine poisoning cases are seldom reported and the other toxins detected in such cases complicate the role of xylazine in the cause of death. Therefore, we carefully reviewed related updated information on xylazine, summarized the knowledge from clinical and forensic perspectives and can thus provide a reference in such cases and throw light on further study in the field of xylazine poisoning.

Keywords: xylazine poisoning; clinical practice; forensic practice; characteristics; future challenge

1. Introduction

Xylazine is a strong agonist of $\alpha 2$ adrenergic receptors that could decrease the release of norepinephrine and dopamine from the brain and thus form a sedative effect [1]. In veterinary practice, it can achieve the required sedative effect used alone or combined with other narcotic drugs including ketamine [2]. The good sedative effect has made it popular in veterinary practice worldwide since the United States introduced it in the 1960s [3]. Generally, it is only legally available for animals with a veterinary prescription, so it is not listed as a controlled substance all over the world [4].

Xylazine can depress the central nervous system and respiratory functions, resulting in bradycardia, hypotension, and transient hyperglycemia; therefore, it has never been approved for human use [5,6]. Indeed, xylazine poisoning in humans was a rare thing at first, whereas, recently, xylazine has been emerging as a street drug [7]. At first, xylazine was used in drug-facilitated crimes such as sexual assault and robbery, but now it is more common in various street drugs across the US and is associated with drug overdose [5–8].

During the last 5 years, the incidence of xylazine involved in drug-related deaths in West Virginia increased by more than six-fold [9]. The increase in xylazine participation in rural states indicated the widespread and emerging public legal issues concerning the misuse and abuse of xylazine in the United States [9]. Recently, a xylazine-related death was first reported in Europe [10]. Indeed, xylazine poisoning is emerging in clinical and forensic practice; whereas most of them are case reports and they are often combined with other toxins or drugs, the pathological changes in xylazine poisoning cases have seldom been reported [10–15]. Therefore, the lethal concentration of xylazine in blood and other organs is far from identified and the underlying mechanism in the cause of death in xylazine-related deaths needs further study to be demonstrated.

Citation: Mai, T.; Zhang, Y.; Zhao, S. Xylazine Poisoning in Clinical and Forensic Practice: Analysis Method, Characteristics, Mechanism and Future Challenges. *Toxics* **2023**, *11*, 1012. https://doi.org/10.3390/toxics11121012

Academic Editor: Youssef Sari

Received: 16 November 2023
Revised: 7 December 2023
Accepted: 9 December 2023
Published: 11 December 2023

To provide reference on xylazine poisoning, we carefully reviewed the latest updated information on it. Blood and urine were easy to acquire in clinical and forensic practice and the concentration of other drugs and toxins in these samples was well established in previous studies. Therefore, we not only summarized the symptoms of the victims in such cases, but also collected the blood and urine concentrations of xylazine in those cases. The articles involved in the present review were published before November 2023 on PubMed and CNKI. Xylazine was a constant keyword, and poisoning was added to focus on clinical and forensic interest. All the reported concentrations of xylazine were converted in ng/mL or ng/g to make a comparison.

2. Xylazine, Structures and Pharmacokinetics

2.1. Xylazine, Structures and Usage

Xylazine is a strong agonist of α2 adrenergic receptors, with a chemical name 2-(2,6-dimethylanilino)-5,6-dihydro-4h-1,3-thiazine. Its molecular formula is C12H16N2S with a 220.33 relative molecular mass. The chemical structure of xylazine is very similar to phenothiazines, tricyclic antidepressants and clonidine [16]. Since it was approved in 1960s, following comprehensive research including into its synthesis process, pharmacological properties, etc., its good sedative effect in veterinary practice has made it popular. It is provided in the form of hydrochloride in a solution containing 20 or 100 mg/mL, indicated by a free base. The routine administration of xylazine is intramuscular, intra-venous or subcutaneous; the intramuscular and subcutaneous injection of xylazine is absorbed quickly in equines. The dosages of xylazine vary from 0.55 to 8.8 mg/kg intramuscularly or 1.1 mg/kg a pound intravenously [17]. This dose could provide analgesia of 15 to 30 min and sedation of 1 to 2 h [18]. Generally, pharmacokinetics information about xylazine in humans is absent.

2.2. Pharmacokinetics

The intramuscular or subcutaneous injection of xylazine is absorbed quickly [19]. It produces a fast onset of action with a short duration. In addition, the intensity and duration of sedative or analgesic effects are proportional to the dose of drug used, and species also plays a role in the process of sedative effects [20]. An ordinary dose in a sheep can keep the animal quiet and asleep for 1~2 h, with analgesia for 15~30 min [18]. Cattle are the most sensitive animals to xylazine, and it is reported that the dose of xylazine used in cattle amounts to 1/10 of that in horses or dogs to achieve the same sedative or analgesic level [20].

The pharmacokinetic parameters of xylazine have been confirmed in various animal species [21]. Generally, the absorbance, metabolism and elimination of xylazine is quite rapid. It absorbs and spreads so quickly that the brain and kidney had the highest concentration of xylazine in organs a few minutes post-intravenous administration. The sedative effect can appear within 5 min and lasts up to 4 h [22,23]. The routine dose of xylazine in animals was 0.5–5.0 mg/kg. The highest concentration of xylazine in plasma occurs 0.2~0.3 h after intramuscular administration of 0.6–1.4 mg/kg in large animals such as sheep [24]. It was reported that less than 1% of the drug is excreted unchanged in the urine in the cow and about 8% in the rat [20].

3. Analytic Aspects

The past decade has provided an important impetus for LC-MS/MS technology and research. Due to the rather high selectivity and sensitivity of LC-MS/MS, better performance than other technology and reliable results it is a perfect method that can achieve the goal of various applications [25–27]. LC-MS/MS was successfully applied in pharmacology and toxicology cases and became an important instrument that cannot be ignored [25]. For one thing, pharmacology is the basis for drug monitoring and finding the correct treatment strategy for patients. Furthermore, LC-MS/MS are the most crucial instrument configurations used for drug identification and illicit drug screening in toxicology and forensic practice and research, providing enormous support to related legal practice. It is no wonder that LC-MS/MS play a critical role in the identification of xylazine.

In 2018, Krongvorakul et al. first established a LC-MS/MS method to detect xylazine in the serum and urine of the victims in drug-facilitated crimes and confirmed the concentration of serum (0.057 μg/mL) at 6 h and urine (0.294 μg/mL) at 8 h after she drank the drink [8]. In 2021, Xu et al. used LC-MS/MS to detect xylazine in the blood of 12 victims in a poisoning case; the minimum detection limit was 0.02 ng/mL and xylazine displayed a good linear relationship in the range of 0.1~200 ng/mL while the correlation coefficient R2 = 0.9978 and the blood concentration of xylazine in the victims ranged from 9.6 to 139.5 ng/mL [28]. In 2023, Yao et al. used LC-MS/MS to detect the potential drugs in 10 victims' consumption of beef and venison; xylazine was detected in the blood and urine of the victims and rather high concentrations of xylazine were detected in the cooked beef and venison [29].

To better understand the current stage in the analysis method of xylazine, we identified several biological analysis methods of xylazine used in the past 10 years and summarize them in Table 1 [28–31]. LC-MS/MS, SPE-HPLC/MS/MS and UPLC-QTOF/MS were the validated methods in the detection of xylazine. Serum, blood and urine were common biological samples in the identification of xylazine as shown in Table 1. The LODa of the reported method was largely affected by the matrix, sample preparation and detection method. With the technology development of MS, the detection method revealed little difference in the identification of xylazine.

Table 1. Some analytic methods of xylazine and metabolites (DMA) used in biological samples.

	Analysis	Matrix (mL)	Sample Preparation	Sample	LODa ng/mL	Linear Range ng/mL	Author, Publication Year
Xylazine	LC-MS/MS	1.0 2.0 g	Liquid—liquid extraction	Blood, urine, beef, venison	Blood 2.5, urine, 2.5, Meat 2.5 μg/kg	1.0~300.0	Yao et al. 2023, [29]
	LC-MS/MS	1.0	Liquid—liquid extraction	Blood	0.02	0.1~200	Xu et al. 2023, [28]
	LC-MS/MS	0.5	Liquid—liquid extraction	Serum, Urine	Serum, 1, Urine 1	Blood 10~750, Urine 10~750	Krongvorakul et al. 2018, [8]
	SPE-HPLC/MS/MS	1.0	SPE	Blood, Urine	Blood 0.4, Urine 0.3	Blood 2~2000, Urine 2~2000	Liu et al. 2017, [30]
	UPLC-QTOF/MS	1.0	Liquid—liquid extraction	Blood, Urine	Blood 0.4, Urine 0.3	Blood 2~2000, Urine 2~2000	Gao et al. 2015, [31]
DMA	SPE-HPLC/MS/MS	1.0	SPE	Blood, Urine	Blood 0.4, Urine 0.3	Blood 2~2000, Urine 2~2000	Liu et al. 2017, [30]
	UPLC-QTOF/MS	1.0	Liquid—liquid extraction	Blood, Urine	Blood 0.5, Urine 0.3	Blood 10~4000, Urine 10~4000	Gao et al. 2015, [31]

Matrix (mL): minimal sample size.

Blood and urine were common in the identification of xylazine in forensic practice. However, other organs such as the liver were rarely involved in such cases, and pre- and postmortem organ distributions of xylazine were seldom involved. Due to the occurrence of food poisoning, the left food was also an ideal sample in such cases to confirm the existence of xylazine.

Table 1 also indicates that the sample preparation of the validated methods was a traditional liquid–liquid extraction, which required large amounts of organic solvent and a complicated operation. The procedures of most methods were time-consuming, qualitative instead of quantitative, and needed a rather large sample size, and thus most of them were not suitable for the rapid detection of xylazine for clinical and forensic purposes. To date, few studies have centered on the tissue distribution of xylazine in the forensic perspective.

Although xylazine is emerging as an adulterant and is most commonly associated with other drugs including fentanyl, the metabolic pathways and major metabolites in humans who suffered from xylazine poisoning are absent. In 2015, Gao et al. established a UPLC-QTOF/MS method to detect the concentrations of xylazine and 2,6-xylidine (DMA) in the blood and urine [31]. In 2019, Cui et al. used the liquid chromatography-quadruped/orbitrap mass spectrometry method to explore the metabolites of xylazine in human urine and concluded that the hydroxylated products, oxidation products, S-oxidation products, etc., were the main metabolites of xylazine, which corresponds to the results of rat urine, rat liver microsomes, and horse urine [32]. The rather similar metabolites of xylazine in human urine in Chinese and Caucasian humans indicated the common metabolic pathways and major metabolites in different populations [33]. Previous studies also confirmed that xylazine undergoes phase I metabolic reactions such as hydroxylation, oxidation, N-dealkylation, and S-oxidation in the human body, and the hydroxylated metabolites then act on glucuronic acid and sulfuric acid to produce phase II metabolic reactions [32–34]. The main metabolite of xylazine, DMA, was a useful biomarker for post-poisoning surveillance.

4. Clinical and Forensic Aspect of Xylazine Poisoning

4.1. Characteristics of Xylazine Poisoning

To identify the characteristics of xylazine poisoning cases in clinical and forensic practice, we searched "Xylazine poisoning" in PubMed and CNKI, excluded the cases without humans involved, and identified 20 papers which included more than 160 cases; the related information is summarized in Tables 2 and 3 [8,10,11,13,15,28–30,34–45]. As shown in Table 2, most of the involved cases were accidental; however, homicide and suicide were also seen in the xylazine poisoning cases. It was reported that it may occur in persons who accidentally ate meat that was injected with xylazine, and due to the rather small dose of xylazine, the victim may be rescued [28,29]. This was associated with mass poisoning incidents, and it may occur in homicide cases. In 2023, Xu et al. reported that a criminal purchased xylazine on the internet and poisoned the breakfast in a hotel; the scene investigation revealed that the criminal had been fired a few ago and he did this for revenge. Recently, injection was a common exposure of xylazine in the USA, and it was also combined with fentanyl or heroin in those illicit drug overdoses [3,4]. However, inhalation was a new approach in the exposure of xylazine and it may become popular as necrotizing skin ulceration has followed the injection. This related skin injury is different from the wound common in injection drug users and can occur at or away from the injection site regardless of the exposure route [3,4].

Table 2. Xylazine poisoning cases in clinic and forensic practice without autopsy details.

Authors and Publication Year	Age/Sex	Cause	Exposure Route	Primary Symptoms	Time Interval between Oral and Symptom	Drug Concentration
Yao et al. 2023, [29]	Unknown, 5 victims	Accident	Oral	Drowsiness, hypokalemia	15–40 min Mean 25 min	Blood 4.8–11.3 ng/mL, Urine 6.7–218.1 ng/mL, Cooked beef 440.0 μg/kg
	Unknown, 5 victims	Accident	Oral	Dry mouth, dizziness, drowsiness	30–60 min Mean 45 min	Blood 8.6–46.8 ng/mL, Venison 193.0 μg/kg Cooked venison 5.0 mg/kg
Xu et al. 2023, [28]	Unknown, 12 victims	Homicide	Oral	Dizzy	Unknown	9.6~139.5

Table 2. *Cont.*

Authors and Publication Year	Age/Sex	Cause	Exposure Route	Primary Symptoms	Time Interval between Oral and Symptom	Drug Concentration
Gill EL et al. 2023, [34]	8/M	Unknown	Unknown	loss of consciousness, hypopneic with pinpoint pupils, headache	Unknown	Xylazine and fentanyl positive
Krongvorakul et al. 2018, [8]	73/F	homicide	Oral	Coma, bradycardia	Unknown	Gastric content, positive. Serum, negative
	71/F	homicide	Oral	Coma, bradycardia	Unknown	Urine, 0.294 µg/mL Serum, 0.0057 µg/mL
	76/M	homicide	Oral	Coma, bradycardia	Unknown	Urine, 0.533 µg/mL
Liu et al. 2017, [30]	unknown	Accident	Oral	Unknown	Unknown	Peripheral blood xylazine and 2,6-Dimethylaniline positive
Forrester MB. 2016, [35]	76 victims, 41M, 33F, mean age 37	49 uninten-tional, 24 were intentional, 1 misuse; 1 adverse reaction, 1 unknown	39 injection, 21 ingestion, 12 dermal route, 11 ocular route, 2 inhalation, and 2 unknown, 9 multiple.	Drowsiness/lethargy [36], Bradycardia [15], Hypotension [8], Hypertension [7] Puncture/wound [6], Slurred speech [6], Coma [5], Ocular irritation/pain [5], Respiratory depression [4]	Unknown	Positive
Wang et al. 2014, [36]	Unknown, 1 male, 4 females	Suicide	Oral	Unknown	Unknown	200–1000 mg, one death
Hou et al. 2013, [37]	Unknown, more than 50 victims	Accident	Oral	Unknown	Unknown	Peripheral blood, cooked meat xylazine positive
Meyer et al. 2013 [38]	14/M	Accident	Injection	Coma, bradycardia	Unknown	Plasma 0.3 mg/L of xylazine and 0.1 mg/L of ketamine
Zhang et al. 2010, [39]	31/M	Accident	Inhaled	Dyspnea	Unknown	Unknown
Liu et al. 2007, [40]	19/M	Accident	Oral	Coma, 35.5 °C, bradycardia	Unknown	Urine 582 mg/L of ketamine, 448 mg/L of norketamine, 745 mg/L of phenobarbital, and 762 mg/L of xylazine
Velez et al. 2006, [13]	38/M	Accident	Unintentional irrigation of both eyes	Bradycardia, hypotension to 90/60 mm Hg, consciousness impairment,	2 h	Unknown
Xia et al. 2006, [41]	7 victims, 1 male, 6 females, 28–48 years old	homicide	Oral	Dizziness, drowsiness, palpitations, impaired consciousness	0.5–2 h	Heart blood 12.86 µg/mL
Elejalde et al.2003, [42]	18/M	Suicide	inhaled	Blood pressure 90/60 mmHg, heart rate 45 bpm, disorientation, dysarthria, dysmetria, ataxia, sinus bradycardia	Unknown	urine for common toxins was negative

Table 2. *Cont.*

Authors and Publication Year	Age/Sex	Cause	Exposure Route	Primary Symptoms	Time Interval between Oral and Symptom	Drug Concentration
Capraro et al. 2001, [15]	16/M	Suicide	inhaled	Coma, bradycardia	Unknown	Blood 0.54 μg/mL perhaps 3 mg of dose
Samanta et al. 1990, [43]	19/M	Accident	Injection	80/60 mm Hg, small pupils	30 min	Unknown

Table 3. Xylazine poisoning cases in forensic practice with forensic autopsy.

Authors and Publication Year	Age/Sex	Cause	Exposure Route	Primary Symptoms	Time Interval between Oral and Symptom	Time Interval between Oral and Death	Autopsy and Pathology Findings	Postmortem Sampled Time	Postmortem Drug Concentration ng/mL
Rock et al. 2023, [10]	43/F	Suicide	Injection	Unknown	Unknown	Unknown	Recent puncture wounds to the groin	Unknown	Blood 38 Urine 135
Zhang et al. 2021, [44]	49/F	Homicide	Injection	Dizzy	30 min	Less than 10.5 h	Bleeding point is seen in the conjunctiva, bleeding was observed on the injection site of right thigh	Unknown	Heart blood 2.4 Injection site positive
Moore et al. 2003, [45]	42/M	Suicide	injection	Unknown	Unknown	Unknown	Unknown	Unknown	Heart blood 2300 ng/mL, peripheral blood 2900, Bile 6300, Kidney 7.8 mg/kg, Liver 6.1 mg/kg, Urine 10
Poklis et al. 1985, [12]	36/M	Suicide	Injection	Unknown	Unknown	Within 4 h	Injection sites on antecubital fossa, congested edematous lung	Unknown	Blood 200, Brain 0.4 mg/kg, Kidney 0.6 mg/kg, Liver 0.9 mg/kg, Lung 1.1 mg/kg, Adipose 0.05 mg/kg, Urine 7000

Coma, bradycardia and hypotension were the primary symptoms in xylazine poisoning cases and they may occur within 30 min and last for several days. Hypokalemia and small pupils were also reported. The time interval between xylazine exposure and the primary symptom in Table 2 was as soon as 15 min or as late as 2 h, and due to the unavailable information on the time interval between xylazine exposure and sample collected, the association link between clinical characteristics and blood or urine concentrations was hard to build. The signs and symptoms of xylazine poisoning are common to various disease states and/or toxicological exposures; therefore, xylazine cannot be detected rapidly in real time and can be a barrier to treatment, and should be added to the differential diagnosis if a toxicological cause is suspected.

Generally, central nervous system depression was common in such cases and respiratory depression was reported in people taking xylazine; the underlying mechanism may be attributed to the increase in the occurrence of respiratory depression induced by opioids. It is a common view that naloxone does work on respiratory depression caused by opioid whereas it does little to xylazine-induced respiratory depression [3,4]. However, xylazine poisoning may mimic an opioid overdose, and due to the prevalence of opioid overdoses, naloxone is still needed in such cases. Additional therapeutic support is required even if naloxone administration has been given. This additional therapeutic support includes keeping control of the airways, supplemental oxygen, and other treatment as needed. To date, there are still no approved drugs that can reverse the effect of xylazine in humans [3,4].

To date, the characteristics of xylazine poisoning have been so obscure that even experienced physicians may make incorrect diagnoses. However, some signs should arouse the attention of clinicians to the possibility of xylazine poisoning. When skin ulcers are seen without known reasons or naloxone has little effect on opioid overdose patients, or people who eat cooked beef and meat purchased from the same shop show coma and bradycardia, the clinicians should keep in mind the possibility of xylazine poisoning to make an early diagnosis. Once confirmed, oxygen-inhaling, naloxone therapy and symptomatic treatments were needed.

4.2. Characteristics in Xylazine-Related Death Cases

Generally, pure xylazine overdose death cases were uncommon in forensic practice. Although numerous studies indicated the increasing incidence of xylazine in forensic practice, most of them just focused on the detection method of xylazine and other characteristics of xylazine-related deaths and the pathological changes in such cases were rarely reported. The reasons may be the non-specific pathological changes in such cases, and xylazine contributes little to the cause of death in most of cases. Table 3 summarizes the forensic autopsy cases of xylazine overdose death. All the involved cases were homicide or suicide, and injection was the only exposure. The groin, right thigh and antecubital fossa were the injection sites; the bleeding injection site, bleeding point in the conjunctiva and congested edematous lung were non-specific and were also seen in other causes of death.

Although Sibbesen et al. retrospectively studied 3292 drug deaths from 2019 to 2021 in West Virginia and identified 117 cases involving xylazine, they found that xylazine-related death always involved other drugs such as opioids, stimulants, benzodiazepines, and antidepressants/antipsychotics. It also indicated that liver diseases were more common in xylazine-related deaths. Injection was still the leading exposure of xylazine in xylazine-related deaths. Other exposure routes include snorting, smoking; oral and inhaling were also reported.

Skin ulcers are a primary health problem in chronic xylazine users that arouse public attention. These skin ulcers can occur far from the injection site, which could distinguish them from other chronic drug users. The underlying mechanisms were attributed to the oxygenation response to xylazine intoxication of the skin, and the chronic use of xylazine may limit the mobility of the limbs and lead to amputation in severe cases [9,46–50]. Generally, the injection site in such cases should be examined carefully and distinguished from iatrogenic injection injuries. Therefore, it is critical to emphasize the scene investigation, case history and toxicological test results.

4.3. Mechanism of Xylazine-Related Death

In 2007, the abuse of xylazine in humans was first reported in Puerto Rico [51,52]. Since then, xylazine abuse has been recorded in other states (mainly in the Northeast), as well as an increasing number of drug-related deaths [53]. It was confirmed as a fentanyl adulterant as it prolonged the duration of fentanyl effects and increased the duration of brain hypoxia [51]. When used with opioids, xylazine has a synergistic effect and it could increase the level of brain hypoxia that may increase the risk of overdose or death [54]. Bradycardia, a depressed central nervous system, and hypotension were the most common side effects associated with human xylazine poisoning. To date, there is no clear definition of xylazine toxic in humans and the lethal concentrations of xylazine in humans have been absent. The role of xylazine in xylazine-related death is far from confirmed as xylazine and fentanyl were the common findings in such cases. The toxic mechanism of xylazine and the combination of xylazine and fentanyl still need further study to be confirmed.

4.4. Blood and Urine

Xylazine Concentrations

The exact contribution of xylazine to the process of death is still unknown, although its pharmacological effects were reported to enhance the respiratory and central nervous system depressive effects of other substances. The toxicity and lethal concentrations of xylazine are still unclear and there is considerable overlap in the blood concentration in patients who suffered xylazine poisoning and xylazine overdose death cases [1]. It is reported that the blood concentration of xylazine ranged from 5 to 49 ng/mL in xylazine overdose death cases, whereas blood or plasma concentrations ranging from 30 to 460 ng/mL were also detected in non-lethal xylazine overdose cases [1]. The concentration of xylazine summarized in Tables 2 and 3 showed that the blood concentration of xylazine can reach 540 ng/mL in non-lethal xylazine overdose cases. Although there is not a significant difference in xylazine concentrations in the number of co-intoxicants, rather higher concentrations of xylazine were seen in decedents who took only one other drug compared to those who took more drugs. Indeed, xylazine plays an important role in the death mechanisms of other co-intoxicants.

Table 3 also shows that the blood concentration of xylazine in homicide cases was higher than in suicide cases, and was much higher in pure xylazine poisoning cases compared to combined with other toxics. To date, only a homicide case was reported and xylazine was the only reported toxic; whether the type of death and the cause of death had a role in the concentration of xylazine in different samples still needs further study. The blood concentration of xylazine was the lowest in most cases, while the urine concentration of xylazine was much higher. The urine concentration of xylazine can reach 533 ng/mL even in non-lethal xylazine overdose cases. The concentration of xylazine in the heart blood was lower than in the peripheral blood; considering the existence of postmortem redistribution in toxic overdose death cases, the heart blood may not be the most suitable sample in such cases. And due to the existence of putrefaction and autolysis, blood samples may be unable to be collected.

4.5. Alternative Samples

The organ and fluid distribution of xylazine in xylazine overdose death cases has rarely been reported in forensic practice. To date, only two research articles published the organ distribution of xylazine in xylazine overdose death cases with paradoxical results. In 1985, Poklis et al. reported that a 36-year-old male injected xylazine to commit suicide and the organ concentration of xylazine was as follows: lung (1100 ng/mL), liver (900 ng/mL), kidney (600 ng/mL), brain (400 ng/mL) [11]. In 2003, Moore et al. reported that a 42-year-old male injected xylazine to commit suicide and the concentration of xylazine in the kidney (780 ng/mL) was higher than in the liver (610 ng/mL) [45]. This may be attributed to the intake dose of xylazine as the intake dose of other drugs revealed the distribution characteristics in vivo are different in various lethal doses. Due to the existence of postmortem dispersion and other factors, the postmortem redistribution of xylazine was an inevitable phenomenon.

The affected factors included diffusion distance, postmortem gavage time and dose. It is prone to disperse into tissues and organs close to the gastrointestinal tract, whereas it did have some impact on the heart blood, bile, and peripheral blood, and never affected the brain and muscles. Therefore, to identify the concentration of xylazine in forensic practice, not only the routine examination materials in forensic toxicological practice such as the heart blood, liver and stomach content should be collected, but also the less affected tissues and organs such as the brain and muscle should be taken for examination.

When it comes to the xylazine overdose death cases whose exposure route was injection, the injection site was the ideal sample for the detection of xylazine. Therefore, it should be carefully examined, taken fixed in formalin for histopathological examination and taken stored at -80 °C for subsequent analysis. Generally, the routine sample for the detection of xylazine was not identified in forensic practice, and it still needs further study to confirm the ideal sample in such cases.

4.6. Other Drugs or Toxins in Such Cases

It has been reported that in almost all xylazine deaths, 98.3%, the patient also took fentanyl, and most of the decedents were young white men [9]. Previous studies have documented this high fentanyl prevalence, consistent with the identification of xylazine as a fentanyl adulterant [46–50]. Xylazine is increasingly involved in drug overdose decedents that ingest more toxicants, and xylazine abuse is an emerging public health problem [46–50].

It has been reported that drug or alcohol abuse history increases the incidence of xylazine-related death [9]. Xylazine-related deaths are more common in individuals who take more than two drugs, including xylazine. In 2022, Sibbesen et al. retrospectively reviewed 3292 drug deaths in West Virginia from 2019 to mid-2021, identified the characteristics of xylazine-related death, found that approximately 90% of the xylazine-related decedents took more than three drugs, and that 40% of the xylazine-related decedents had taken more than five drugs [9]. Opioids, stimulants, benzodiazepines, and antidepressants/antipsychotics are commonly associated with xylazine-related deaths. Fentanyl analogues, heroin and other opioids are found in about a quarter of xylazine-related deaths and nearly one in five of non-xylazine deaths.

Few studies focus on the interaction between xylazine and fentanyl. In 2023, Smith et al. used C57BL/6 mice which were 42 days of age to quantify the lethal effects of fentanyl and xylazine, found that 56 mg/kg fentanyl administration could shift the LD50 of xylazine from 157.2 mg/kg to 32.0 mg/kg and 100 mg/kg xylazine administration could shift the LD50 of fentanyl from 131.3 mg/kg to 1.27 mg/kg, suggesting the synergistic manner in their interaction and that the combination of them can lead to more rapid death than alone [54]. Choi et al. used adult male Long–Evans rats that weighed 450 ± 50 g to evaluate the neuro effects of xylazine alone or combined with fentanyl and heroin and found that xylazine alone not only can decrease temperature and locomotor activity, but also prolong the lower oxygen levels in the brain [55].

The existence of other drugs complicates the role of xylazine in xylazine-related death in forensic practice. The concentration of other drugs may be far from the lethal concentration, but xylazine may enhance the toxic effect of them, and result in unexpected death. However, few studies focused on the interactions between them, and other drugs may affect the metabolism of xylazine, which needs further study.

5. Future Challenges

Due to the characteristics of xylazine and the influence of the internet, xylazine abuse is increasing in clinical and forensic practice, and the identification of such cases is still a challenge. Therefore, we summarized the updated information about xylazine and xylazine poisoning cases and thus provide a reference and give advice on future research.

Firstly, we summarized the characteristics of xylazine poisoning cases in clinical and forensic practice and thus provided reference in clinical and forensic practice. Secondly, the detecting method of xylazine and samples (tissues and liquid) were also listed to update toxic information about xylazine and thus throw light on further study. Thirdly, other drugs or toxicants may also be detected in xylazine poisoning cases, which may complicate such cases in clinical and forensic practice and need further study to make clear their interactions.

Lastly, the frequency of illicit drug users who use xylazine has exceeded the scope of accidental drug outbreaks. It is speculated that it may be attributed to the spread of internet usage in illicit drug users, and the ease of obtaining the drug from the internet. It is a common view that xylazine can be used as a veterinary anesthetic. The over-the-counter availability and low price of xylazine provides the victims with a convenient way to obtain it. Therefore, public health agencies should consider focus on the education of xylazine poisoning and monitor the real-time information on xylazine poisoning [56].

6. Conclusions

Above all, although we update the latest information about xylazine poisoning, the exact lethal concentration of xylazine is still vague; many factors, such as the age of the victim, pre-existing morbidity and co-exposure, complicate the critical role of xylazine in xylazine poisoning cases. Therefore, the concentration of xylazine and its role in the symptoms of xylazine poisoning cases still need further study to confirm. Furthermore, the combined drugs in xylazine-related poisoning cases complicate the role of xylazine in clinical and forensic practice and the interaction between xylazine and other drugs was seldom involved in the present studies. And lastly, the organ distribution and toxic mechanism of xylazine poisoning still needs further study to be confirmed and we hope more clinical and forensic pathologists and toxicologists can join us and make it clear.

Funding: This research was funded by the Fundamental Research Funds for the Central Universities (2023KYPT18).

Data Availability Statement: Not applicable.

Conflicts of Interest: The authors declare no conflict of interest.

References

1. Ruiz-Colón, K.; Chavez-Arias, C.; Díaz-Alcalá, J.E.; Martínez, M.A. Xylazine intoxication in humans and its importance as an emerging adulterant in abused drugs: A comprehensive review of the literature. *Forensic Sci. Int.* **2014**, *240*, 1–8. [CrossRef] [PubMed]
2. Huang, L.; Yang, G. Repeated exposure to ketamine-xylazine during early development impairs motor learning-dependent dendritic spine plasticity in adulthood. *Anesthesiology* **2015**, *122*, 821–831. [CrossRef]
3. Rubin, R. Here's What to Know About Xylazine, aka Tranq, the Animal Tranquilizer Increasingly Found in Illicit Fentanyl Samples. *JAMA* **2023**, *329*, 1904–1906. [CrossRef]
4. Gupta, R.; Holtgrave, D.R.; Ashburn, M.A. Xylazine—Medical and Public Health Imperatives. *N. Engl. J. Med.* **2023**, *388*, 2209–2212. [CrossRef] [PubMed]
5. Greene, S.A.; Thurmon, J.C. Xylazine—A review of its pharmacology and use in veterinary medicine. *J. Vet. Pharmacol. Ther.* **1988**, *11*, 295–313. [CrossRef]
6. Spoerke, D.G.; Hall, A.H.; Grimes, M.J.; Honea, B.N., 3rd; Rumack, B.H. Human overdose with the veterinary tranquilizer xylazine. *Am. J. Emerg. Med.* **1986**, *4*, 222–224. [CrossRef] [PubMed]
7. Alexander, R.S.; Canver, B.R.; Sue, K.L.; Morford, K.L. Xylazine and Overdoses: Trends, Concerns, and Recommendations. *Am. J. Public. Health* **2022**, *112*, 1212–1216. [CrossRef] [PubMed]
8. Krongvorakul, J.; Auparakkitanon, S.; Trakulsrichai, S.; Sanguanwit, P.; Sueajai, J.; Noumjad, N.; Wananukul, W. Use of Xylazine in Drug-Facilitated Crimes. *J. Forensic Sci.* **2018**, *63*, 1325–1330. [CrossRef] [PubMed]
9. Sibbesen, J.; Abate, M.A.; Dai, Z.; Smith, G.S.; Lundstrom, E.; Kraner, J.C.; Mock, A.R. Characteristics of xylazine-related deaths in West Virginia-Xylazine-related deaths. *Am. J. Addict.* **2023**, *32*, 309–313. [CrossRef] [PubMed]
10. Rock, K.L.; Lawson, A.J.; Duffy, J.; Mellor, A.; Treble, R.; Copeland, C.S. The first drug-related death associated with xylazine use in the UK and Europe. *J. Forensic Leg. Med.* **2023**, *97*, 102542. [CrossRef]
11. Hoffmann, U.; Meister, C.M.; Golle, K.; Zschiesche, M. Severe intoxication with the veterinary tranquilizer xylazine in humans. *J. Anal. Toxicol.* **2001**, *25*, 245–249. [CrossRef] [PubMed]
12. Poklis, A.; Mackell, M.A.; Case, M.E. Xylazine in human tissue and fluids in a case of fatal drug abuse. *J. Anal. Toxicol.* **1985**, *9*, 234–236. [CrossRef]
13. Velez, L.I.; Shepherd, G.; Mills, L.D.; Rivera, W. Systemic toxicity after an ocular exposure to xylazine hydrochloride. *J. Emerg. Med.* **2006**, *30*, 407–410. [CrossRef] [PubMed]
14. Cooper, G. The rise and rise of fentanyl in postmortem casework. *J. Forensic Sci.* **2023**, *68*, 1675–1685. [CrossRef] [PubMed]
15. Capraro, A.J.; Wiley, J.F., 2nd; Tucker, J.R. Severe intoxication from xylazine inhalation. *Pediatr. Emerg. Care* **2001**, *17*, 447–448. [CrossRef] [PubMed]

16. Stillwell, M.E. A reported case involving impaired driving following self-administration of xylazine. *Forensic Sci. Int.* **2003**, *134*, 25–28. [CrossRef] [PubMed]
17. U.S. Food and Drug Administration, Animal & Veterinary, Animal Drugs@FDA. Available online: http://www.accessdata.fda.gov/scripts/animaldrugsatfda/ (accessed on 30 August 2023).
18. Paddleford, R.R.; Harvey, R.C. Alpha 2 agonists and antagonists. *Vet. Clin. North. Am. Small Anim. Pract.* **1999**, *29*, 737–745. [CrossRef] [PubMed]
19. Spyridaki, M.H.; Lyris, E.; Georgoulakis, I.; Kouretas, D.; Konstantinidou, M.; Georgakopoulos, C.G. Determination of xylazine and its metabolites by GC-MS in equine urine for doping analysis. *J. Pharm. Biomed. Anal.* **2004**, *35*, 107–116. [CrossRef]
20. Garcia-Villar, R.; Toutain, P.L.; Alvinerie, M.; Ruckebusch, Y. The pharmacokinetics of xylazine hydrochloride: An interspecific study. *J. Vet. Pharmacol. Ther.* **1981**, *4*, 87–92. [CrossRef]
21. Knych, H.K.; Stanley, S.D.; McKemie, D.S.; Arthur, R.M.; Kass, P.H. Pharmacokinetic and pharmacodynamics of xylazine administered to exercised thoroughbred horses. *Drug Test Anal.* **2017**, *9*, 713–720. [CrossRef] [PubMed]
22. De Carvalho, L.L.; Nishimura, L.T.; Borges, L.P.; Cerejo, S.A.; Villela, I.O.; Auckburally, A.; de Mattos-Junior, E. Sedative and cardiopulmonary effects of xylazine alone or in combination with methadone, morphine or tramadol in sheep. *Vet Anaesth. Analg.* **2016**, *43*, 179–188. [CrossRef] [PubMed]
23. Gallanosa, A.G.; Spyker, D.A.; Shipe, J.R.; Morris, D.L. Human xylazine overdose: A comparative review with clonidine, phenothiazines, and tricyclic antidepressants. *Clin. Toxicol.* **1981**, *18*, 663–678. [CrossRef] [PubMed]
24. Adam, M.; Lindén, J.; Raekallio, M.; Abu-Shahba, A.; Mannerström, B.; Seppänen-Kaijansinkko, R.; Meller, A.; Salla, K. Concentrations of vatinoxan and xylazine in plasma, cerebrospinal fluid and brain tissue following intravenous administration in sheep. *Vet. Anaesth. Analg.* **2021**, *48*, 900–905. [CrossRef] [PubMed]
25. D'Ovidio, C.; Locatelli, M.; Perrucci, M.; Ciriolo, L.; Furton, K.G.; Gazioglu, I.; Kabir, A.; Merone, G.M.; de Grazia, U.; Ali, I.; et al. LC-MS/MS Application in Pharmacotoxicological Field: Current State and New Applications. *Molecules* **2023**, *28*, 2127. [CrossRef]
26. Campêlo, J.M.; Rodrigues, T.B.; Costa, J.L.; Santos, J.M. Optimization of QuEChERS extraction for detection and quantification of 20 antidepressants in postmortem blood samples by LC-MS/MS. *Forensic Sci. Int.* **2021**, *319*, 110660. [CrossRef] [PubMed]
27. Rubicondo, J.; Scuffi, L.; Pietrosemoli, L.; Mineo, M.; Terranova, F.; Bartucca, M.; Trignano, C.; Bertol, E.; Vaiano, F. A New Multi-Analyte LC-MS-MS Screening Method for the Detection of 120 NPSs and 49 Drugs in Hair. *J. Anal. Toxicol.* **2023**, *46*, e262–e273. [CrossRef]
28. Xu, J.J.; Lu, Y.M. Xylazine detected in a poisoning case: A case report. *Guang Dong Gong An Ke Ji* **2023**, *31*, 78–80.
29. Yao, Y.; Ma, M.Y.; Liu, F.; Xu, F. Detection of two food poisoning cases caused by xylazine residual using liquid chromatography-mass spectrometry. *Chin. J. Health Lab. Tec.* **2023**, *33*, 1540–1544.
30. Liu, W.W.; Li, J.; Xu, J.P.; Zheng, J.; Wang, J.W. Detection of xylazine and 2,6-xylidine in blood and urine by SPE-HPLC/MS/MS. *Chin. J. Forensic Med.* **2017**, *32*, 182–185. [CrossRef]
31. Gao, X.; Guo, H.; Du, Y.; Gu, C. Simultaneous Determination of Xylazine and 2,6-Xylidine in Blood and Urine by Auto Solid-Phase Extraction and Ultra High Performance Liquid Chromatography Coupled with Quadrupole-Time of Flight Mass Spectrometry. *J. Anal. Toxicol.* **2015**, *39*, 444–450. [CrossRef]
32. Cui, J.J.; Sun, Y.Y.; Shen, B.H.; Liu, W.H.; Xiang, P. Identification of the Metabolites of Xylazine in Human Urine by Liquid Chromatography-Quadrupde/Orbitrap Mass Spectrometry. *J. Chin. Mass. Spectrom. Soc.* **2019**, *40*, 280–288.
33. Meyer, G.M.; Maurer, H.H. Qualitative metabolism assessment and toxicological detection of xylazine, a veterinary tranquilizer and drug of abuse, in rat and human urine using GC-MS, LC-MSn, and LC-HR-MSn. *Anal. Bioanal. Chem.* **2013**, *405*, 9779–9789. [CrossRef]
34. Gill, E.L.; Mack, E.A.; Osterhoudt, K.C.; Shaw, L.M.; Milone, M.C. An Epidemic within a Pandemic: An 8-Year-Old Boy with Xylazine-Complicating Fentanyl Poisoning, and Drug Detection Challenges. *J. Appl. Lab. Med.* **2022**, *7*, 1492–1495. [CrossRef] [PubMed]
35. Forrester, M.B. Xylazine Exposures Reported to Texas Poison Centers. *J. Emerg. Med.* **2016**, *51*, 389–393. [CrossRef] [PubMed]
36. Wang, Y.Y. Study on treatment effect of xylazine intoxication by hemoperfusion combined with plasma exchange. *Chin. J. Trauma. Disabil. Med.* **2014**, *22*, 28.
37. Hou, L.; Zhang, X.Y. Analysis study of veterinary anesthetic poisoning. *Forensic Sci. Technol.* **2013**, *5*, 56–57. [CrossRef]
38. Meyer, G.M.; Meyer, M.R.; Mischo, B.; Schofer, O.; Maurer, H.H. Case report of accidental poisoning with the tranquilizer xylazine and the anesthetic ketamine confirmed by qualitative and quantitative toxicological analysis using GC-MS and LC-MS(n.). *Drug Test. Anal.* **2013**, *5*, 785–789. [CrossRef] [PubMed]
39. Zhang, S.Q.; Li, P.; Song, Q.M.; Zhang, F.D. Xylazine poisoning: A case report. *China Mod. Dr.* **2010**, *48*, 100.
40. Liu, C.M.; Chiu, M.J.; Fang, C.C.; Chen, W.J. Xylazine abuse: A rare cause of syncope. *Clin. Toxicol.* **2007**, *45*, 309–311. [CrossRef]
41. Xia, S.X.; Jiao, A.F. Xylazine poisoning: Seven case reports. *Clin. Focus* **2006**, *2*, 84.
42. Elejalde, J.I.; Louis, C.J.; Elcuaz, R.; Pinillos, M.A. Drug abuse with inhalated xylazine. *Eur. J. Emerg. Med.* **2003**, *10*, 252–253. [CrossRef] [PubMed]
43. Samanta, A.; Roffe, C.; Woods, K.L. Accidental self administration of xylazine in a veterinary nurse. *Postgrad. Med. J.* **1990**, *66*, 244–245. [CrossRef] [PubMed]
44. Zhang, Y.; Li, X.F.; Ji, Y.Q.; Wang, G.J. Death induced by xylazine injection: A case report. *J. Forensic Med.* **2020**, *36*, 874–876.

45. Moore, K.A.; Ripple, M.G.; Sakinedzad, S.; Levine, B.; Fowler, D.R. Tissue distribution of xylazine in a suicide by hanging. *J. Anal. Toxicol.* **2003**, *27*, 110–112. [CrossRef] [PubMed]
46. Delcher, C.; Anthony, N.; Mir, M. Xylazine-involved fatal overdoses and localized geographic clustering: Cook County, IL, 2019–2022. *Drug Alcohol Depend.* **2023**, *249*, 110833. [CrossRef] [PubMed]
47. Kariisa, M.; Patel, P.; Smith, H.; Bitting, J. Notes from the Field: Xylazine Detection and Involvement in Drug Overdose Deaths—United States, 2019. *MMWR Morb. Mortal. Wkly. Rep.* **2021**, *70*, 1300–1302. [CrossRef] [PubMed]
48. Thangada, S.; Clinton, H.A.; Ali, S.; Nunez, J.; Gill, J.R.; Lawlor, R.F.; Logan, S.B. Notes from the Field: Xylazine, a Veterinary Tranquilizer, Identified as an Emerging Novel Substance in Drug Overdose Deaths—Connecticut, 2019-2020. *MMWR Morb. Mortal. Wkly. Rep.* **2021**, *70*, 1303–1304. [CrossRef] [PubMed]
49. Nunez, J.; DeJoseph, M.E.; Gill, J.R. Xylazine, a Veterinary Tranquilizer, Detected in 42 Accidental Fentanyl Intoxication Deaths. *Am. J. Forensic Med. Pathol.* **2021**, *42*, 9–11. [CrossRef] [PubMed]
50. Dai, Z.; Abate, M.A.; Groth, C.P.; Rucker, T.; Kraner, J.C.; Mock, A.R.; Smith, G.S. Fentanyl and other opioid involvement in methamphetamine-related deaths. *Am. J. Drug Alcohol Abus.* **2022**, *48*, 226–234. [CrossRef] [PubMed]
51. Reyes, J.C.; Negrón, J.L.; Colón, H.M.; Padilla, A.M.; Millán, M.Y.; Matos, T.D.; Robles, R.R. The emerging of xylazine as a new drug of abuse and its health consequences among drug users in Puerto Rico. *J. Urban. Health* **2012**, *89*, 519–526. [CrossRef]
52. Johnson, J.; Pizzicato, L.; Johnson, C.; Viner, K. Increasing presence of xylazine in heroin and/or fentanyl deaths, Philadelphia, Pennsylvania, 2010–2019. *Inj Prev.* **2021**, *27*, 395–398. [CrossRef] [PubMed]
53. Friedman, J.; Montero, F.; Bourgois, P.; Wahbi, R.; Dye, D.; Goodman-Meza, D.; Shover, C. Xylazine spreads across the US: A growing component of the increasingly synthetic and polysubstance overdose crisis. *Drug Alcohol. Depend.* **2022**, *233*, 109380. [CrossRef] [PubMed]
54. Smith, M.A.; Biancorosso, S.L.; Camp, J.D.; Hailu, S.H.; Johansen, A.N.; Morris, M.H.; Carlson, H.N. "Tranq-dope" overdose and mortality: Lethality induced by fentanyl and xylazine. *Front. Pharmacol.* **2023**, *14*, 1280289. [CrossRef] [PubMed]
55. Choi, S.; Irwin, M.R.; Kiyatkin, E.A. Xylazine effects on opioid-induced brain hypoxia. *Psychopharmacology* **2023**, *240*, 1561–1571. [CrossRef]
56. Zagorski, C.M.; Hosey, R.A.; Moraff, C.; Ferguson, A.; Figgatt, M.; Aronowitz, S.; Stahl, N.E.; Hill, L.G.; McElligott, Z.; Dasgupta, N. Reducing the harms of xylazine: Clinical approaches, research deficits, and public health context. *Harm Reduct. J.* **2023**, *20*, 141. [CrossRef]

MDPI
St. Alban-Anlage 66
4052 Basel
Switzerland
www.mdpi.com

Toxics Editorial Office
E-mail: toxics@mdpi.com
www.mdpi.com/journal/toxics

www.ingramcontent.com/pod-product-compliance
Lightning Source LLC
Chambersburg PA
CBHW061612210926
51850CB00009B/2296
* 9 7 8 3 7 2 5 8 0 4 4 2 9 *